Sustaining the World's Wetlands

Richard C. Smardon

Sustaining the World's Wetlands

Setting Policy and Resolving Conflicts

 Springer

Richard C. Smardon
College of Environmental Science
    and Forestry
State University of New York
Syracuse, NY 13210
USA
rsmardon@esf.edu

ISBN 978-1-4899-8481-4   ISBN 978-0-387-49429-6 (eBook)
DOI 10.1007/978-0-387-49429-6
Springer Dordrecht Heidelberg London New York

Printed on acid-free paper

Springer is part of Springer Science+Business Media (www.springer.com)

Foreword

Wetlands throughout the world, including those described in this book are among the most sensitive and vulnerable ecosystems. They are critical habitats to the world's migratory birds and a broad range of endangered mammal, reptile, amphibian, and plant species. They provide a broad range of flood storage, pollution control, water supply, ecotourism functions to indigenous peoples and country populations as a whole.

They are also at the center of severe land and water use conflicts. These are conflicts between counties where wetland resources or the water supplies required for such resources involve more than one country. These are conflicts in use such as conflicts between habitat protection and charcoal production in mangroves. These are conflicts between groups of peoples such as indigenous peoples and hydropower advocates. Many wetlands have already been destroyed by water extractions, dams, levees, channelization, and fills. Others have been degraded by water pollution, overfishing and overhunting, timber harvest, and a host of other activities.

This book describes these conflicts and international policies and institutions developed to protect and manage wetland resources. Most of the broader literature and other books on wetlands focuses on wildlife. Wildlife is described in the case studies, which follow. But, Richard Smardon provides us with more. He traces the history of conflicts and the development of policies and institutions to protect and manage wetland resources.

Richard has patiently prepared the book. It has been several decades in the making. During this time, Richard and his colleagues and students have not only investigated but also participated in efforts to protect and manage wetland resources domestically and internationally such as his work in the Yucatan Peninsula. Richard has throughout his career been interested in the role of local people in resource management.

The case studies which follow will be of interest to anyone wishing to protect wetland ecosystems. They will be of interest to teachers wishing students to understand the complexities of natural resource policy making. They will be of interest to NGOs and governments wishing to reduce conflicts and better manage and restore wetlands.

The case studies are illustrated with many fine figures and photographs and abundant references for anyone seeking more information.

My colleagues and I have had the pleasure of working with Richard and his students for many years. This includes lively discussions on the case study wetlands described in this book. During this time, Richard has given freely his time to aid wetland protection and restoration efforts at all levels of government and by NGOs. This work is much appreciated.

We hope you enjoy the book and find it useful. I thank Richard for preparing the book and sharing his insights with us.

Berne, New York   Jon Kusler, PhD
Association of State Wetland Managers

Preface

Wetland assessment and management continues to be a major policy issue around the world especially with CEC environmental putting pressure on European countries, with the Rio Conference and Ramsar Treaty stressing wetland protection, and with the continuing debate about wetland management in North America. There is an international audience as witnessed by continuing interest/attendance at international meetings and conferences on Ramsar Treaty, IUCN Biosphere Reserve management, sustainable development implementation, and eco/nature tourism. There is a strong academic interest in wetland policy and management conflicts as examples of resource conflict, sustainable development, local equity, and decision making. So, the book could be used as a textbook for departments of environmental studies, ecology, human ecology, natural resource management, environmental science, geography, applied anthropology, international policy, and conflict resolution at the upper undergraduate and graduate levels. Key themes that will be treated in almost all the eight case studies are as follows: (1) trade-offs between sustainable use of wetlands for food, fuel, and fiber vs. protection of ecosystem diversity and stability and (2) respective roles of Big International Non-government Organizations (BINGOs), national and regional government, and local community-based organizations when faced with wetland management issues. Developed countries/regions and developing countries are facing equally challenging but different wetland management issues.

With the advent of global warming and resultant regional climate change, effective wetland management strategies are urgently needed. This book focuses on the roles of different actors in different contexts as both developed and developing countries strive for sustainable wetland use and management.

Syracuse, New York   Richard C. Smardon

Acknowledgments

The author wishes to thank the many wetland experts around the world who encouraged him to keep working on this book and reinforced the basic themes. Specific acknowledgements for all those who made individual contributions to the development of specific case study chapters are given below.

For review comments on the Wadden Sea wetlands chapter, I sincerely thank Professor Dr. H.L.F. Saeijs, chief engineer, director of Delta Area and professor of Water Quality Policy and Sustainability, Erasmus University in Rotterdam, the Netherlands.

For the Axios River Delta comments the author wishes to thank Costas Cassios of the Athens National Technical University and Topiotechniki whose project allowed him to actually see the Axios Delta area. Special thanks to Chrysoula Athanasiou of the WWF-Red Alert Project who provided detailed comments on an earlier draft plus valuable updates. Additional material was provided by Professor P.A. Gerakis of the Gaulandris Museum of Natural History, Greek Biotype/Wetland Centre, Thermi, Greece and Professor J. Szijj, University of Geasamthochschule-Essen, Germany.

For the Kafue Flats, Zambia River chapter, many thanks for the commentary provided by Bernard Kamweneshe of the Zambia Department of National Parks and Wildlife, Dr. Charles Namafe, School of Education, University of Zambia, and Ms, Monica Chundama, WWF Program Officer on an earlier working draft of this chapter.

For east Kolkata wetland, coauthors plus the primary book author wishes to acknowledge the efforts of Dr. N.C. Landi, deputy director of the Zoological Survey of India, New Alipore, Calcutta, for his careful editing and rewrites of this chapter.

Many thanks to George Archibald who agreed to let ICF cooperate with the author on the Tram Chim Nature Reserve, Vietnam, study and to Jeb Brazen, ICF, for letting the author interview him in depth and to Rich Beilfuss, ICF, for his detailed comments on the first chapter draft. Thanh Vo, a doctoral student at SUNY/ESF, provided recent updates plus photos of Tram Chim.

The author is deeply grateful for the interview time with Stephen Brown, Cam Davis, Gail Gruenwald, and Nancy Patterson of Environment Canada for the chapter on Great Lakes wetlands.

The author is indebted to early review comments from Joann Andrews (past president of PRONATURA) and Dr. John Frazier (formerly of CINVESTAV and now with Smithsonian Institution) plus key feedback from Drs. Betty Faust, Julia Fraga, and Jorge Euan of CINVESTAV, Merida', Mexico, for Ria Celestun and Ria Lagartos wetlands chapter. Special contributions were provided by Scott Moan (former landscape architecture masters student) for his work on Ria Lagartos and to Gabriela Canamar (another landscape architecture student) who did translation work plus an ecotourism survey of Celestun.

For Mankote mangrove chapter the author acknowledges the collaboration of Matius Burt, Yves Renard, Allan Smith, and the local mangrove producers whom he interviewed plus the review comments of Allan Smith on an earlier draft of this chapter.

Contents

Chapter 1
International Wetland Policy and Management Issues

Introduction

Wetlands are among the world's most important environmental resources; yet remain among the world's least understood and most seriously abused assets. Of all global systems, wetlands are the source of some of today's most contentious, difficult, and politically sensitive environmental and social questions. Increasingly, in both the developed world and developing world, the future of wetlands seems to depend on economic, social, and political development trends and the outcomes of litigation, and legislative/administrative debate rather than natural processes. Yet natural processes result in ecosystem functions that have real economic value to society which can be expressed in terms of yield over time, such as fisheries production, maintenance of water quality, and flood damage aversion. The purpose of this book is to examine the international environmental policy implications of wetland use and management conflicts.

Wetlands occupy the transitional zone between permanently wet and generally dry environments (Finlayson and Moser 1991, p. 8) and generally have some form of temporary flooding, saturated soils, and resultant plant communities that have adapted to these conditions. There are different forms of both freshwater and saltwater or brackish wetlands including marshes, swamps, peatlands, floodplain wetlands; mangroves, nipa and tidal freshwater swamp forest; lake edge wetlands, estuaries and lagoons, and even man-made wetlands. A number of authors have addressed wetland types and their occurrence around the world (Finlayson and Moser 1991, Kusler and Opheim 1996, Mitsch et al. 1991, Mitsch and Gosselink 2000, Whigham et al. 1993).

In the traditional view, wetlands are wastelands (Maltby 1986, p. 1, Mitsch and Gosselink 2000, p. 13). Words like *marsh swamp, bog and fen* imply little more than dampness, disease, difficulty, and danger. Such wasted lands can be put to good use, however, if they are "reclaimed" for agriculture and building. From mythology, we have the view that wetlands were bogs and swamps inhabited with creatures, pixies, heathens, and monsters. This mythology was transported from Europe to North America and probably is still with us, explaining part of the negative attitude toward wetlands (Smardon 1983).

R.C. Smardon, *Sustaining the World's Wetlands*,
DOI 10.1007/978-0-387-49429-6_1, © Springer Science+Business Media, LLC 2009

But historically, far from being wastelands, wetlands are among the most fertile and productive ecosystems of the world. They are essential life-support systems, play a vital role in controlling water cycles, and help to clean up our environment as biofilters. Some wetlands produce up to eight times as much plant matter as an average wheat field, promising higher crop yields if the fertility of the wetland soil can be harnessed and the ecosystem managed for sustained production. Actually wetlands were the mainstay of the ancient Mayan food production system, which was able to maintain multiple cities 2000 years ago (Smardon 2006). Many, if not all of the world's great civilizations were born in wetland regions, such as in the floodplains of the Nile, the Tigrus–Euprates, and the Indus Rivers and in the Yucatan peninsula (Maltby 1986, Mitsch and Gosselink 2000, p. 8). Wetlands traditionally are known for their value for biodiversity and as habitat for plant animal and fish species (Verhoeven et al. 2006, Bobblink et al. 2006, Mitsch and Gosselink 2000). More recently a number of environmental services or functions from wetlands are being recognized, such as

- natural protection against extreme floods and storm surges;
- storage of freshwater to be used for drinking water or for irrigation;
- water quality enhancement if located along streams, rivers, and lakes;
- spawning habitat for fish if located along rivers, shallow lakes, and coastal wetlands;
- long-term net carbon storage regionally and globally (Verhoeven et al. 2006, Bobblink et al. 2006, Mitsch and Gosselink 2000).

Within the Summary of the Millennium Ecosystem Assessment (2005) "Ecosystems and Human Well-being: Wetlands and water" states that wetland ecosystems including rivers, lakes, marshes, rice fields, and coastal areas provide many services that contribute to human well-being and poverty alleviation (Millennium 2005, p. 1). Specific wetland functions that can be linked to human well-being include the following:

Inland fisheries, especially important for protein supply for developing countries (Verhoeven et al. 2006).
Principal supply of renewable freshwater for human use comes from inland wetlands including lakes, rivers, swamps, and shallow groundwater aquifers (Mitsch and Gosselink 2000, Verhoeven et al. 2006).
Water purification and detoxification of wastes (Mitsch and Gosselink 2000, Verhoeven et al. 2006).
Climate regulation through sequestering and releasing fixed carbon in the atmosphere (Verhoeven et al. 2006).
Mitigation of climate change by wetlands such as mangroves and floodplains in reduction of storm surges.
Cultural services: wetlands provide significant aesthetic. Educational, cultural and spiritual benefits including recreation and tourism activities (Smardon 2003) (Table 1.1).

Wetlands are found on every continent except Antarctica and in every clime from the tropics to the tundra (Mitsch and Gosselink 2000, p. 35). Estimation of

Table 1.1 Ecosystem services provided by/derived from wetlands (Millennium 2005)

Services	Comments and examples
Provisions	
Food	Production of fish, wild game, fruits and grains
Freshwater	Storage/retention of water for domestic, industrial, agricultural use
Fiber and fuel	Production of logs, fuel wood, peat, fodder
Biochemical	Extraction of medicines and other biotic materials
Genetic materials	Genes for resistance to plant pathogens
Regulatory	
Climate regulation	Source of/sink for greenhouse gases; influence on local and regional temperature, precipitation, etc.
Water regulation	Groundwater recharge/discharge
Water purification/ treatment	Retention, recovery, removal of excess nutrients/pollutants
Natural hazard regulation	Flood control and storm protection
Erosion regulation	Retention of soils and sediments
Pollination	Habitat for pollinators
Cultural	
Spiritual and inspirational	Source of inspiration – attach spiritual/religious values to wetland ecosystems
Recreational	Opportunities for recreational activities, e.g., fishing
Aesthetic	Finding beauty/aesthetic value in wetland ecosystems
Educational	Opportunities for formal and informal education and training
Supporting	
Soil formation	Sediment retention and accumulation of organic matter
Nutrient cycling	Storage, recycling, processing, and acquisition of nutrients

the worldwide extent of wetlands is difficult because of the variation in definitions of wetland cover types and the fact that wetlands constantly change in area with variation in water levels. Based on several estimates, the extent of the world's wetlands is thought to be from 7 to 9 million km^2 or about 4–6% of the world's land surface (Mitsch and Gosselink 2000, p. 35). According to the Millennium Assessment (2005) it is in excess of 1,280 million ha (1.2 million km^2) but wetlands everywhere are under threat and/or stress.

Wetland Stress and Loss

Despite the strength of the early association of wetlands with human communities (Maltby 1986, Coles and Coles 1989, Mitsch and Gosselink 2000, Smardon 2006) the historical trends worldwide have been to modify or change wetlands so they can be used for non-wetland purposes or suffer cumulative stress resulting in an ecologically degraded condition.

It is estimated that more than 50% of specific types of wetlands in parts of North America, Europe, Australia, and New Zealand were converted during the twentieth century (Millennium 2005, p. 3). Coastal wetland ecosystems are under extreme pressure and it is estimated that about 35% of mangrove (from countries with multiple year data) have been lost during the last two decades because of agricultural development, deforestation, and freshwater diversion (Baldock 1984). Major causes of wetland loss include the following.

Clearing and drainage for agricultural expansion and increased withdrawal of freshwater.

By 1985, an estimated 56–65% of inland and coastal marshes had been drained for intensive agriculture in Europe and North America, 27% in Asia, 6% in South America, and 2% in Africa. Practices, which stress or degrade wetland ecosystems include the following:

- Agricultural practices such as extensive use of water for irrigation and excessive nutrient loading from use of nitrogen and phosphorous in fertilizers.
- Introduction of invasive alien species causing local extinction of native freshwater species.
- Freshwater diversion from estuaries causing less delivery of water and sediment to nursery areas and fishing grounds.
- Disruption and fragmentation of coastal wetlands important as migration routes for waterfowl and other birds.

According to the Millennium Wetland Assessment: "The primary direct driver of the loss or degradation of coastal wetlands, including saltwater marshes, mangrove, sea grass meadows, and coral reefs, has been conversion to other land uses. Other direct drivers affecting coastal wetlands include diversion of freshwater flows, nitrogen loading, over harvesting, siltation, changes in water temperatures, and species invasions. The primary indirect drivers of change have been the growth of human populations in coastal areas coupled with growing economic activities" (Millennium 2005, p. 6).

In addition to the aforementioned factors global climate change may well exacerbate the loss and degradation of wetlands by

- changes in coastal wetlands due to sea-level rise, increased storm and tidal surges, changes in storm intensity and frequency, and subsequent changes in river flow regime, and sediment transport;
- changes in the distribution of coastal wintering shorebirds and other waterfowl as well as habitat loss;
- incidence of vector-borne diseases such as malaria and dengue and of water-borne diseases such as cholera (Millennium 2005, p. 7).

The association of wetlands with diseases such as malaria, schistosomiasis and in the northeastern US, eastern equine encephalitis has been a strong and emotive factor in drainage or heavy use of insecticides. Recent outbreaks of West Nile virus in bird populations in the United States and the fear of wild waterfowl transferring avian flu in Asia and Europe have intensified these fears in recent years. This encourages wetland loss throughout the world and is an argument used to support development plans in lesser developed countries (LDCs). The papyrus swamp in the Hula Valley in Israel was drained in the 1950s at least in part as a malaria eradication measure.

Increasing competition of the land occupied by wetlands – which is often ideally located for water supply systems and for riverine, estuarine, or coastal access – is an inevitable consequence of present urban and industrial expansion. Development of water oriented recreation and irrigated agriculture. This land, which is characterized by flat terrain, was historically avoided because of the expense and the technical difficulties of development. The availability of foreign investment and development aid has hitherto pristine and/or little modified wetlands in the tropics and subtropics to increasing threats. We have seen this phenomena occurring recently in southern Mexico, where wetlands are being converted for rice production, cattle ranching, or oil production (Smardon 2006).

Primarily, because the historical losses have been the smallest in the developing nations of the world – it is in these countries that future wetlands losses will be the greatest. These losses are especially acute in coastal wetlands because of land use conversion, erosion, and coastal pollution as well as natural erosion and sea level rise due to climate change (Baca and Clark 1988). One type of coastal wetland, under extreme stress throughout the world is the tropical mangrove, including those in Africa (see Brinson and Lugo 1989, Dugan 1988, Nelson et al. 1989), India and Bangladesh (see Azarth et al. 1988, Haq 1989, Ambasht and Srivastava 1994), Southeast Asia (see Dugan 1988, Maltby 1986), Australia (see Finlayson 1989, Nelson et al. 1989), and Central/South America (see Quesada and Jimenz 1988, Toledo et al. 1989). Many of the interior riverine wetland systems are also under extreme duress particularly in South America, Africa, India, and Southeast Asia (Nelson et al. 1989, Dugan 1988). It is these regions where the traditional association between human communities and wetland ecosystems has been retained most firmly and intimately, and that the subsequent loss of human livelihoods and values will be the greatest.

Development of International Wetland Policy

An understanding of the value of wetlands as well as the nature of wetland stress/loss is necessary to set the stage for discussion of wetland policy. What is being done? At what levels? And how effective is such policies?

Wetlands are the only ecosystem type that have their own international convention, the Ramsar Convention. The convention on Wetlands of International Importance – usually known as the Ramsar Convention after the Iranian City on the Caspian Sea where it was adopted in 1971 – is the principal global instrument for international cooperation on wetland conservation. At present 152 countries are signatories to the convention. These countries include states from the industrialized world together with a growing number of developing countries in Africa, Asia, and South/Central America (IUCN 1980, 1984, IUCN/ IWRB 1980, 1984, Mathews 1993, Navid 1988) (see Figs. 1.1, 1.2, 1.3, and 1.4):

Fig. 1.1 Countries that are Ramsar Treaty contracting partners. Contracting parties are shaded in *dark gray*. Figure drawn by Samuel Gordon and adapted from Ramsar Convention for Wetlands: http://www.ramsar.org

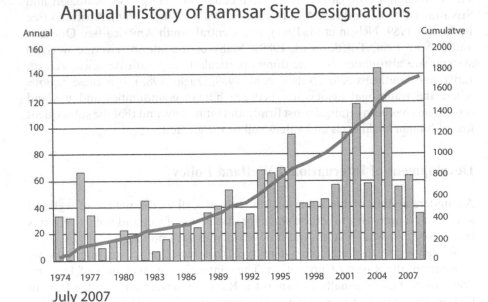

Fig. 1.2 Annual history of Ramsar site designations. The first Ramsar site was designated in 1974. In this graphic, the annual level of Ramsar site designation is depicted along with a cumulative trend line. Figure drawn by Samuel Gordon and adapted from Ramsar Convention for Wetlands: http://www.ramsar.org

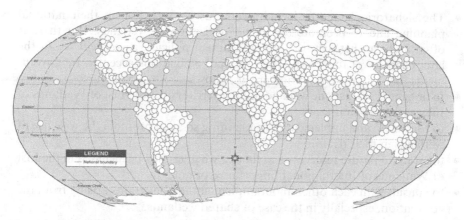

Fig. 1.3 Figure illustrating approximate distribution of Ramsar sites around the world. Figure drawn from Samuel Gordon and adapted from Ramsar Convention for Wetlands: http://www.ramsar.org

Total Designated Ramsar Site Area/Region (hectares)

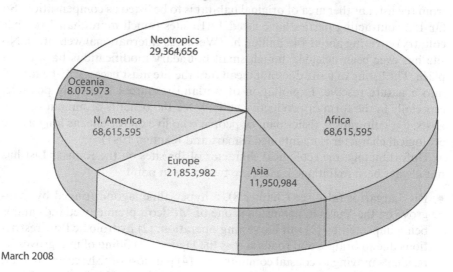

Neotropics
29,364,656

Oceania
8.075,973

N. America
68,615,595

Europe
21,853,982

Asia
11,950,984

Africa
68,615,595

March 2008

Fig. 1.4 Total designated Ramsar site area/region (ha). Ramsar Contracting Parties come from one of six administrative regions. This chart depicts the approximate current total designated area of Ramsar sites from each region. The cumulative area of Ramsar sites worldwide now totals approximately 159,551,478 ha designated in 1,721 sites. Figure redrawn by Samuel Gordon and adapted from Ramsar Convention for Wetlands: http://www.ramsar.org

- The signatories agree to include wetlands conservation in their national planning and to promote the sound utilization of wetlands. But there is often a chasm between rhetoric and actual policies, as we will see in the later chapters of this book. The principal obligations accepted by governments that join the convention are as follows:
- to designate at least one wetland in their territory for the List of Wetlands of International Importance;
- to formulate and implement their planning so as to promote the wise use of wetlands in their territory;
- to establish nature reserves as wetlands, whether they be included in the List of Wetlands of International Importance or not;
- to consult with each other about implementing obligations arising from the convention, especially in the case of shared wetlands.

During the initial years of the convention, most of the parties' attention was devoted to developing the List of Wetlands of International Importance (Ramsar Convention 1984). Designation of a wetland for the list means the area's ecological character is to be maintained, and notification of any change in ecological character, actual or potential, has to be given to the Ramsar Bureau in Gland, Switzerland. If, as a result of urgent national interest, a site is removed from the list, another area of original habitat is to be listed as compensation. So far 152 contracting parties have listed 1,615 sites (much more than I site per country) covering about 145 million ha (Wetlands International web site[1]). No site has ever been deleted, though small boundary modifications have taken place. The listing of a site does not mean that the site must necessarily be turned into a nature reserve. Exploitation of wetland resources is entirely possible, especially in the form of productive activity for the benefits of human pastoralists, agriculturalists, fisherman, or people who live on the site, as long as the ecological character is maintained (Smart and Kanters 1991).

Unfortunately, the ecological character of the sites on the Ramsar List has not always been maintained. Here are two cases in point:

- Ria Largartos (also see Chapter 8), a hypersaline lagoon fringed by mangrove on the Yucatan peninsula is one of Mexico's premier wetlands and is being impacted by (1) salt harvesting operation, (2) hydrologic flow restrictions due to bridges and roads across the lagoon, (3) filling of mangroves by residents moving to coastal communities, (4) plus loss of fish rearing habitat plus overfishing (Rosete et al. 1991, Faust and Sinton 1991, Smardon 2006).
- In Europe, the Axios River Delta in Greece (see Chapter 3), another Ramsar listed wetland, is being impacted by intensive rice farming, overgrazing on the floodplain, and water pollution from upstream sources. Its ecological character is also not being maintained (IUCN 1990, Tsiouris and Gerakis 1991).

---

[1] Wetlands International web site: http://www.wetlands.org/RSDB/_COP9Directory/

The third conference of the Ramsar Parties, held in Regina, Saskatchewan, Canada in 1987 produced a very important document, which was annexed to the convention recommendations (Ramsar Convention Bureau 1988). This document provided not only amended criteria on how to identify a "Wetland of International Importance" but also provided a definition of "wise use" and guidelines for achieving wise use:

- The *wise use* of wetlands is defined as "their sustainable utilization for the benefit of human kind in a way compatible with the maintenance of the natural properties of the ecosystem".
- *Sustainable utilization* is defined as "human use of a wetland, so it may yield the greatest continuous benefit to present generations while maintaining its potential to meet the needs and aspirations of future generations".
- The *natural properties of an ecosystem* are defined as "those physical, biological, or chemical components such as soil, water, plants, animals and nutrients, and their interaction between them".

The Regina Conference also established a Working Group on Criteria and Wise Use, which was charged with examining how to elaborate the criteria and how to apply the wise use provisions. The Working Group's report was circulated to the contracting parties, in anticipation of the conference of the parties held in June/July 1990 in Montreux, Switzerland.

The Working Group's report developed the Regina guidelines further. It recognized that the elaboration of national wetland policies would be a long-term process, and that immediate action should be taken to stimulate wise use. The revised guidelines therefore include both short- and long-term elements that are divided into (1) actions that establish national wetland policies; (2) priority actions at the national level; and (3) priority actions at particular wetland sites.

In order to increase knowledge and awareness of the importance of wetlands, the benefits and values of wetlands were listed at the Regina Conference. The listing was taken from Adamus and Stockwell (1983) and Adamus et al. (1987), which provided a codification of wetlands' functions. This is of special interest because of its focus on functions of importance to people – and thus on the possibilities for human participation and wise use including

> groundwater recharge;
> groundwater discharge;
> flood storage and desynchronization;
> shoreline anchoring and dissipation of erosive forces;
> sediment trapping;
> nutrient retention and removal;
> food chain support;
> habitat for fisheries;
> habitat for wildlife;
> active recreation;
> passive recreation and heritage value.

The overriding concern at the Leiden Conference on the People's Role in Wetland Management (Marchand and Udo de Haes 1990, Marchand and Udo de Haes 1991) was with the application of the wise use guidelines. Examples were needed that could be adopted or developed by other contracting parties. The papers presented at the workshop in Leiden did offer a number of such examples and gave interesting insight into the attitudes of both countries that were contracting parties and those that were not. One of the key issues is how the convention's work can be extended beyond the conservation of waterfowl habitat, in order to give greater weight to all aspects of wetlands and to develop the north–south dialogue for full consideration of wetland-dependent livelihoods. Other legal and technical limitations of the Ramsar Convention will be covered in subsequent chapters.

Since 1971, the Ramsar Convention parties have held nine major meetings: the fifth meeting in Kushiro, Japan, in 1993, the sixth meeting in Brisbane, Australia, in 1996, the seventh meeting in San Jose, Costa Rica, in 1999, the eighth meeting in Valencia, Spain, in 2002 and the ninth meeting in Kampala in 2005. These conferences of the parties have resulted in a number of programs and at least 200 official decisions (117 resolutions and 83 recommendations) (see http://www.ramsar.org).

In addition to the Ramsar Convention, there are other forms of international wetland recognition. Some wetlands are given regional or national recognition (Carp 1980). Others are recognized and protected by being UNESCO biosphere reserves or parks such as the Ria Lagartos Biosphere Reserve in Mexico (Chapter 8 this volume and Smardon and Faust 2006) or the Trebon Basin in the Czech Republic (Kvet et al. 2002). Within biosphere reserves, nature reserves and parks – the usual management or protection device is zonation – where uses and activities are increasing restricted as one moves closer to critical habitat areas. The implementation of zoning has been problematic in multiple-purpose biosphere reserves, where there is a range of activities undertaken by local people living within the biosphere reserve. This is related to the problem of local enforcement by government agencies or non-governmental groups (NGOs) that manage the biosphere reserve. Or more importantly, it may be due to lack of participatory processes in development of management plans for these biosphere reserves (see Smardon and Faust 2006).

Wetlands under private ownership pose special management problems stemming from the difficulties of maintaining ecological integrity if the economic use of the wetlands is not restricted. There may be resultant resource conflict as well as upstream or upper watershed uses that are not compatible with downstream ecological integrity. We also have the example of the United States and Canada trying to restrict the inappropriate use of privately owned wetlands through permit and review systems. This raises the taking issue of economic loss of property rights without giving appropriate compensation as well as due process and delay in decision making. Future chapters will address some of these basic management issues that affect decision making within the context of public versus private land ownership, governmental versus non-governmental management with attendant legal and economic issues.

National Wetland Policy Developments Around the World

There is a striking similarity in the wealth and diversity of the wetlands in different parts of the world. There is also a worldwide similarity in the need for nations to cooperate with local communities, if their wetlands are to be conserved. However, realization of programs for protection and wise use is difficult – not only because of the lack of insight into wetland functioning but also because of lack of current detailed data about sustainability versus exploitation. But there are some interesting developments taking place in unexpected places.

There has been tremendous variation in wetland protection policy, especially if we compare North America to Europe and other developing countries. The following section provides some highlights of this variation. For instance, for North America we have the United States, Canada, and Mexico included within this volume (Chapters 7 and 8).

In the United States we have had a history of government programs, which supported conversion of wetlands to other uses until the 1970s (Mitsch et al. 1994, Mitsch and Gosselink 2000, Vileisis 1997). Policies within agencies such as the US Army Corps of Engineers, the Soil Conservation Service, and the Bureau of Reclamation encouraged the destruction of wetlands, while US Department of Interior's Fish and Wildlife Service encouraged their protection (Mitsch et al. 1994, World Wildlife Fund 1992). In 1987 a National Wetlands Policy Consortium convened by the Conservation Foundation, at US EPA's bequest, recommended a "no net loss" policy (The Conservation Foundation 1988), which was subsequently adopted by George W. Bush I and Bill Clinton's administration.

Even with a national wetland regulatory program implemented by US EPA and the US Corps of Engineers (Section 404 of the US Clean Water Act), plus many state programs, there were contentious legal issues. One of these is the "taking issue" regarding regulating private wetland property by federal or state agencies. Should the private property landowner be compensated if their property is declared federal or state wetland jurisdiction and development is restricted (Mitsch et al. 1994)?

The other major issue with the Section 404 of the Clean Water Act is what wetland areas fall under the acts jurisdiction? From 2001 to 2006 US Federal Courts issued thirty-seven decisions regarding scope of CWA jurisdiction (Kusler et al. 2006). There were three major Supreme Court cases of note:

- In US vs. Riverside Bayview Homes, Inc. 474 US 121 (Sup. Ct. 1985) the court unanimously upheld the Corps jurisdiction over wetlands adjacent to navigable-in-fact waterways.
- A 5-4 divided court in Solid Waste Agency of Northern Cook County vs. Army Corps of Engineers SWANCC 531 US 159 (Sup Ct. 2002) held a series of ponds in northern Illinois was not subject to CWA jurisdiction solely based on their use by migratory birds. The court distinguished but did not override Riverside Bay view (Kusler et al. 2006).

- Rapanos vs. US 126 S. Ct. 2208 (Sup Ct. 2006), the third case, did not override either Riverside Bayview or SWANCC (Kusler et al. 2006). The case vacated to lower appellate court decisions upholding CWA jurisdiction for wetland which were separated from ditches or drains leading into navigable waters by a berm and for wetlands linked to navigable waters through a system of drainage ways and ditches (Kusler et al. 2006).

The US Corps of Engineers has to sort all this out in terms of the jurisdictional issue and its "about as clear as mud"!

In Canada, problems with the wise use of wetlands are concentrated at the borders of major urban areas, especially around the Great Lakes and the St. Lawrence River (see Chapter 7). In this region a great deal of money and attention has been given to the wise use of wetlands, especially to passive and outdoor recreation and other non-consumptive uses. There was activity in the early 1990s by the bi-national Great Lakes Wetlands Policy Consortium (an NGO) that has complied some 50 recommendations to pressure both Canadian and US agencies to do more with wetlands protection, management and even creation to offset previous wetland losses and impacts (see Brown 1990, Gruenwald 1990, Loftus et al. 2004). In addition there is the North American Waterfowl Management Plan, whose objective is the protection of existing wetlands and the creation of additional wetlands to ensure adequate habitat for migratory waterfowl along the North Atlantic Flyway in the United States, Canada, and Mexico (see Lambertson 1990, Loftus et al. 2004, Rubec 1994).

Interesting developments in Europe include (1) recognition of the importance of small wetlands, especially marshes, (2) the need for a census of small remaining wetlands, (3) the need for modern cost–benefit analyses concerning modern cropping of rice paddies, (4) the need for cooperation between farmers, recreationalists, and other participants to make conservation of small wetlands economically viable, and (5) the need for approval at the national level of a law for the protection of national wetlands (Maltby 1986, Williams 1990).

In the Netherlands we can see the results from the Dutch Society for the Preservation of the Wadden Sea in Chapter 2 of this volume over a 25-year history. The Netherlands part of the Wadden Sea has many values and is particularly appropriate, given the issue of wise use inside a large Ramsar site. Emphasis is given to the important role of policies that directly affect the inhabitants of the area including (1) decreasing the intensity of farming on the Wadden Isles, (2) giving more attention to nature-oriented recreation, and (3) preventing reclamation of salt marshes and mudflats.

Other European countries may or may not have specific regulatory programs protecting wetlands. For instance, Sweden investigated national wetland protection laws (Leander and de Mare 1994) but did not pass such legislation. Sweden has several environmental laws that require landowners to preserve or not pollute existing wetlands with appropriate economic compensation (Leander and de Mare 1994). This is an issue that affects much of Europe according to

Turner and Jones (1990), which includes market failure case studies for the United Kingdom, France, and Spain.

There is hope that ecotourism and other uses will be useful in sustaining more compatible usage of many wetlands throughout the world, particularly in Central and South America (Rosete et al. 1991, Smardon 2006). If tourism-generated revenues are returned to local peoples and/or community-based organizations (CBOs) these funds can be used to maintain wetland-dependent livelihoods or at least divert land use activities that would have a deleterious impact on the wetland ecosystem. However, it remains to be seen whether ecotourism or nature tourism is sustainable.

Wetland conversion and wetland drainage goes ahead despite the possibility that greater benefits might come from more carefully considered management and exploitation. Developed countries have apparently not learned from their centuries of experience. The Irish Peat Board, for instance, argues that any ecological damage brought about by peat mining in Ireland is a small price to pay for reduced import bills and an improved standard of living.

In the People's Republic of China, multiple use ecosystems have been established via measures adopted for local management of wetlands. In Vietnam a very alert and adaptable approach to wetlands use has resulted in a change from intensive cultivation of rice on recently drained grounds to the cultivation of less-intensive crops, which are more appropriate to the principle of wise use and sustainability. But, at the same time, there is no legal recognition of jurisdiction of wetlands for government or private ownership in Vietnam.

The same mistakes are being exported to the developing world, where many of the biggest wetland conversion projects are being carried out with foreign aid. The Netherlands, which has a longer history of expertise in land drainage than most other countries, has financed drainage surveys in Zambia and Jamaica. Swedish and Finish funding has supported a prospective peat mining project in Jamaica, and Japanese money went into a plan to drain Jamaican wetlands for agriculture (Maltby 1986).

In the late 1970s the World Bank financed feasibility studies and the preparation of plans to drain and divert to agriculture 570,000 ha of wetlands in south Sumatra and central Kalimantan. Between 1981 and 1984 the World Bank loaned $87 million for two Indonesian swamp reclamation projects, which together resulted in the drainage of 39,000 ha of wetlands for agricultural use and resettlement. The hydrological disruption, peat subsidence, and acid sulfate soil problems that have resulted from this reclamation have caused major ecological and environmental degradation. In some cases the land has been abandoned, but the extent of damage to water quality, fisheries, and wildlife habitats remains largely unknown.

In the mid-1970's the Inter-American Development Bank (IDB) partly financed the drainage and conversion of agriculture of 2,000 ha of Jamaica's Black River Upper Morass. Earlier, in the same decade, the IBD loaned $50 million toward agricultural conversion of 165,000 ha of marshland in Mato Grosso State in Brazil and $95 million for drainage and irrigation of 81,000 ha of wetlands in Guyana. Finally we have the grandfather of all development schemes in the early 1990s, which was the drainage and proposed canal system, which would cut through the Pantanal – which is the largest wetland region in South America – covering land area in three countries.

Some of this practice has been slightly reversed in the 1990s but the world's poorer countries are often caught in a conflict of interest. Some LDC "environmentalists" pull no punches and argue that there is no vested interest in a constancy that does not serve well the aspirations or needs of the people. They further argue that some destructive development must be allowed, and that the environment is basically resilient and tolerant of a certain degree of impact. Furthermore, they argue that improvement in the living standards and national wealth of the LDCs is urgently needed. The industrialized nations of the world gained their wealth at the expense of earlier wetland destruction are difficult to counteract or argue with.

Efforts of the developed world must concentrate not only on means of enhancing wetland management to optimize their sustainable utilization but also on means of preventing uncontrolled financing of schemes that will lead to wetland destruction. The following chapters will pinpoint areas where such successful and unsuccessful efforts have occurred and what we should learn from them.

Non-governmental organizations (NGOs) have a major role to play in lobbying for the wise use of development funds and for careful management of wetlands. In 1985, the World Wildlife Fund (WWF), the International Union of the Conservation of Nature (IUCN) launched a campaign at promoting better public awareness about wetlands and their importance. A major goal of the campaign is to ensure that wetland development goes ahead only when all the implications are understood and when plans have been made to ensure that negative environmental impact is minimized. This campaign has been followed by subsequent campaigns along similar themes.

Sound management and conservation of wetlands are very important in the developed world, where so little of the original wetlands area remains and where concerns for environmental quality remain high. Wetlands are, however, crucial in the developing world, where survival of people as well as ecological and genetic resources is linked inextricably with wetland functioning.

Although there are major structural differences in the management requirements for wetlands in different parts of the world, it would be naïve, to separate entirely the issues of wetland protection and management in developed nations from those of developing nations. International development and technology aid in the role of funding agencies in remote wealthy nations are important factors influencing the survival of wetlands in developing nations.

The climate for action is still positive. This is reflected in the current high profile of wetland scientific research, in the specific activities of national governments, and by the increasing influence of NGOs such as WWF, IUCN, Wetlands International, and the Ramsar Bureau. In Europe, the CEC's Directorate General XI (Environment) has taken major initiatives in investigating the problems of the management and protection of the coastal wetlands around the Mediterranean Sea. Politicians worldwide have often exploited the "green" label – in some cases with substantive results and media coverage of environmental issues is fairly constant, especially with environmental and economic implications of regional climate change and the roles of wetlands. So this is the opportunity for pushing the importance of effective wetland management worldwide.

But, before we take action, we should be aware of the respective roles of government, international, national, and regional NGOs, local community-based organizations (CBOs), and specific livelihood linkages to wetland resources locally. It is author's thesis that effective wetlands management is strongly linked to the interchange of government, NGO, and CBO roles plus local residents wise use of wetlands internationally. Only when we can understand how these roles and linkages work we can have lessons to impart about effective wetland management policy. There have been very good international guidance provided by the Ramsar Bureau, Wetlands International, IUCN, and various authors (Dugan 1990, Maltby 1991, Rubec 1989).

Most recently the Millennium Ecosystem Assessment has specifically addressed "Ecosystems and Well Being: Wetlands and Water" (Millennium Assessment 2005). Within the summary for decision makers, the authors of the report stress wetland services and human well-being including services for "those living near wetlands [that] are highly dependent on those services and are directly harmed by their degradation" (Millennium Assessment 2005, p. 1). Other services listed include water purification and detoxification of waste, climate regulation, mitigation of climate change and cultural services. The report also reviews status and trends of wetlands, causes of wetland loss and degradation, explore four possible scenarios for plausible futures as well as potential responses to these alternate futures. The drivers of wetland stress and change are very much as we have already seen in this introductory chapter. The new emphasis in the Millennium Assessment (2005) is (1) the sustainable possibilities and tradeoffs for groups utilizing wetlands for food fiber and fuel and (2) wetlands role in climate change amelioration and resultant stress on wetland systems from climate change.

We need to take a closer look at cases where all the actors and linkages are at play and we can attempt to identify what is working or not with sustainable wetlands management.

The case studies were chosen as being roughly geographic representation of major wetlands systems in Europe, Africa, Asia, North America, and Latin America/Caribbean. They were also chosen because there was significant NGO involvement and there was substantial access to background information on wetland management history, The author has direct knowledge of the Axios River Delta in Greece, Great Lakes Coastal Wetlands in the United States and Canada, Ria Celestun and Ria Lagartos estuarine lagoons in Mexico and Mankote mangrove in St. Lucia. Interviews were done with major participants for the Tran Trim Nature Preserve in the Mekong Delta, Vietnam, Great lakes coastal wetlands in Canada and the US, and the Mankote mangrove in St. Lucia. Major local actors were asked to review earlier versions of the case studies. Within each case study there is usually a regional policy context, a history documenting how changes have occurred to the respective wetland, detailed listing of dominant flora and fauna to show how the wetland has changed as well as documentation of the respective roles of key individuals, organizations, and other stakeholders affecting wetland management decisions.

Plan of the Book

This is the purpose of the middle eight chapters within this book – to look at specific case studies around the world – to see whether we can draw some inferences about sustainable wetlands management from environmental, economic, and social perspectives. The final chapter will be a summary of lessons learned. The following outlines some of the highlights of succeeding chapters.

Chapter 2 presents the development of the Tripartite Management Plan for the Wadden Sea in the Netherlands, Germany, and Denmark. This case study illustrates the power of the grass routes friends of the Wadden Sea in the Netherlands and how three countries can come together to manage a large regional productive wetland with many uses and functions.

Chapter 3 illustrates the plight of a highly stressed Axios River Delta on the Mediterranean in Greece. The case study follows the development of the WWF-IUCN-University of Thessaloniki Action Plan, what has worked and what has not and some implications for Mediterranean wetlands in general.

Chapter 4 covers the tragic demise of the Kafue Flats in Zambia and how hydroelectric development has forever changed the ecological character of these riverine grasslands and the linkages to local livelihoods. There are some glimmers of hope for local cooperative management of a changed ecosystem.

Chapter 5 presents the East Kolkata Lagoon System for water treatment, which is very innovative from economic and social perspectives. Also included will be the use of this same system for urban ecotourism and aquaculture.

Chapter 6 covers the creation of the Tram Trim Nature Reserve in the Mekong River Delta in Vietnam. In this case the International Crane Foundation works with Vietnamese national and local authorities to negotiate a management plan that balances crane habitat protection vs. rice and fisheries production as well as recreates the hydrology of the Mekong.

Chapter 7 reviews the creation of the bi-national Great Lakes Wetlands Policy Consortium and the outcomes of such in the United States and Canada. It also reviews common wetland management issues in North America.

Chapter 8 is a presentation of wetland management issues of the Yucatan Mexico coastal lagoons of Ria Celestun and Ria Largartos. These coastal hypersaline and mangrove fringed lagoons are highly stressed and even Ramsar recognition and biosphere reserve status does little to reduce the stress. You will be surprised as one local fishing village literally takes the "bull by the horns" to solve their own resource dilemma.

Chapter 9 finishes the case studies with the story of a small mangrove wetland in St. Lucia typical of mangroves throughout Latin America and the Caribbean. This is a story of local innovation for sustainable charcoal production while maintaining key habitat areas in the mangrove wetland.

Chapter 10 summarizes international, regional, and site-specific issues presented in the case studies and also summarizes effective innovations or major barriers to sustainable wetland management.

Acronyms

CBO: community-based organization
EC: European Commission
CWA: US Clean Water Act
IDB: Inter-American Development Bank
IUCN: International Union for the Conservation of Nature
IWRB: IWRB
LDCs: lesser development countries
NGO: non-government organization
SWANCC: Solid Waste Agency of Northern Cook County vs. US Army Corps of Engineers
WWF: Worldwide Fund for Nature-World Wildlife Fund

References

Adamus, P. R. and L. T. Stockwell. 1983. *A Method of Wetland Functional Assessment.* US Federal Highway Administration, Washington, DC, Vol. 1 Report FHWA IP-82-83, Vol. II, Report FHWA-IP-82-24.

Adamus, P. R., E. Clairain, E. J. Smith, and R. E. Young. 1987. *Wetland Evaluation Technique (WET)*, Vol. 2: Methodology, Operational Draft. Vicksburg, MI, US Army Corps of Engineers Waterways Experiment Station.

Ambasht, R. S. and N. K. Srivastava. 1994. Restoration strategies for the degrading Rihand River and Reservoir ecosystems in India. In W. J. Mitsch (ed.) *Global Wetlands: Old World and New*, pp. 725–728. Amsterdam: Elsevier.

Azarth, J., P. Banth, H. Azarth, and V. Selvan. 1988. Impact of urbanization on the status of mangrove swamps in Madras. In D. D. Cook et al. (eds.) *The Ecology and Management of Wetlands, Volume 2: Management Use and Value of Wetlands*, pp. 225–233. Portland, Oregon: Timber Press.

Baca, B. J. and J. R. Clark. 1988. Coastal management practices for prevention of future impacts on wetlands. In D.D. Cook et al. (eds.) *The Ecology and Management of Wetlands Volume 2: Management Use and Value of Wetlands*, pp. 28–44. Portland, Oregon: Timber Press.

Baldock, D. 1984. *Wetland Drainage in Europe.* London: IIED/IEEP.

Bobblink, R., D. F. Whigham, B. Beltman, and J. T. A. Verhoeven. 2006. Wetland functioning in relation to biodiversity and restoration in Wetlands. In R. Bobblink, B. Beltman,

J. T .A. Verhoeven, and D. F. Whigham (eds.) *Functioning, Biodiversity Conservation and Restoration, Ecological Studies Volume 191*, pp. 1–12. New York: Springer.

Brinson, M. M. and A. E. Lugo. 1989. Tropical wetlands: An overview of their distribution and diversity. In J. A. Kusler and S. Daly (eds.) *Wetlands and River Corridor Management*, pp. 83–89. Berne, New York: Association of State Wetland Managers.

Brown, S. 1990. Preserving Great Lakes wetlands: An environmental agenda. In J. Kusler and R. Smardon (eds.) *Wetlands of the Great Lakes*, pp. 319–331. Berne, New York: Association of State Wetland Managers.

Carp, E. 1980. *Directory of Wetlands of International Importance in the Western Palearctic.* Gland, Switzerland: International Union of the Conservation of Nature.

Coles, B. J. and J. M. Coles. 1989. *People and Wetlands, Bogs, Bodies and lake Dwellers.* London: Thames and Hudson.

Dugan, P. J. 1988. The importance of rural communities in wetlands conservation and development. In D.D. Cook et al. (eds.) *The Ecology and Management of Wetlands Volume 2: Management Use and Value of Wetlands*, pp. 3–11. Portland, Oregon: Timber Press.

Dugan, P. J. 1990. *Wetland Conservation: A Review of Current Issues and required Action.* Gland, Switzerland: IUCN The World Conservation Union.

Faust, B. and J. Sinton. 1991. Let's dynamite the salt factory: Communications, coalitions and sustainable use among users of a biosphere reserve. In J.A. Kusler (eds.) *Ecotourism and Resource Conservation*, pp. 602–624, Vol. II. Madison WI: Omni press.

Finlayson, M. 1989. Plant ecology and management of an internationally important wetland in monsoonal Australia. In J. A. Kusler and S. Daly (eds.) *Wetlands and River Corridor Management*, pp. 90–98. Berne, New York: Association of State Wetland Managers.

Finlayson, M. and M. Moser. 1991. *Wetlands.* Gloucester, UK: International Waterfowl and Wetland Research Bureau (IWWRB).

Gruenwald, G. 1990. Recommendations of the Great Lakes Wetlands Policy Consortium. In J. Kusler and R. Smardon (eds.) *Wetlands of the Great Lakes*, pp. 17–18. Berne, New York: Association of State Wetland Managers.

Haq, S. 1989. Protection of Gangetic wetlands in Bangladesh. In J.A. Kusler and S. Daly (eds.) *Wetlands and River Corridor Management*, pp. 116–118. Berne, New York: Association of State Wetland Managers.

IUCN. 1985. *Wetlands Conservation Programme.* Gland, Switzerland: IUCN.

IUCN. 1980. *The Ramsar Convention: A Legal Review, Conference on the Convention of Wetlands for International Importance Especially as Waterfowl Habitat.* Conf. 5, Cagliari, Italy, Nov. 24–29, 1980. Gland, Switzerland: IUCN.

IUCN. 1990. *Conservation and Management of Greek Wetlands*, Thessalanki Workshop Proceedings. Gland, Switzerland: IUCN.

IUCN/IWRB.1980. *The Ramsar Convention: A Technical Review: Conference on the Convention of Wetlands of International Importance Especially as Waterfowl Habitat.* Conf. 4, Cagliari, Italy, Nov. 24–29, 1980. Gland, Switzerland: IUCN.

IUCN/IWRB. 1984. *Overview of National Reports Submitted by Contracting Parties and Review of Developments Since the First Conference of the Parties.* Doc. C2.6, Groningen, Netherlands May 7–12, 1984.

Kusler, J., P. Parenteau, and E. A. Thomas. 2006. "Significant Nexus" and Clean Water Act Jurisdiction, Draft discussion paper, 26 pp.

Kusler, J. and T. Opheim. 1996. *Our National Wetland Heritage: A Protection Guide.* Environmental law Institute, Washington, DC, 149 pp.

Kvet, J., J. Jenik, and L. Soukupova. 2002. *Freshwater Wetlands and Their Sustainable Future: A Case Study of the Trebon Basin Biosphere Reserve, Czech Republic.* UNESCO and the Parthenon Publishing Group and CRC Boca Raton, 495 pp.

Lambertson, R. 1990. Fish and Wildlife Service Initiative and the North American Waterfowl Management Plan. In J. Kusler and R. Smardon (eds.) *Wetlands of the Great Lakes*, pp. 10–13. Berne, New York: Association of State Wetland Managers.

Leander, B. and L. de Mare. 1994. Management of wetlands in Sweden: legal prerequisites and constraints. In W. J. Mitsch (ed.) *Global Wetlands: Old World and New*, pp. 625–636. Amsterdam: Elsevier.

Loftus, K.K., R. C. Smardon, and B. Potter. 2004. Strategies for the stewardship and conservation of Great lakes coastal wetlands. *Aquatic Ecosystem Health and Management*, 7(2): 305–330.

Maltby, E. 1991. Wetland management goals, wise use and conservation. *Landscape and Urban Planning*, 20(1–3): 9–18.

Maltby, E. 1986. *Waterlogged Wealth: Why Waste the World's Wet Places*. London, UK: Earthscan.

Marchand, M. and H. A. Udo de Haes (eds.). 1990. *The Peoples Role in Wetland Management*. Leiden, The Netherlands: Center for Environmental Studies, Leiden University.

Marchand, M. and H. A. Udo de Haes (eds.). 1991. The Peoples Role in Wetland Management: Wetlands Special Issue. *Landscape and Urban Planning*, 20(1–3): 1–276.

Mathews, G. V. T. 1993. *The Ramsar Convention on Wetlands: Its History and Development*. Gland, Switzerland: Ramsar Convention Bureau, 120 pp.

Millennium Ecosystem Assessment. 2005. *Ecosystems and Human Well Being: Wetlands and Water Synthesis*. Washington, DC: Water Resources Institute, 70 pp.

Mitsch, W. J. and J. G. Gosselink. 2000. *Wetlands*. 3rd ed., New York: John Wiley and Sons, 920 pp.

Mitsch, W. J., R. H. Mitsch, and R. E. Turner. 1994. Wetlands of the Old and new Worlds: Ecology and management. In Mitsch, W. J. (ed.) *Global Wetlands: Old World and New*, pp. 1–55. Amsterdam: Elsevier Science.

Navid, D. 1988. Developments in the Ramsar Convention. In D. D. Cook et al. (eds.) *The Ecology and Management of Wetlands Volume 2: Management Use and Value of Wetlands*, pp. 21–27. Portland, Oregon: Timber Press.

Nelson, R. W., R. S. Ambasht, C. Ameros, G. W. Begg, A. A. Bonetto, I. R. Wais, E. Dister, E. Wenger, C. M. Finlayson, J. K. Handoo, A. K. Pandit, K. M. Mavuti, D. Parish, and P. Savey. 1989. River Floodplain and Delta Management Team: A project of the Worlds Wetlands Project. In J. A. Kusler and S. Daley (eds.) *Wetlands and River Corridor Management*, pp. 75–82. Berne, New York: Association of State Wetland Managers.

Quesada, C. A. and J. A. Jimenz. 1988. Watershed management and a wetlands conservation strategy: The need for a cross-sectoral approach. In D. D. Cook et al. (eds.) *The Ecology and management of Wetlands Volume 2: Management Use and value of Wetlands*, pp. 12–20. Portland, Oregon: Timber Press.

Ramsar Convention Bureau. 1984. *Proceedings of the Second Conference of the Contracting Parties*. Groningen Netherlands, May 7–112 1984. Gland, Switzerland: Ramsar Convention Bureau.

Ramsar Convention Bureau. 1988. *Proceedings of the Third conference of the Contracting Parties*. Regina Canada, May 27–June 5 1987. Gland, Switzerland: Ramsar Convention Bureau.

Rosete, R. M., R. C. Smardon, and S. Moan. 1991. Developing principles of natural and human ecological carrying capacity and natural disaster risk vulnerability for application to ecotourism development in the Yucatan, Peninsula. In J. A. Kusler (ed.) *Ecotourism and Resource Conservation*, pp. 740–752, Vol. II. Madison, WI: Omni press.

Rubec, C. D. A. 1989. Wetland science networking and coordination of international concerns. In J. A. Kusler and S. Daly (eds.) *Wetlands and River Corridor Management*, pp. 32–36. Berne, New York: Association of State Wetlands Managers.

Rubec, C. D. A. 1994. Canada's Federal policy on wetland conservation: A global model. In W. J. Mitsch (ed.) *Global Wetlands: Old World and New*, pp. 537–554. Amsterdam: Elsevier.

Smardon, R. C. (ed.). 1983. *The Future of Wetlands: Assessing Visual-Cultural Values*. Totowa, NJ: Allenheld-Osmun Press.

Smardon, R. C. 2003. The role of nongovernmental organizations for sustaining wetland heritage values. In M. Gavari-Barbas and S. Guihard-Anguis (eds.) *Regards Croises sur le*

Patrimoine dans le Monde a L'Aube du XXXI Siecle, pp. 785–815. Presses de L'Universite de Paris-Sorbonne.

Smardon, R. C. 2006. Heritage values and functions of wetlands in Southern Mexico. *Landscape and Urban Planning*, 74(3–4): 296–312.

Smardon, R. C. and B. B. Faust. 2006. Introduction: international policy in the biosphere reserves of Mexico's Yucatan peninsula. *Landscape and Urban Planning*, 74(3–4): 160–192.

Smart, M. and K. J. Kanters. 1991. Ramsar participation and wise use. *Landscape and Urban Planning*, 20(1–3): 269–274.

The Conservation Foundation. 1988. *Protecting America's Wetlands: An Action Agenda: Final Report to the National Wetlands Policy Forum*. Washington, DC: The Conservation Foundation.

Toledo, A., A. V. Botello, M. Herzog, and F. Contreiss. 1989. Environmental Studies on wetlands of Coatzacoalcos, Veracruz state, Mexico. In J.A. Kusler and S. Daly (eds.) *Wetlands and River Corridor Management*, pp. 102–107. Berne, New York: Association of State Wetland managers.

Tsiouris, S. E. and P. A. Gerakis. 1991. *Wetlands of Greece: Values, Alterations, Conservation*. Thessalanki: WWF Laboratory of Ecology and Environmental Protection, Faculty of Agriculture, Aristotelian University of Thessalanki IUCN.

Turner, K. and T. Jones (eds.). 1990. *Wetlands: Market and Intervention Failures*. London: Earthscan Publications Ltd, 202 pp.

Verhoeven, J. T. A., B. Beltman, F. F. Whigham, and R. Bobblink. 2006. Wetland functioning in a changing world: Implications for natural resource management. In Verhoeven, J. T. A., B. Beltman, R. Bobblink, and D. F. Whigham (eds.) *Wetlands and Natural Resource Management, Ecological Studies Vol. 190*, New York: Springer, pp. 1–12.

Vileisis, A. 1997. *Discovering the Unknown Landscape: A History of America's Wetlands*. Washington, DC: Island Press.

Whigham, D. F., D. Dyjova, and S. Hejny. 1993. *Wetlands of the World: Inventory, Ecology and Management*. Dordrecht, The Netherlands: Kluwer, 768 pp.

Williams, M. (ed.). 1990. *Wetlands: A Threatened Landscape*, Oxford, UK: Basil Blackwell, 419 pp.

World Wildlife Fund. 1992. *Statewide Wetland Strategies: A Guide to Protecting and Managing the Resource*. Washington, DC: Island Press.

Cited International Wetland Web Sites

Conservation International: http://www.conservation.org

International Corporate Wetlands Restoration Project (ICWRP): http://www.icwrp.org

Ramsar Convention for Wetlands: http://www.ramsar.org

US Fish and Wildlife Service, Wildlife without Borders: http://www.fws.gov/international/ramsar/facts_sheet.htm

Wetlands International: http://www.wetlands.org

Wetlands and Water Resources Program – IUCN: http://www.iucn.org/themes/wetlands/

WWF's Climate Change Program: http://www.panda.org/news_facts/newsroom/index.cfm

Chapter 2
The Wadden Sea Wetlands:
A Multi-jurisdictional Challenge

Introduction

This is a story of a very large coastal wetland complex bordering the North Sea plus three countries in Europe. Intertwined with a multi-jurisdictional management issues is the role of several NGOs most notably the Society for the Preservation of the Wadden Sea. This case study will present the wetland resource, the various threats to the resource, a three-country institutional context, and finally the role and history of the NGOs involved.

The Wadden Sea covers an area of 8,000 km², half of which is tideland and an additional 1,000 km² made up of the Wadden Islands. More than half (60%) of the tideland found between Europe and North Africa to the mangrove coasts is situated in the Wadden Sea. The sea is bounded by three countries: the Netherlands, Germany, and Denmark and sits between Den Holder in the Netherlands and Esberg in Denmark (see Fig. 2.1).

Historical Overview

Humans have interacted with the Wadden Sea since its origin 7,500 years ago. Exploitation, habitat alteration, and pollution have strongly increased since the Middle Ages, affecting abundance and distribution of many marine mammals, birds, fish invertebrates, and plants. Large whales and some large birds disappeared more than 500 years ago. Most small whales, seals, birds, large fish, and oysters were severely reduced by the late nineteenth and early twentieth centuries, leading to the collapse of several traditional fisheries (Lotze 2005).

Since 1600 the surface area of the Dutch Wadden Sea has decreased by successive reclamation of salt marshes. In 1933 the Zuiderzee (3,200 km²) was closed off from the Wadden Sea causing an increase in tidal range and current velocities in the remaining parts. In 1969 the Lauwerzee (91 km²) was closed off and turned into a freshwater lake. Dredging in harbors and shipping routes as well as extraction of sand and shells became common practice and contributed to turbidity of the Wadden Sea. Discharge of nitrogen and phosphorus into the western Wadden Sea increased manifold since 1950 causing an increase in

R.C. Smardon, *Sustaining the World's Wetlands*,
DOI 10.1007/978-0-387-49429-6_2, © Springer Science+Business Media, LLC 2009

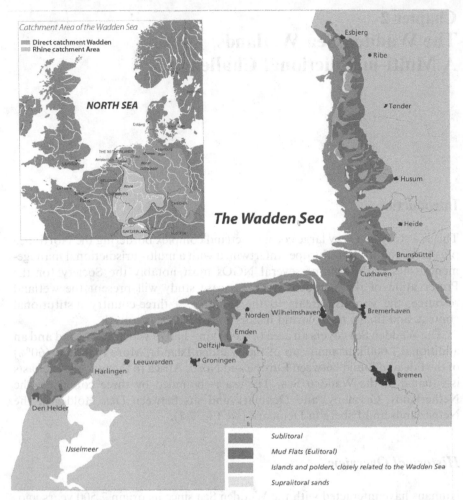

Fig. 2.1 (a) Catchment area of the Wadden Sea and The Wadden Sea: Sublittoral, mudflats, islands, and supralittoral areas redrawn by Samuel Gordon. Sources: Adapted from Common Wadden Sea Secretariat, Undated. *The Wadden Sea: A Shared Nature Area*, p. 3, and WWF, 1991. *The Common Future of the Wadden Sea*, p. 57

phytoplankton production, duration of phytoplankton blooms, and intertidal macrozoobenthic biomass (de Jonge et al. 1993, Swennen 1989).

Fisheries changed drastically since the 1930s. Fishing in the Zuiderzee herring came to an end shortly after closing off the Zuiderzee. The anchovy fishery ceased in 1960 and that of the flounder in 1983 (de Jonge 1993). Under-sized brown shrimps were fished until 1971 and selective shrimp trawls and sorting devices with flushing seawater were introduced to reduce mortality among young flatfish and shrimp. Oysters became extinct in the 1960s due to over-exploitation of natural beds. Production of mussels increased more than

Fig. 2.1b and c Aerial photo of part of the Danish Wadden. Source: Common Wadden Sea Secretariat, Undated. *The Wadden Sea: A Shared Nature Area*, p. 1

10 times between 1950 and 1961 due to "culturing", and catches of cockles increased slowly between 1955 and 1984. Whelks were fished until 1970 (de Jonge et al. 1993).

The most important changes in the biotic system of the Wadden Sea (de Jonge et al. 1993) were increased production of microalgae and intertidal macrozoobenthos which can be attributed to increased nutrient loads. Eutrophication provided ample food supply for mussels, which were harvested mainly by man and eider duck, and may have caused increased growth rates in juvenile plaice. Increased turbidity may have impaired life conditions for adult dab and assisted in recovery of substantial eelgrass beds after their disappearance in the 1930s (de Jonge et al. 1993, Swennen 1989).

Increased turbidity in the Wadden Sea is probably caused by the closing of the Zuiderzee in 1931 by a significant increase of dredge spoil disposal near Hoek van Holland between 1970 and 1983 and by more than a 10-fold increase in mussel culturing since 1950 (de Jonge et al. 1993). Stocks of several bird species breeding in the Wadden Sea area suffered great losses in the early 1960s due to pesticides. Most of the populations have recovered.

The Wadden Sea Physical Environment

One of the key characteristics of the area is the great tidal variation from 1.36 m in Den Helde in the Netherlands to 3.43 m in Husun in Germany. This variation in conditions is instrumental to the exceptional diversity and wealth of flora and fauna. The coastal landscapes and dunes rank among Europe's most beautiful places. On the islands, more than 900 different plant species, 300 moss species, 350 species of lichen, and 650 species of fungi occur.

There is also an abundance of birds and the area plays a vital part in the survival of about 50 different species, originating from the larger part of the northern hemisphere: from northeastern Canada, Greenland, and Spitzbergen up to central Siberia. Estimates have shown that there are 9.3 million herbivorous water birds utilizing the area for foraging and migratory rest stop. Species that can be seen include barnacle geese, osprey, spoonbills, sheldrake, avocet, sandwich terns, sandpipers, bar-tailed godwits, and oyster catchers (Smit 1989). One of the current issues is the conflict between commercial shellfish fishing and shellfish-dependent birds (Kees 2001, Verhulst et al. 2004, van Eerden et al. 2005, Goss-Custard et al. 2004, van Berkel and Revier 1991).

Then there are the fish. Estimates indicate that the Wadden Sea has an average fish density of one fish per square meter, which means billions of fish. The catch taken in the North Sea is considerable. Eighty percent of all plaice and 50% of all sole caught in the North Sea grew up in the Wadden Sea, representing annual turnover of many hundreds of millions in dollars (de Jonge et al. 1993, Swennen 1989).

At the incoming tide large shoals of fish, mainly flat fish, like plaice, flounder, dab and sole, spread over the inundated sandbars to look for food. They mainly feed on the smaller shellfish, worms, shrimps, and crabs. The fish in the Wadden Sea can be classified into several groups: sedentary fish such as eelpout, butterfish, and scorpion fish spend their lives in the Wadden Sea. Migrants, including flounder, garfish, and gray mullet, visit the mud flats only in a certain period, mostly in the summer. Several species of fish find themselves as occasional visitors to the North Sea (de Jonge et al. 2006, Swennen 1989).

The fisheries in the Wadden Sea concentrate on mussels, cockles, and shrimps. Mussels are cultivated in the western part of the Wadden Sea. A management problem is that mussel and cockle fishing seriously disturbs wildlife in the area (Goss-Custard et al. 2004, Verhulst et al. 2004). Natural mussel beds have vanished with the removal of the mussel seed. The cockle fishers cause disturbance of benthic species on the bottom. As a result 25% of the flats in the Wadden Sea have been closed for mussel and cockle fisheries, while effects of shrimp fishing are still being investigated. Fishing licenses have been limited and there is some minor fishing activity for eel, sole, gray mullet, and smelt.

In terms of biodiversity of the Wadden Sea we have a few definitive studies. Wolff (2000) examined various causes of expiration of marine and estuarine species within the Wadden Sea and their relative importance. He obtained data from geological, archeological, historical, and biological publications. According to Wolff (2000) at least 10 species of algae, 10 invertebrates, 13 fish, 5 birds, and 4 marine mammals became extinct during the past 2,000 years. Habitat destruction played a part in 26 cases, over-exploitation in at least 17 cases, and pollution in at least three cases. According to Fog et al. (1996) eight species of amphibians and four species of reptiles are threatened in at least one subregion of the Wadden Sea. Of these, seven species of amphibians and all four species of reptiles are threatened for the entire area and are therefore placed on the International IUCN Red list.

The only mammal left in the Dutch coastal waters is the seal (see Fig. 2.2). Its reason for staying in the Wadden Sea is also the abundant food stocks, the peace, and the space still to be found there. In summer the females have their young on the high exposed sand bars. They also use these sand bars as places to rest. In the 1950s there were still about 2,500 seals in the Dutch Wadden Sea, but their number rapidly declined as a result of hunting and human disturbance, and later water pollution. After reaching a low of 350 animals in 1975, their number increased to about 1,000 in 1988. In that same year, a virus disease attached to the colony and in combination with water pollution decimated the animals to 350 in 1989. The seal has become an indicator of environmental quality and its numbers have increased to almost 1,200 in 1994.

In 1962 the Netherlands prohibited seal hunting. Germany and Denmark followed the lead in 1973 and 1976, respectively. The places where seals used to

Fig. 2.2 Seals on an offshore shoal. Source: Common Wadden Sea Secretariat, Undated. *The Wadden Sea: A Shared Nature Area*, p. 3

rest are under additional protection. The Netherlands has two resorts for seals: one on the island of Texel, in the research center of EcoMare, and the other a village of Pieterburen. Germany has one in Norden and one in Fridrichskoog, and Denmark one in Esbjerg.

The major landscape features heading back from the waters edge are salt marshes, islands, dunes, and embankments around dike edges (see Fig. 2.3). Much of the original marsh was destroyed by reclamation, but new salt marsh has also been created due to natural siltation and accretion processes. The salt marshes are extremely productive or fertile and are valuable as pastures for farmers at the seaside. These same farmers have been trying to stimulate the formation of salt marshes and these methods vary from country to country. In 1930 the Netherlands took over the Schleswig–Holstein method, which implies the stimulation of silt deposit by ditches and osier dams. When the deposit had become high enough a dike was constructed and so a new plodder had been created. Now and again a newly "reclaimed" salt marsh was protected against further influence of the sea by a low dike. A similar salt marsh is called a "summer Plodder". In the 1960s the reclamation of the Dutch Wadden area was stopped. Only maintenance of reclamation works is kept up. There is also experimentation with different species such as *Juncus* and *Phragmites* for brackish marsh creation (Bakker et al. 1993, Huiskes 1988) (Fig. 2.4).

Fig. 2.3 Typical estuarine pattern. Source: Common Wadden Sea Secretariat, Undated. *The Wadden Sea: A Shared Nature Area*, p. 7

The long chain of islands and high sandbars, approximately 50 in number, characterizes the European Wadden area. Most of these islands were formed after the last ice age from the beach ridges along the coast. Windblown sand made these ridges higher and the spreading vegetation settled the newly

Fig. 2.4 Coastal dunes. Source: Common Wadden Sea Secretariat, Undated. *The Wadden Sea: A Shared Nature Area*, p. 12

developed dunes. Not all islands developed in this way. The Halligen in the Schleswig–Holstein area are remains of an extensive area of salt marshes. The Danish Wadden Islands were also formed on sandbars; the wide beaches are the result of the enormous transport of sand in this part of the Wadden Sea.

Dunes are not only formed by the wind piling up loose sand,but the sand may also be blown away again, unless plants hold it. Sometimes the sea washes away large parts of the dunes during a gale. In the Netherlands, this is a real problem on the island of Texel. When the dunes protect the land from the sea, much effort is being made to keep them as they are.

Marram is planted and reed mats are put up to prevent the dunes from eroding. Longitudinal dikes are also built across the beach to ward off the current. This was done on Vieland. So the beach holds its initial width and the waves can only wash dunes during extremely heavy storms. On Texel the beach is raised with new layers of sand to protect the dunes. The west sides of most of the islands of Lower Saxony have been "embedded in concrete" by heavy dikes. In the Netherlands, by contrast, it is possible to keep the coastal strip more dynamic. The key is that dune land variation in lime, lime limited, wet and dry creates the variation and diversity in vegetation. Dunes also function as fresh water collection devices.

For centuries embankment of land outside the dikes was common practice in the Wadden area. Creeks were cut off in order to improve the protection of the hinterland. It also made more soil available for farming and cattle breeding and more recently for industrial and military activities. The land outside the dikes, however, is also of great importance for wildlife as it provides many bird species with grounds to feed, rest, and breed. In recent years the motives for embankment of new land outside the dikes came under great pressure, because there is increasing demand for farmland, and embankments demand great financial sacrifices.

Due to the efforts of conservationist's organizations like the Wadden Society the general public became aware of the natural values of the land outside the dikes. In the course of time this had a political effect in conjunction of large-scale plans for embankments along the coastline of North Holland (Balgzand), Friesland (North Friesland outside the dikes), Groningen (North Groningen and the Dollart), and Germany (Dollart, Tumlauer Bucht).

Specific Wetland Resource Management Issues and Threats

The following sections outline the major events affecting the Wadden Sea wetlands for the Dutch Wadden including the Lauwersmeer and Ens-Dollart area, the German Wadden including Lower Saxon, Elbe, and Schleswig–Holstein areas, and the Danish Wadden including Skallinger, Varde A, Romo and Fano, and Esbjerg. Major source material for this section is from Wadden Society (1994) and WWF (1991).

The Dutch Wadden

The western and eastern parts of the Dutch Wadden area show great differences. The Wadden Sea between Den Helder, Vieland, and Harlingen is much deeper than the eastern part. Therefore the surface of the sandbars being uncovered in the eastern Wadden is larger. At low tide the ferries heading for Ameland and Schiermonnikoog sail in gullies between the emerging sand bars.

Because of the peace, space and the beautiful landscape in the islands are ideal holiday resorts. The problems involved with recreation on the Wadden Islands are numerous. Vulnerable dune land had to be closed to the public. The flow of tourists created the need for all kinds of additional facilities, including housing, water supply, waste removal, and transport. The Wadden Society is of the opinion that recreation should not expand, but should be stabilized at the present level. Fortunately this same view is held by most of the people on the Wadden Islands.

The first inhabitants of the Wadden area could only maintain themselves by building their houses on man-made mounds. By the beginning of the second century the first dikes were built when the connecting roads between the mounds and the walls of the salt marshes were leveled. The monasteries in the area have always played an important part in dike construction. In the Frisian Wadden the Portuguese landlord Caspar di Robles took the initiative to improve dike maintenance during the Eighty Year's War. The delta project drawn up after the tragic Zeeland flood in 1953 also included raising the dikes in the Wadden area.

In the course of this century recreation on the Wadden Islands, where many landscapes are combined (beach, dune land, salt marsh, woodland, and plodder), has developed into an important means of existence. This led to drastic changes. Several farmers decided to close their farms and become "recreational farmers". Water sports have also expanded enormously in the Wadden Sea. During high season the islands harbor so many tourists that the total number of inhabitants is increased 10-fold. On the one hand, recreation affects nature and landscape, on the other hand it has focused the attention on the Wadden area with favorable effect on conservation and protection. In recent years there is a tendency to expand the season in order to reduce the flow of tourists during the high season.

The Wadden Sea is very attractive for water sportsmen. The number of yachts is still increasing, and several Wadden Islands decided to enlarge their marinas and the effects are not all positive. Careless water sportsman can seriously disturb natural areas at critical times. Seals are very vulnerable in summer when their young are born. This also applies for breeding, roosting, and foraging birds. Since 1981 the number of areas coming under the Nature Conservation Act has largely been extended. Some parts may not be entered by boat or otherwise and sometimes entering is only allowed for nature study or research.

Lauwersmeer

In 1969 the Lauwersmeer was separated from the Wadden Sea by a dike with the intention of improving the drainage of the provinces of Groningen and Friesland. This dike has been provided with a lock and a drainage sluice. The result was the Lauwersmeer, a hinterland consisting of land and water. The Lauwersmeer area is important for all kinds of migratory birds, such as geese. There are many kinds of recreational facilities, especially for water sports. The military exercise ground that has been established there does not fit in with areas so near the Wadden Sea and near recreational activity.

In 1965 a plan was launched to connect the island of Ameland with the mainland by means of two dams. This plan was the impetus to set up the Dutch Society for the Preservation of the Wadden Sea. The Wadden Society succeeded in preventing the plan from being realized. However, reclamation of an area of 4,000 ha along the Frisian coast, the so-called Noord-Friesland Buitendijks (North Friesland outside the dikes), was still being pursued. But in the last instance the Wadden Society also blocked this plan. Conservationists do not support future plans for reclamation of parts of the Dutch Wadden area. In the German Wadden, however, such plans are still an issue.

The Dutch Wadden area is also used for military purposes. Especially the western part is extremely popular with the Ministry of Defense. Military activities take place near the city of Den Helder and the island of Texel and on and around the island of Vieland. In addition there is a route for low-flying military aircraft over the eastern part, and the Lauwersmeer has an exercise ground and shooting range. The German Wadden area is also disturbed by military activities affecting human and wildlife activities.

Currents transport great quantities of polluted water from the European rivers (Rhine, Schekdt, Meuse, Elbe, and Ems) into the Wadden Sea. The atmosphere, the IJesselmeer, dumping in the North Sea, and discharges from the Wadden Sea itself all add to the pollution. Agriculture and shipping oil (oil spills) are also to be blamed for the pollution of the Wadden Sea. The polluting substances penetrate into the food chain via plankton. As a result seals are weakened, become infertile, and are susceptible to virus diseases. The number of fish diseases in the North Sea and Wadden Sea still increases. Man is also part of the food chain and recently found susceptible to the long-term subtle affects of toxics. Fortunately the flow of polluting substances from the large rivers is decreasing lately.

The bottom of the Wadden Sea holds natural gas in some locations. Oil companies are constantly searching for these gas fields. On the island of Ameland and the western part of the Wadden Sea exploitation has already started. But there are more sites where the presence of natural gas has been established. Exploitation of natural gas disturbs the ambient environment, seriously affects the landscape for a long time, and leads to settlement. As a result, vulnerable dune land and marshes are submerged, and the areas appropriate for foraging birds decrease in size. The Wadden Society resisted the new plans the oil concerns made to put new drilling rigs in the Wadden Sea from 1994. Exploitation of natural gas is not accepted within the context of the Wadden Sea as a nature reserve.

Ems–Dollart Area/the Netherlands and Germany

The Dollart is a deep bay in the Ems estuary between the Netherlands and Germany. Its natural value is very high. It is a sheltered area, and so the smallest particles of silt can settle down in the Dollart. The soft layer of silt, which is formed in this way, is very attractive for certain birds like the avocet. The Dollart is a brackish tideland. The water becomes brackish because the sea saltwater blends with the freshwater of the rivers Ems and Westerwoldse A. These conditions create unique vegetation: the Dollart area is famous for its high bushes of sea asters.

Many chemical industries are concentrated in the Ems–Dollart area, near the cities of Delfzijl and Emden. As a result chemicals continuously affect soil, water, and air. The German plans to establish a large-scale industrial harbor in this area were not realized, but the area is under pressure from new plans and proposed ventures. In the 1970s the digging of a channel outside the dikes through the Dollart did not happen. A large part of the Dollart is a national nature reserve and is administered by the Stichting Groningen Landschap (Foundation of the Groningen Landscape) and the Vereniging tot Behoud van Natuurmonumenten (Society for the Preservation of Nature Reserves).

Lower Saxony – Germany

The Wadden area of Lower Saxony extends between the estuaries of the Rivers Ems and Elbe. Between the Ems estuary and the Jadebusen lies the East Frisian Wadden area, which closely resembles the Dutch Wadden. More to the east, between the Jadebusen and the Elbe, the Wadden has developed in a slightly different way under the influence of currents and estuaries. Seven inhabited and two uninhabited islands and sandbars bound the East Frisian Wadden area. Between the Jadebusen and the Weser estuary lie the Wadden of the Hohe Weg. Between the rivers Weser and Elbe are the Wurster Watt, the Wadden area of Knechtsand, and the Neuwerker Watt.

During the last 200 years the coast of lower Saxony showed drastic changes. Several estuaries were formed such as the Dollart, Leybucht, and Jadebusen. Many of them were embanked in the course of the centuries. The Jadebusen did not change any more after it was formed. Afterward it partly silted up, and during this process high moor peat was deposited. In the eleventh century a dike ran from the city of Wilhelmshaven to the present Jadebusen, along the peninsula of Butjadingen on the river Weser. The dike was swept away by storm tides taking the settlements with them. Only part of the peat moor has survived in the nature reserve called Das Schwimmende Moor (the Floating Moor).

Germany was the first to discover the recreational value of the Wadden area, far before the Netherlands did. The island of Ameland had its first "bathing establishment" in 1850, while the German island of Norderney had known recreation for 300 years already. In the nineteenth century the islands were considered resorts where one could restore one's health. High-rise blocks and promenades have affected the original character of Norderney and the island of Borkum nearby. The other islands of Lower Saxony, especially Spiekeroog and Baltrum, have retained their own character.

The Elbe – Germany

This swiftly flowing river has always influenced the Wadden area at the mouth of the River Elbe. The sand coming in from the west is checked by the Elbe; so a high slack water was formed at Neuwerk and Scharhorn. Horse and wagon can easily reach Neuwerk. The new man-made island of Nigehorn near Scharhorn has been created for the birds. The Elbe is responsible for the flow of great quantities of polluted water to the Wadden Sea. This water comes from the industries in and around Hamburg.

Schleswig-Holstein – Germany

This part of the Wadden is quite different from those in Lower Saxony and the Netherlands. The islands were formed in a different way and were called "Halligen". They are partly remains of the salt marshes that were washed away by a fierce storm tide in 1634. In the course of time these islands grew at the side of the mainland, while parts of them at the seaside were washed away. The salt marshes were already inhabited before 1634. On the Halligen are mounds, called Warften, on which one or more farmhouses were built. Some Halligen are connected to the mainland by means of a dam. After the storm tide of 1962 most Halligen have been provided with summer dikes. A unique feature of this Wadden area is the "Wanderdunen" on Sylt, a bare dune land. This conservation area is subject to continuous erosion. Along the coast of Schleswig-Holstein active reclamation is still common practice.

In the last 50 years tens of thousands of hectares of biologically valuable ground outside the dikes were lost. In the Schleswig-Holstein Wadden a new dike was constructed in the Nordstrander Bucht, which resulted in the loss of 90 km^2 of the Wadden area. Elsewhere an area of 570 ha is threatened by embankment.

The Wadden area of Schleswig-Holstein is a very popular recreation area, especially the island of Sylt with its ample facilities. High-rise blocks dominate the capital of Westerland. Sylt is connected with the mainland by a dam. The train running across the dam takes hundreds of thousands of tourists with their cars to the island yearly. The islands of Pellworm and Amrum also attract many tourists. On most of the Halligen recreation is still a small-scale affair.

Some years ago Lower Saxony, Hamburg, and Schleswig-Holstein have designated "their" Wadden areas as National Parks. Unlike the Netherlands, the federal governments disposed of legal tools enabling them to take drastic protective measures. Germany has taken advantage of this possibility by creating special zones. In some of these zones, nature has absolute priority over all human activities. In other zones some activities are permitted. And there are buffer zones and zones where nature has no priority at all. The criteria for zoning are different in both federal states concerned. Unlike the Netherlands, Germany has not coordinated the administration of these zones.

The German Wadden area is exposed to several threats. Recreation is much more intensive than in the Netherlands, also because many islands are easily accessible by dams. Large-scale embankment projects were carried out in the Leybucht and the Norstander Bucht. Oil exploration takes place near the bird island of Trishen. Military activities are still expanded in the area. Large industrial centers are established near Emden, Bremerhaven, Wilhelmshaven, and Cruxhaven, involving contamination risks for soil, water, and air. A large nuclear power plant is situated near Esenshamm in the Weser area, as is the case at Bokdorf on the Elbe. Hamburg harbors many metallurgical and chemical concerns. The Rivers Weser and Elbe are permanent sources of pollution for the Wadden Sea. But Germany is also beginning to realize the importance of protecting nature reserves such as the Wadden Sea.

The Danish Wadden

Two fixed points determine the shape of the Danish Wadden; the Horns Rev nears the coast of Blavandshuk (the most western point of Denmark) and the Rote Kliff on Sylt. The Danish Wadden is very dynamic. The coastline changes visibly every year as a result of the enormous quantities of sand supplied by the sea. So the exceptionally wide beach was created on the islands of Romo, Mando, and Fano. Except for the inhabited islands of Romo, Mando, and Fano, the Danish Wadden area comprises the uninhabited islets of Jordsand and Langli, the peninsula of Skallingen, and a few large sandbars. The coastline of the Danish Wadden is also greatly determined by dikes.

Skallingen

The Danish Ministry for the Environment bought Skallingen as a conservation area in 1976. This peninsula, which has a length of 13 km, came into being as a result of the transport of sand that formed a whole with the beach ridges. It consists of a row of dunes at the backside of which is an extensive salt marsh bordering the Ho Bugt and transacted by many channels. The salt marsh measures about 700 ha. Human activities have also marked the landscape of Skallingen. The erosion of the dunes is partly blamed on recreation, and intensive grazing causes the harm done to the salt marsh. Besides, many dikes of dry sand have been put up and ditches dug. The south point suffers from serious erosion.

Varde A

It is quite exceptional, especially in the Wadden Sea, that man allows rivers to flow freely into the sea without taking precautionary measures in his effort to check its stream by dikes and locks. The Varde A is such an exception. That is why such an unusual landscape has been created in and near its estuary. The extraordinary variety of its vegetation is a result that the freshwater river blends with seawater. At extremely high water levels and stormy weather the saltwater can penetrate a few kilometers into the riverbed. These "annoying" inundations have been resisted everywhere else but they provide beautiful landscape in the Varde A area.

Romo and Fano

Romo has a very wide beach, which at some places reaches a width of 4 km. Primary dune formation takes place on this beach. The old dunes on a large part of the island are overgrown with heather. At the side of the mud flats is a small strip of salt marsh. Since 1947 Romo is connected with the mainland by a dam, which divides the Danish Wadden area into two parts. As a result the island is under great pressure by recreation. The structure of Fano slightly resembles that of Romo. The beach is not as wide but the dunes and the strip of salt marsh are similar. Parts of the dunes are covered with woods. Fano has also reached the limits of its recreational possibilities.

Esbjerg

Due to the relatively low population density the Danish Wadden area is less disturbed than the Dutch and German parts. Esberg is the only large town in this neighborhood. The fish processing industries in this large fishing harbor are mainly responsible for the discharge of large quantities of wastewater. Moreover, the sewers of the city discharge into the Wadden Sea and the rubbish dump is situated near the beach. This is why organic matter mainly pollutes this part of the Wadden Sea. The coastal area north of Esberg along the Ho Bugt has a steep coast.

Protection and Management To Date

Major source material for the following section includes work by Bachest (1991), Dettmann and Enemark (2004), Hergreen (1991), and de Jong and Siderius (1995). Nienhuys (1990), Revier (1995), Swennen (1989), Waddensea Secretariat (1997), Walters (1990), van Zutphen (1989), van der Zwiep (1990), and van der Zwiep and Backes (1994). The Netherlands, Germany, and Denmark have all taken measures to protect the (remaining) ecological, cultural, and scenic values of the Wadden area or parts of it. Key or important steps were taken in the mid-1960s. At this time there was relentless pressure for more economic exploitation of the area, including recreation. At the same time there was pressure for extensive reclamation and embankments, which were engineered beyond protection goals.

In those days, nature conservation was almost exclusively concerned with the protection of rare species of birds, and numerous sanctuaries were designated for this purpose. The protection of the area in Denmark goes back to the 1930s at which time one can find the first implementation of preservation regulations, as laid down in accordance with the Nature Conservation Act of 1917 (Swennen 1989).

From this early time of habitat and species preservation, the situation changed when the general public became increasingly aware of the ecological and scenic values at stake, and of activities which threatened these same values with fast decline or total destruction. The usual response was regulations (laws), which purport to restrict certain uses of the Wadden Sea areas. However, there was further decline of the area due to loopholes in the regulations, insufficient attention to particular values of the area, and lack of clear quality requirements. Finally a set of regulations was issued in the three states that truly did not reflect the ecological relations and connections characteristic of the area. Furthermore, these regulations were the result of numerous political compromises. Many of the regulations shared no connection to one another, and were issued by competing legislative bodies and competing authorities. There was no relation between the three countries bordering the Wadden Sea.

At the same time there were similar shared concerns. In the Netherlands and Germany there were people who argued in favor of valuation and description of the Wadden Sea as an ecological entity, recognizing protection as being of national significance. In the Netherlands, this led to proposals for a special Wadden act, and in Germany for a National Park Act for the Wadden area. Politicians were not ready for such institutional mechanisms and presented their own proposals. In Denmark, the most important Nature Conservation Act came into force as a result of the discussion about proposed dams in the 1960s. A final solution is yet to be found and regulations concerning the affected area have been and are still developing.

Before moving on to institutional mechanisms existing in each of the three countries, we should at least acknowledge three major NGOs that have focused public attention on the issues mentioned and in some cases forcing action. In 1965 the Dutch Society for the Preservation of the Wadden Sea was established. The Wadden Society goals include optimal conservation of the natural and historical–cultural values of the Wadden area. Several working groups in the Wadden Society engage in diverse issues such as water, military use, recreation, industrialization, and management. All legal means, which might lead to a favorable policy review, are applied such as

- consultation, objections, publicity, political pressure;
- information and advice;
- stimulating alternatives;
- mobilization of all environment-minded Dutchman.

The society has approximately 60,000 members, 300 of them active. The members receive the "Wadden Bulletin", a periodical with many activities about landscape and nature in the international Wadden area and interviews with people working and living in the area. Activities of the society are also given much attention.

In Germany, the Schutzstation Wattenmeer and the World Wildlife Fund (WWF) Wattenmeerstelle are active in Schleswig-Holstein. Not only do they engage in campaigns against embankment plans and nuclear plants but they also give information. Several islands have information centers, which also publish a newsletter (Informationsbrief).

In 1977 a Danish Wadden Group was established which at the time resisted reclamation plans and the increasing facilities for water sports. The Fishing Museum in Esbjerg has brought out quite a lot of publications on the Danish Wadden. These three groups have been the major NGO actors for preservation and ecosystem management of the Wadden Sea wetlands. The following sections will outline existing institutional protection measures for the three countries followed by international treaties and provisions.

The Netherlands

The Wadden policy in the Netherlands is based on the Physical Planning Act and the Nature Conservation Act. These two regulations support a complicated system that tries to make use compatible with protection. At the same time efforts have been made to solve the problem of coordinating competing powers of national, regional, and local authorities, and those of numerous other departments and institutions. The Nature Conservation Act grants the status of nature reserve by means of a designation with all concomitant legal consequences. The physical planning key decision (PKB), which is based on the Physical Planning Act, regulated the various forms of exploitation and coordination of administrative aspects.

The combining of the two regulations was necessary because the Nature Conservation Act cannot do justice to both the ecological and social functions of such a large area. On the other hand, the legal status of the PKB was too unstable and judicially weak to serve as a basis of integration for protection and use of the area.

In the Dutch system, these values are first described in the PKB. By doing so, the advantages of the physical planning law as the most favorable instrument to weigh all interests at issue, including the interests of nature, could be used. In this respect, use of a new instrument like the PKB based on the Physical Planning Act can be supported. It provides an opportunity to straighten out the rather complicated relations between ecological and social interests. By combining both instruments, the Dutch government made a lot of concessions to the Nature Conservation Act and the values and interests that the law is supposed to protect, and thus to the ecological values as well.

The policy established in this combination of regulations is based on conservation, protection, and recovery of the Wadden Area. Human use is not excluded. The PKB indicates what forms of use are meant, and how these are to be fitted to actual situations, for example, by granting permits under the Nature Conservation Act so as to cause as little damage as possible to the ecological value of the area. The PKB further indicates to which geographic area it applies. The area is limited to the actual marine part between the dikes on the mainland and the southern part of the islands, and some of the uninhabited parts of the islands.

The Dutch set of regulations shows severe shortcomings according to van der Zweip and Backes (1994). Although PKB allows the environment in the area some preponderance, it is vague. The description of the interests that are considered acceptable and admissible is so noncommittal, and shows so many loopholes, that almost all activities can be allowed: industrialization, military activities, traffic and transport, recreation, fishing, etc. The only chance of restricting these is actually to be found in the roles for granting permits under the Nature Conservation Act or any other sector law at issue.

The boundaries of the nature reserve designated under the Nature Conservation Act do not correspond with those mentioned in the PKB. Since November 1993, the Nature Conservation Act had dealt with 90% of the area. That means that a more or less wide region the necessary junction between the PKB and the Nature Conservation Act can be laid. The protection of the other 10% of the Dutch Wadden Sea will remain incomplete. For this remaining part of the area, the rule applies that implementation of the PKB policy depends on weighing the pros and cons outlined in the sector law concerned. In other words, the PKB policy is dependent on the weighing described in sector laws. In those sector laws, the interests of nature can be omitted or sector interests can be predominant.

In the future, according to van der Zweip and Backes (1994) the link between the Nature Conservation Act and the PKB may become problematic. According to the new Nature Conservation Act, currently being discussed in the Dutch Parliament, the provinces will be largely qualified to implement this law. Furthermore the PKB is not legally binding in relation to lower administrative bodies. This would make the current legal instruments even more unclear and fragmentary (Fig. 2.5).

Fig. 2.5 Agricultural use of Wadden Sea marshes. Source: Common Wadden Sea Secretariat, Undated. *The Wadden Sea: A Shared Nature Area*, p. 8

Germany

In Germany, there are a lot of various instruments, which together form the judicial basis for the protection of the Wadden Sea. The legal system is rather complicated for several reasons. In the first place, because of the federal form of government, which means that the federal states (Lande) are the first to be responsible for nature conservation, while the federal government acts only as a coordinating body. All efforts to change this situation by a constitutional amendment failed in the early 1970s. The lack of unity among the authorities issuing regulations has been an important stumbling block in the development of the protection of the Wadden area in Germany.

This protection is mainly based on the Nature Conservation Acts and especially on the regulations for national parks. The Nature Conservation Act of the federation defines what this protection should comprise and the federal states have to work out the details of the regulation. As the German part of the Wadden area extends over four federal states (Lower Saxony, Bremen, Hamburg, and Schleswig-Holstein) the legal powers required for protection of the area as a total entity are dissipated. For example, as early as 1974 large parts of the area were already designated as "wetlands of international importance" under the Ramsar Convention. Other parts were not designated. The area of Schleswig-Holstein was designated as a national park in 1985, the area of Lower Saxony in 1986, and the Hamburg area in 1990. Bremen was left out. Through the City of Bremerton, the city-state of Bremen borders the Wadden area, is a party to the trilateral Wadden consultants at a governmental level, though it has no Wadden territory of its own, and therefore no specific Wadden regulations of its own.

The three existing regulations of the national parks are not only different from one another in a substantial way as to their form (laws in Hamburg and Schleswig-Holstein, a regulation/bylaw in Lower Saxony) but also in their territorial scope (with or without islands, salt marshes, forelands, and/or dikes). Furthermore, their degree of effectiveness is quite different. Nevertheless, the regulation of national parks in Germany were the only chance of realizing wider-ranging protection of the area than was possible under the already existing regulations for nature conservation, as these were haphazardly applied. It also offered the opportunity to create an administration infrastructure (national park administration) to manage the national parks as entities and to provide funds for their management.

These national park settlements embrace nature conservation aims, and formulate the acceptable and admissible social uses. All this is expressed in a set of local, periodic bans, orders, and exemptions. All regulations in the Lander are based on splitting up the area into zones, which apply different forms of protection. Roughly, the area is divided up into three zones: zone I in which the interests of nature are predominant and human use principle is excluded; zone II in which human use is not excluded but where important protection measures are taken; and zone III which includes all remaining areas, above all recreation areas. A system of prohibitions applying to these zones has to guarantee that use is compatible with protection, i.e., that human activities do not harm the natural values.

Although protection of the Wadden area in Germany is mainly based on the Nature Conservation Acts, the Physical Planning Act also plays an important point. The systems, however, are not directly linked. The purposes established through town and country planning are necessary additions. The hierarchical structure of this instrument for planning is one of the reasons why certain uses, including environmental uses, can be weighed and established at an adminis-trative level. Though these uses can only be roughly described, they have a highly standardized effect. In this way, various forms of exploitation have been defined by zoning. The pros and cons of relevant interferences (some indicated in the planned purpose) are meticulously weighed, both with respect to protec-tion and to various other functions.

Lastly, sector law should be mentioned. On the basis of the constitutional distribution of legislative power, some activities are exclusively regulated by sector law (shipping on the Wadden Sea, for example). The nature conservation laws of the federal states and the regulations concerning the national park settlements adopted under them may not include any restrictions with regard to these activities. Restriction for the sake of nature protection can be enforced only under the sector law concerned. In actual practice, the authorities some-times fail to do this, or if they do, the result is unsatisfactory. This can be considered a weak spot in the German legal system for the protection of the Wadden Sea.

Denmark

The Danish set of instruments for the protection of the Wadden Sea has been highly refined in recent years. This applies to the legal foundations support-ing the protective measures as well. In 1992, the various nature protection laws were streamlined and integrated into the new Act on Nature Protection. The former designations under these nature protection laws were combined in 1985 to form one designation of large parts of the Wadden Sea as a nature preserve. However, the protection of the Danish Wadden Sea is not fully integrated into one regulation under the nature protection laws. In addition to the general conservation rule according to the Nature Protection Act that covers the whole region, there are special territorial laws applying to specific areas (for the reclaimed Margrethe Kog and the Tonder Marsh salt marshes).

Protection under the nature protection laws is complemented by protection on the basis of town and country planning. The Danish physical planning laws have also been drastically revised in recent years, especially with respect to the integration of the various physical planning laws into the new 1992 Planning Act. Just as in Germany, Danish physical planning is hierarchical. The regional plans, which are drawn up by the two counties in the Wadden Sea region, are of special importance. After weighing all interests playing a part in the area concerned, the counties decide on the functions and the possibilities for development of the space concerned. The county and municipal councils will strive to implement the guidelines of the regional plan. Their planning and development activities may not contradict the regional plans. In most cases, the county or municipal councils operate at first instance; hence the physical plans have a significant practical importance.

International regulations have a large effect on the implementation of the above-mentioned regulations in Denmark, especially the Ramsar Convention and the EC Bird Directive, and in the future, the Habitat Directive. The Danish Wadden Sea was designated as a wetland of international importance under the Ramsar Convention in 1987, and earlier, in 1983, as a special protection zone in accordance with Section 4 of the Bird Directive. The provisions from international agreements and EC directives are in principle not directly binding in Denmark, but are first to be transformed into national law. Nevertheless, the Nature Complaint Board, in particular, uses the provisions from the Ramsar Convention and the EC Bird Directive for judicial review even without a clear national foundation. In real practice, those international agreements and regulations are therefore of utmost importance for the protection of the Wadden Sea, at least as far as the jurisdiction of the Nature Complaint Board is concerned.

Besides this, specific decisions such as conservation decisions, which are proposed by the nature conservancy boards in accordance with the rules laid down in the Nature Protection Act, are important. Such specific conservation decisions (e.g., regarding air traffic, marinas, water catchments, road projects, or management measures for special areas) are brought to the Nature Protection Board of Appeal. The power of this board as an administrative appeals board is based on such cases, which are viewed as administrative decisions based on the Nature Protection Act. These specific conservation decisions constitute another important environmental instrument for the protection of the Danish part of the Wadden Sea.

International Rules and Implementation

In 1982 the Danish, German, and Dutch governments agreed on a "Joint Declaration on the Protection of the Wadden Sea". They declared they would "...consult each other in order to coordinate their activities and measures to implement the international legal instruments with regard to the comprehensive protection of the Wadden Sea region as a whole". The legal instruments separately mentioned are as follows:

- The Convention on Wetlands of International Importance especially as Waterfowl Habitat (Ramsar February 2, 1972)
- The Convention on the Conservation of Migratory Species of Wild Animals (Bonn, June 23, 1979)
- The Convention on the Conservation of European Wildlife and Natural Habitats (Bern, September 19, 1979)
- The relevant EC Council Directives, especially the one issued on April 2, 1979, on the Protection of Wild Birds (79/409 EC); EC Bird Directive which is linked to
- Council Directive (92/43/EC) on the Conservation of Natural Habitats of Wild Fauna and Flora (May 21, 1992); EC Habitat Directive

Ramsar Convention

The Ramsar Convention aims at the protection and conservation of wetlands (as discussed in Chapter 1), which means protection of the whole biotype rather than only species. An area satisfying the criteria established by the convention can be presented on the list of wetlands. At present, almost the whole Wadden area has been designated as such and appears on the list.

Once an area has been designated, the contracting party is obliged under Section 3.1 to "...promote the conservation of wetlands included in the list". Under Section 4.1 there is also an obligation to promote the conservation of wetlands by establishing nature reserves whether or not these wetlands appear on the list. Denmark drew the conclusion that the convention does oblige the designation of areas satisfying the criteria, and consequently the protection of them by means of the national regulations (nature reserves). The Netherlands took the line that designation can only be realized and effective if the area is already protected under national regulations and inclusion on the List of Wetlands of International Importance only sets "the seal of protection". It is interesting, given this background that the Netherlands did not designate all parts of the Wadden area, which are already fully protected under national regulations. Though the whole PKB area is designated as a Ramsar area, not included are important parts of the islands, a strip of the North Sea, and parts of the mainland, which are also part of the Wadden Ecosystem. The German government has established new nature reserves or extended existing reserves on listed sites since for their inclusion on the list.

The Ramsar Convention has a strong influence on the protective measures for the Wadden area. In all these countries, the area, or large parts of it, is protected under national laws. This is also due to international supervision or observance of the convention through the permanent secretariat, periodic Conferences of the Contacting Parties, and access of NGOs to various events and processes. Through this public exposure, obligations though not directly binding, become morally binding. In Denmark, this is expressed by the fact that the convention is used as a direct judicial criterion for assessment of human activities in the area. Public pressure made Germany observe the principle of compensation set down in Section 4.2 of the convention. The Dollarthafen project could only be carried out if 2,000 ha of nearby grassland was inundated by way of compensation for the loss of the listed wetland of the Ostfrisische Wattenmeer and the Dollart.

Yet there are differences both in the dimensions of the designated area and in the extent to which the various obligations of the convention are met. As a rule, the Dutch islands do not come under the designation. In Germany and Denmark, only parts of them do. Denmark has also included a belt of the North Sea in the designation; Germany and the Netherlands have not. The Netherlands does not recognize the principles of compensation set down in Section 4.2 of the convention, whereas Germany does. There are several interpretations of the concept of "wise use" (Section 3.1 of the convention). However at the Sixth Trilateral Government Wadden Sea Conference in 1991, the three countries agreed to fill these gaps.

World Heritage Commission

At the Sixth Trilateral Governmental Wadden Sea Conference in 1991, the three countries agreed on presenting a joint proposal that the international Wadden areas should be put on the World Heritage List of the Convention of Cultural and Natural Heritage. The convention concerning the Protection of World Cultural and Natural Heritage was adopted in 1972 in Paris and came into force in 1975. On the basis of the convention, area of "outstanding universal value" can be put forward for inclusion in the World Heritage List. If the area is put on the list, then the contracting parties are legally bound to conserve and protect it. The provisions concerned (in Sections 4 and 5) are more imperative than those of the Ramsar Convention. World Heritage List Convention offers financial and technical opportunities to support conservation and protection, but there is debate whether designation on the World Heritage List offers more effective protection.

Bonn Convention

The Bonn Convention especially aims at the protection of wild migratory animal species. The contracting parties bind themselves to conserve the habitats of the migratory species. The convention is a framework treaty on the basis of which regional agreements can be concluded. In 1988, the Netherlands, Germany, and Denmark concluded an agreement on the Conservation of Seals of the Wadden Sea. This agreement came into force on October 1, 1991, and is the first regional agreement under the Bonn Convention. So far the agreement has had a favorable effect on scientific research, the monitoring of the seal population, the designation of special resting areas and public information. But as far as the protection of habitat of the seals against the threats of water pollution is concerned, neither the convention nor the Trilateral Seal Agreement has had any perceptible effect.

The convention can play an important part as an instrument for the coordination of the protection of migratory wild animals (especially birds) and their habitats on their long trek between north Siberia and Africa. The Wadden area has the important function of being the "intermediate station" for these birds. In this context, attention should be given to the paper prepared for the International Union for Conservation of Nature and Natural Resources (IUCN): *Elements of an agreement on the Conservation of Western Palearctic Migratory Species of Wild Animals.* This document states that "...the development of cooperative links with the governing bodies and secretariats established under the other international instruments dealing with certain Western Palearctic migratory species or their habitats is necessary". Furthermore, the document concludes that although none of the instruments (Ramsar, Bonn, Bern, EC Directives) covers the full range of all Western Palearctic species listed in the appendices of the Bonn Convention, they all deal with at least some of these species over the whole or part of their range. It is therefore essential that coordination mechanisms be established to avoid duplication of effort and ensure effective implementation of conservation and management measures.

EC Directives

Because of their binding effect, the most important international protective measures are those taken by the European Commission (EC). Among these measures are the Bird Directive and the new Habitat Directive, which have the most forward reaching consequences for the protection of the area. The Habitat Directive takes over and reinforces the function and the legal consequences of the Bird Directive as far as the designated areas under this directive are concerned, and also the provisions and obligations resulting from the Bern Convention. The Habitat Directive has been enforced in the national regulations of the three Wadden countries.

So far, the Bird Directive has inadequately been implemented with respect to the Wadden area, both as an obligation to designate protected areas and to the observance of the protective measures required. Both the Netherlands and Germany have been reproved for this several times by both the European Commission and the European Court of Justice.

One of these cases is the judgment made by the European Court of Justice in the Leybucht case regarding the structure of the new Habitat Directive. The judgment was that it is justified that the Habitat Directive be more strictly implemented and applied. The directive's implementation does not only involve designation of areas to be protected but also judicial consequences. The provision in Section 63 prohibits, to a certain extent, permission being given to carry out projects and plans, which may be detrimental to the natural characteristics of designated areas. The granting of permission has been made dependent on environmental research and "imperative reasons for overriding public interest". This provision imposes considerable restrictions on the various authorities in their scope of policy-making. There are many questions concerning further implementation of the Habitat Directive.

Section 3 explicitly states that the areas of the Bird Directive are automatically listed as Special Areas of Conservation. The Netherlands designated the area as area for bird protection (according to PKB boundaries). Germany also designated important parts of the area especially in Lower Saxony. Denmark designated this area as an area under Section 4 of the Bird Directive in 1983.

The Beginning of Cooperative Management of the Wadden Sea

Historically, the protection of the Wadden Sea was set according to a series of national initiatives in the late 1970s and during the 1980s starting with the establishment of the Wildlife and Nature Reserve in the Danish part in 1979/1982, the Wadden Sea Memorandum and Nature Reserve in the Dutch part in 1980/1981, and the three national parks in the German part from 1985 on. The Wadden Sea, from Esberg in Denmark in the north to Den Helder in the Netherlands in the west, is now covered by an almost unbroken stretch of nature reserves and national parks. Parallel talks between the three governments were initiated with the aim of achieving a comprehensive protection of the Wadden Sea as a shared ecosystem, which resulted in the first Trilateral Governmental Conference for the protection of the Wadden Sea in 1978. At the Third Governmental Conference in Copenhagen in 1982, the three governments formalized the cooperation by adopting the "Joint Declaration on the Protection of the Wadden Sea". To extend and strengthen the cooperation, the Common Wadden Sea Secretariat was established in 1987, following a decision at the Fourth Governmental Conference in 1985 (Dettmann and Enemark 2004).

The area of the tri-national cooperation of the Netherlands, Germany, and Denmark is 13,500 km^2 large. The transition zone to the North Sea covers about 4,000 km^2, the islands about 1,000 km^2, the tidal area some 7,500 km^2, the salt marshes and summer plodders some 350 km^2. The four estuaries, the Varde A, the Elbe, the Weser, and the Ems, have a total surface area of 260 km^2. Also some areas on the mainland, which are important for birds, are part of the cooperation area and cover about 250 km^2.

Trilateral Wadden Sea Cooperation

During the years that followed the initial cooperation, the three governments were reluctant to engage in agreements, which contained elements of international legally binding arrangements. The breakthrough in the cooperation came with the adoption of the Joint Declaration in 1982. The Ramsar Convention played an essential role in bridging the formal differences and expresses the political commitment to cooperate in the protection of the Wadden Sea (Dettmann and Enemark 2004). The three countries had ratified the Ramsar Convention and were legally committed to implement its provisions. If in accordance with Article 5, the Wadden Sea countries would consult on a coordinated implementation of the Ramsar Convention with respect to the Wadden Sea – this greatly contributed to comprehensive protection.

According to the Joint Declaration, the governments declared their intention to consult with each other in order to coordinate their activities and measures to implement a number of international legal instruments with regard to the comprehensive protection of the Wadden Sea region as a whole. The international legal instruments, as mentioned previously, are the Ramsar Convention on Wetlands, the Bonn Convention on the Conservation of migratory species, the Bren Convention on the conservation of European wildlife and natural habitats, and the relevant EC directives, in particular the EC-Bird Directive.

The Joint Declaration resolved a dilemma. It is a declaration of intent, stating the political commitment to work toward a common goal, but it includes a number of legally binding international instruments. It was the intention of the parties that counts, rather than the legal character of the instrument. The Joint Declaration served as a catalyst in the period after 1982, and in conjunction with the establishment of the common secretariat in 1987, the Trilateral Wadden Sea Cooperation was intensified and extended (Dettmann and Enemark 2004). The Trilateral Wadden Sea Plan, which was adopted at the Eighth Wadden Sea Environmental Ministers Conference in 1997, entails a comprehensive common policy and management of the Wadden Sea (see Waddensea Secretariat 1997)

The Trilateral Wadden Sea Plan – Key Elements

The Wadden Sea Plan entails policies, measures, projects, and actions, which have been agreed upon by three countries. The plan is a framework for the overall Wadden Sea management and will be revised at regular intervals. It is a statement on how the three countries envisage the future coordinated and integrated management of the Wadden Sea area as well as the projects and actions that must be carried out to achieve the targets.

The plan is a political agreement and will be implemented by the three countries in cooperation, and individually, by the various authorities on the basis of existing legislation and through the participation of interest groups. The implementation of the plan shall not interfere with legislation regarding marine navigation, management of marine navigation routes, harbor management, disaster control, sea rescue services, and other aspects of internal and external security (Waddensea Secretariat 1997).

The Wadden Sea Plan entails a number of critical decisions with regard to the delimitation of the common management area, the shared principles, and action to implement the targets.

Delimitation

The geographic range of the Wadden Sea Plan is the Trilateral Wadden Sea Cooperation Wadden Sea area, which is

- the area seaward of the main dike, or where the main dike is absent, the spring-high-tide-water line, and in the rivers, the brackish-water limit;
- an offshore zone 3 nautical miles from the baseline;
- the corresponding inland areas to the designated Ramsar and/or EC Bird Directive areas;
- the islands.

The trilateral conservation area is situated within the Wadden Sea, and consists of the following:

- In the Netherlands. the areas under the Wadden Sea Memorandum including the Dollard
- In Germany, the Wadden Sea national Parks and protection areas under the existing Nature Conservation Act seaward of the main dike and the brackish water limit including the Dollard
- In Denmark, the Wildlife and Nature Reserve Wadden Sea

It is recognized that within the Wadden Sea area, there are areas where human use has priority. The delimitation of the Wadden Sea area attempts to bridge the formal differences in jurisdiction between the three countries. The Wadden Sea area is a common management area and not a protection area, which allows for the implementation of trilateral agreements, measures, and actions by the application of a wide range of national instruments.

Shared Principles

The Guiding Principle of the Trilateral Wadden Sea Policy is "to achieve, as far as possible, a natural and sustainable ecosystem in which natural processes proceed in an undisturbed way". The principal is directed toward the protection of the tidal area, salt marshes, beaches, and dunes.

In addition, seven management principles have been adopted which are fundamental to decisions concerning the protection and management within the Wadden Sea area (see Waddensea Secretariat 1997):

- The Principle of Careful Decision Making, i.e., to make decisions on the basis of the best available information
- The Principle of Avoidance, i.e., activities which are potentially damaging to the Wadden Sea should be avoided

- The Precautionary Principle, i.e., to take action to avoid activities which are assumed to have significant damaging impact on the environment, even where there is no sufficient scientific evidence to prove a causal link between activities and their impact
- The Principle of Translocation, i.e., to translocate activities which are harmful to the Wadden Sea environment to areas where they will cause less environmental impact
- The Principle of Compensation, i.e., that the harmful effect of activities which cannot be avoided, must be balanced by compensatory measures, in those parts of the Wadden Sea, where the principle has not yet been implemented, compensatory measures will be aimed for
- The Principle of Restoration, i.e., that, where possible, parts of the Wadden Sea should be restored if it can be demonstrated by reference studies that the actual situation is not optimal, and that the original state is likely to be re-established
- The Principles of Best Available Techniques and Best Environmental Practice, as defined by the Paris Commission

Unreasonable impairments of the interests of the local population and its traditional uses in the Wadden Sea area have to be avoided. Any user interests have to be weighed on a fair and equitable basis in light of the purpose of protection in general and in the particular case concerned.

Targets

The trilateral conservation policy and management is directed toward achieving the full scale of habitat types, which belong to a natural and dynamic Wadden Sea. Each of these habitats needs a certain quality (natural dynamics, absence of disturbance, absence of pollution), which can be reached by proper conservation and management. The quality of habitats shall be maintained or improved by working toward achieving targets, which have been agreed upon for six habitat types. Targets on the quality of water and sediment are valid for all habitats. Supplementary targets on birds and marine mammals have been adopted, as well as targets on landscape and cultural aspects (see Waddensea Secretariat 1997) (Table 2.1).

Policy and Management

The key element of the Wadden Sea Plan is the common policy and management (see Fig. 2.6 below). For each target category, trilateral policy, management, and proposals for trilateral projects and actions necessary for the implementation of the targets have been developed.

Table 2.1 Wadden Sea Plan targets

Targets on habitat and species

Salt marshes

The habitat type for salt marsh includes all mainland and island salt marshes, including the pioneer zone. Also the brackish marshes in the estuaries are considered part of this habitat type.

The following targets apply to slat marshes:

- An increased area of natural slat marsh;
- An increased natural morphology and dynamics, including natural drainage patterns, of artificial salt marshes, under the condition that the present surface area is not reduced;
- An improved natural vegetation structure, including the pioneer zone, of artificial salt marshes.

Tidal areas

The tidal area covers all tidal flats and subtidal areas. The border to the North Sea side is determined by an artificial line between the tips of the islands. The borders of the estuaries are determined by the average 10% isohaline at high water in the winter situation.

The following targets are valid:

- A natural dynamic situation in the tidal area;
- An increased area of geomorphologically and biologically undisturbed tidal flats and subtidal areas;
- An increased area, and more natural distribution and development of natural mussel beds, *Sabellaria* reefs and *Zostera* fields;
- Viable stocks and natural reproduction capacity, including juvenile survival, of the common seal and gray seal;
- Favorable conditions for migrating and breeding birds;

= a favorable food availability;
= a natural breeding success;
= sufficiently large undisturbed roosting and molting areas;
= natural flight distances.

Estuaries

Estuaries include the estuaries of the rivers with a natural water exchange with the Wadden Sea. On the landward side, the mean-brackish-water line delimits estuaries. On the seaward side, the border is the average 10% isohaline at high water in the winter situation.

Estuaries will be protected and the riverbanks will remain, and as far as possible, be restored to a natural state.

Beaches and dunes

Beaches and dunes include beaches, primary dunes, beach plains, primary dune valleys, secondary dunes, and heath land behind the dunes.

The following targets apply

- Increased natural dynamics of beaches, primary dunes, beach planes, and primary dune valleys in connection with the offshore zone;
- An increased presence of a complete natural vegetation succession;
- Favorable conditions for migrating and breeding birds.

Offshore zone

The offshore zone ranges from the 3-sea-mile line to an artificial line connecting the outer tips of the islands. The border between the offshore zone and the beaches on the islands is determined by the average low-tide watermark.

Table 2.1 (continued)

The following targets apply to the offshore zone:

- An increased natural morphology, including the outer deltas between the islands;
- Favorable food availability for birds;
- Viable stocks and a natural reproduction capacity of the common seal, gray seal, and harbor porpoise.

Rural area

The rural area includes meadows and arable land on the islands and on the mainland where there is a strong ecological relationship with the Wadden Sea.

The following target applies

Favorable conditions for flora and fauna, especially migrating and breeding birds.

Targets on the quality of water and sediment

Nutrients

A Wadden Sea, which can be regarded as an eutrophication non-problem area.

Natural micropollutants

Background concentrations in water, sediment, and indicator species.

Man-made substances

Concentrations as resulting from zero discharges.

Source: Dettmann and Enemark (2004).

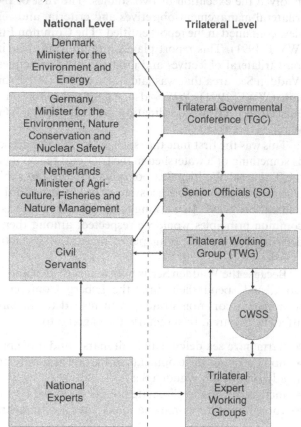

Fig. 2.6 Trilateral wetland governance. Source: Common Wadden Sea Secretariat, Undated. *The Wadden Sea: A Shared Nature Area*, p. 3

Summary and Missing Links

Given the different legal approaches to protection and management of the Wadden Sea wetland area by the three countries plus the different interpretations and implementation of international treaties such as the Ramsar Convention and EC Directives, we can see the linkage, coordination, and management integrative problems. They are myriad, but there have been substantial accomplishments as well. Before delving into the most critical future management issues, we should examine the role of the NGOs in Wadden Sea management.

The Role of the Wadden Sea NGOs

Events in all three Wadden Sea countries illustrate clearly the need to continue work on the development of better methods for the protection of the Wadden Sea. This was apparent in the early 1980s when environmental NGOs from all the Wadden Sea states started working together on the problem. With financial support of the World Wildlife Fund (WWF), they developed a program that involved the execution of two studies. The first of these focused on uniform, trilateral management objectives and criteria, and resulted in the managerial view contained in the report entitled "The Common Future of the Wadden Sea" (WWF 1991). This report played an important role in the formulation of the joint trilateral objectives and joint common principles for management of the Wadden Sea area that was laid down in the 1991 Esbjerg Ministerial Declaration (WWF 1991). WWF also set up coordination stations in Bremen and Husceu as well as coordination with up to 50 NGOs in the international Wadden Sea area.

This was the first time that such an approach had been taken and can be seen as something of a watershed in the trilateral sea cooperation. It was agreed that the Trilateral Wadden Sea policy would aim "to achieve, as far as possible, a natural and sustainable ecosystem in which natural processes proceed in an undisturbed way". In working toward this goal it was agreed that a number of common principles would be respected, among them the principle of careful decision making, the principle of precautionary action, and the principle of translocation.

Because the Wadden Sea is both a nature area and an area where people live, work, and spend their time, the Esberg Conference found it necessary to formulate a common strategy with regard to the variety of human activities affecting the area. In summary, they agreed to

- harmonize sea defense and salt marsh and dune protection;
- no new major developments of harbor and industrial facilities immediately adjacent to the Wadden Sea;
- increase efforts to eliminate pollution caused by shipping;
- cooperate in developing national criteria with regard to dredging operations;

- avoid exploration and exploitation of oil and gas until 1994;
- avoid (in principle) the construction of new pipelines;
- prohibit wind turbines in the Wadden Sea;
- limit the extraction of sand;
- limit the negative ecological impact of the mussel and cockle fishery;
- protect the recreational values of the Wadden Sea;
- reduce the disturbance to wildlife caused by hunting;
- limit the impact of civil air traffic on the Wadden Sea;
- reduce the impact of military activities;
- reduce the impact to the Wadden Sea of polycyclic aromatic hydrocarbons and organic compounds;
- express joint concern about climatic changes and sea level rise;
- develop plans for restoring parts of the Wadden Sea and for the reintroduction of species;
- develop a "Red List" of endangered marine and coastal species and biotopes in the Wadden Sea area;
- conserve seals;
- ensure adequate wardening;
- harmonize environmental impact assessments;
- cooperate in the field of public information;
- cooperate in international flora where the Wadden Sea was an issue (Ramsar Convention, World Heritage Convention, Flyway Cooperation, European Community, and North Sea Conferences).

Although the adoption in 1991 of the common principles and objectives was a significant step in the right direction, there were still major shortcomings in the trilateral cooperation. The principles and objectives were formulated in a way that allows individual countries considerable freedom for interpretation. Consequently, to some they can follow their own course while still complying in terms of the trilateral agreement.

Furthermore, the ministerial declarations are not legally binding. When a participating country does not comply with the adopted principles and objectives the other countries can only react at a political level and there are no significant sanctions. For these reasons and others previously covered in the review of existing management instruments, the NGOs sponsored a second study, which concerned the legal component of the managerial view. The intention was to study how the rules applied to the three countries could be harmonized. This study resulted in the publication entitled "Integrated System for Conservation of Marine Environments – Pilot Study: Wadden Sea" (Zweip and Backes 1994). Hans Revier, director of the Dutch Wadden Society, pointed out the need for such a study

- to document comparisons between the various national laws;
- to look at the principle of unity for ecosystem management and its implications;

- to develop a "level playing field" throughout the whole Wadden Area so as to avoid parties taking advantage of inconsistencies and undue development pressure for some areas;
- to insure NGO knowledge of each country's administrative and legal structure similarities and differences.

The study itself focuses on the legal structure of the instruments, and where and why this structure is not functioning optimally. The study is divided into two parts: first, there is an analysis of national legal frameworks for the protection of the Wadden Sea, showing their strong and weak points and including recommendations for change and second, an attempt is made to further develop the umbrella of legal instruments covering all three Wadden states, i.e., the international and European laws for the protection of the Wadden Sea.

Of most interest are the possible routes, which could be taken to achieve a collective formulation of preconditions and criteria to affect unified ecosystem management of the whole Wadden Sea. One of these is part of EC law, the other of international law. In respect to the former, a special EC Wadden Directive could be established. The study prefers all parts of the area to be designated collectively and in a coordinated manner within the framework of the Habitat Directive, which would be much easier to accomplish.

As for international law, the study authors favor the establishment of a Trilateral Wadden Sea Treaty. If this were formulated in a sufficiently concrete fashion, then it would have a significant influence on national legal systems, and ensure that gaps in the various protective instruments would be filled.

Aside from the legal management structure needed for ecosystem management of the Wadden Sea there are still several outstanding physical–chemical problems that need attention. These are given below:

- Water quality targets especially for total discharge, nutrients, heavy metals, and organic micropollutants
- The need for fisheries management and biodiversity
- The need for an ecosystem management plan for the whole Wadden area
- Credible agreement on oil and gas exploration and production in the area
- Protection of special at-risk populations of seals and dolphins

The Role of Wetland Science in Monitoring, Modeling, and Future Impacts

There has been long-term monitoring of shorebirds on the Wadden Sea over 20 years (Smit 1989). Shorebird surveys in the Wadden Sea have not only revealed the extremely large importance of the area, especially for wading birds, but also show that different areas are exploited by shorebirds in different ways. They have also provided data on changes in bird numbers throughout the year but there still need to be improvements in how the counts are conducted.

In addition there has been 70 years of vegetation plot research in the Netherlands, including the Wadden Sea wetlands (Smits et al. 2002). The database provides insight into vegetation succession, fluctuations within plant communities over time, and the effects of changes of the environment on vegetation.

There have been calls for ecosystem models to quantify material flows to reveal imbalances, which then may indicate the direction of ecosystem change (Reise 1995). More specific models have been proposed for habitat suitability of restoration of *Zostera marina* shellfish beds (van Katwijk et al. 2000).

In terms of monitoring pollutants in the Wadden Sea, Van der Brink and Kater (2006) have used chemical measurements and bioassays to evaluate marine sediments for four groups, including heavy metals, PAHs, chlorinated aromatic compounds, and tin compounds. Measurements were taken at 16 locations in the Wadden Sea, the Netherlands. Principal component analysis indicated that the response to the Microtox Solid Phase bioassay had a positive significant relationship with the levels of PAHs and organic compounds in the marine sediment. These compounds may still be stressors for aquatic invertebrates in the Wadden Sea (Van der Brink and Kater 2006).

Besides pollutants in wetland sediment the other major concern with the health of wetland communities is the relative contribution of sediment and nutrients for maintaining or even building coastal marsh. According to Bakker et al. (1993) the area of salt marsh along the Netherlands Wadden Sea coast no longer increases. Recent erosion rates coincide with a rise in MHT level in the last 25 years. Despite the decrease in area, sedimentation continues, especially in the lower salt marsh, which acts as a sink for nitrogen. Assimilation and mineralization of nitrogen are in balance in most communities along the gradient from lower to higher salt marsh, whereas the above ground production and mean content of plants decreases. Sedimentation on main land marshes can compensate for the expected sea level rise, but this is not the case for island salt marshes. The stability of remaining coastal flats with a rising sea level scenario is also of concern to Danish Wadden Sea scientists as well (Christiansen and Aagaard 2004).

This brings us to the effect of future climate change on the Wadden Sea wetlands vegetation, fish, and wildlife. According to Brouns (1992) one of the main concerns is the rise of sea level and that the sedimentation rates will be insufficient to maintain salt marshes on the barrier islands as stated above. The marshes on the mainland coast will be impoverished, as high and low marshes are not expected to coexist in the same locations. As sediment supply to the Wadden Sea is sufficient to compensate for sea level rise, the estuarine character of the Wadden Sea, with sand and mud flats, is expected to remain largely unchanged (Brouns 1992).

The resultant impacts to wetland-dependent species have been studied for climatic change on Western Palearctic migratory birds by Meekes (1992). He concludes that many migratory bird species will be influenced by climate change, leading to adaptation in the bird's annual cycle. The biggest problems may arise for those birds, which depend on wetlands, because many of these wetlands may desiccate (Meekes 1992).

Summary

So in essence, the role of Wadden Sea NGOs evolved from early protection of species to campaigns against specific development proposals and management activities, to international diplomacy and influence of policy determined at the Trilateral Government Wadden Sea Conferences. The strategy for the 1991, 1994, and 1997 meetings is that of the agenda setters; preparation of major policy documents designed to have maximum impact on policy decision makers. NGOs concerned with the Wadden Sea continue their monitoring role, especially with the international conventions such as Ramsar, Bonn Convention, and EC Bird/Habitat Directives. Above all the Wadden Sea NGOs do an incredible job with education through use of newsletters and other media to keep members and concerned citizens informed. The role of wetland science is also critical in monitoring and reporting on the health and direction of change of ecosystems habitat and specific species as reported above in the previous section

Acronyms

EC: European Commission
IUCN: International Union for the Conservation of Nature
PKB: Physical Planning Act

References

Bachest, S. 1991. Acceptance of national parks and participation of local people in the decision making process. In M. Marchel and N.A. Udo des Has (eds.) Special Issue Wetlands. *Landscape and Urban Planning*, 20: 239–243.

Bakker, J. P., J. de Leeuw, K. S. Dijkema, P. C. Leendertse, H. H. Prins, and J. Rozema. 1993. Salt marshes along the coast of the Netherlands. *Hydrobiologia*, 265(1–3): 73–95.

van Berkel, B. M. and J. M. Revier. 1991. Mussel fishery in the international Wadden Sea, consistent with "wise use"? In M. Marchel and N. A. Udo des Has (eds.) Special Issue Wetlands. *Landscape and Urban Planning*, 20: 27–32.

Brouns, J. J. 1992. Climatic change and the Wadden Sea, The Netherlands. *Wetlands Ecology and Management*, 2(1–2): 23–29.

Christiansen, C. and T. Aagaard. 2004. Wadden Sea research: An introduction. Geografix Tidsskrift, *Danish Journal of Geography*, 104(1): 1–3.

Dettmann, C. and J. Enemark. 2004. Workshop D – The Wadden Sea Plan – common management of a shared wetland. In *5th European Regional Meeting on the Implementation and Effectiveness of the Ramsar Convention*, 4–8th December 2004, Yerevan, Armenia.

Fog, K., R. Podloucky, U. Dierking, and A. H. P. Stumpel. 1996. Red list of amphibians and reptiles of the Wadden Sea area. *Hegoland Marine Research*, 50(Suppl. 1): 107–112.

Goss-Custard, J. D., R. A. Stillman, A. D. West, R.W. G. Caldow, P. Triplet, S. E. A. le V. dit Durell, and S. McGroty. 2004. When enough is not enough: Shorebirds and Shell fishing. *Proceedings of the Royal Society B: Biological Sciences*, 271(1536): 233–237.

Hergreen, R. 1991. Towards a new vision on the development of the Wadden Sea. In M. Marchel and N. A. Udo des Has (eds.) Special Issue Wetlands. *Landscape and Urban Planning*, 20: 168–176.

Huiskes, H. L. 1988. The salt marshes of the Westerschelde and their role in the estuarine ecosystem. *Aquatic Ecology*, 22(1): 57–63.

de Jong, F. and K. Siderius. 1995. The Wadden Sea: Our compassion and concern. *North Sea Monitor*, Dec.: 15–16.

De Jonge, V. N., K. Essink, and R. Boddeke. 1993. The Dutch Wadden Sea: A changed ecosystem. *Hydrobiologia* 265(1–3): 45–71.

Kees, T. P., T. Piersma, and K. Comphuysen. 2001. What peak mortalities of Eiders tell us about the Dutch Wadden Ecosystem. *Wadden Sea Newsletter*, 1: 42–45.

Lotze, H. K. 2005. Radical changes in the Wadden Sea fauna and flora over the last 2000 years. *Hegoland Marine Research*, 59(1): 71–83.

Meekes, H. T. H. M. 1992. An inventory of the possible effects of climate change on Western Palearctic migratory birds. *Wetlands Ecology and Management*, 2(1–2): 31–36.

Nienhuys, K. 1990. The interaction between a Dutch NGO and the local people in the promotion of wise management of the Wadden Sea Area. In M. Marchand and H. A. Udo ed Haes (eds.) *The Peoples Role in Wetland Management, Proceedings of an International Conference* [Leiden, The Netherlands, June 5–8 1989], pp. 766–771. Centre for Environmental Studies, Leiden University.

Reise, K. 1995. Predictive ecosystem research in the Wadden Sea. *Hegoland Marine Research*, 49(1–4): 495–505.

Revier, H. 1995. An integrated system for the protection of the Wadden Sea: A reality or something to work for? *North Sea Monitor*, Dec.: 8–11.

Smit, C. J. 1989. Perspectives in using shorebird counts for accessing long-term changes in wader numbers in the Wadden Sea. *Hegoland Marine Research*, 43(3–4): 367–383.

Smits, N. A. C., J. H. J. Schaminee, and L. van Durren. 2002. 70 years of permanent plot research in the Netherlands. *Applied Vegetation Science*, 5(1): 121–126.

Swennen, C. 1989. Wadden Seas are rare, hospitable and productive. In M. Smart (ed.) *International Conference on the Conservation of Wetlands and Waterfowl* [Heiligenhafen, Federal republic of Germany, 2–6 Dec. 1974], pp. 184–198. Slimbridge, U.K.: International Waterfowl Research Bureau.

Van der Brink, P. J. and B. J. Kater. 2006. Chemical and biological evaluation of sediments from the Wadden Sea, The Netherlands. *Ecotoxicology*, 15(5): 451–460.

Van der Perk, J. and R. de Groot. Undated. *Working Paper 12: Case Study Critical Natural Capital: Coastal Wetlands: The Dutch Wadden Sea*, as part of the CRITINC Project: Making Sustainability Operational: Critical Natural Capital and the Implications of a Strong Sustainability Criterion. Environmental System Analysis Group, Wageningen University & Research Center, 43 pp.

Van Eerden, M. R., R. H. Drent, J. Stohl, and J. P. Bakke. 2005. Connecting seas: Western Palearctic continental flyway for water birds in the perspective of changing land use and climate. *Global Change Biology*, 11(6): 894–908.

Van Katwijk, M. M., D. C. R. Hermus, D. J. de Jong, R. M. Asmus, and V. N. de Jong. 2000. Habitat suitability of the Wadden Sea for restoration of *Zostera marina* beds. *Hegoland Marine Research*, 54(2–3): 117–128.

Verhulst, S., K. Oosterbeck, A. L. Rutton, and B. J. Ens. 2004. Shellfish fishery severely reduces condition and survival of oystercatchers despite creation of large marine protected areas. *Ecology and Society*, 9(1): 17.

Waddensea Secretariat. 1997. *Trilateral Sea Plan: Eight Trilateral Governmental: Conference on the Protection of the Wadden Sea.* Stade, Germany: Wadden Sea Secretariat, http://www.waddensea-secretariat.org/tgc/Wsp/

Wadden Society. 1994. *Be careful of the Wadden Sea,* Harlingen: Wadden Society, 30 pp. brochure.

Walters, A. R. 1990. Top-down approach to protecting wetlands: A Dutch experience. In M. Marchel and N. A. Udo des Has (eds.) Special Issue Wetlands. *Landscape and Urban Planning*, 20: 623–635.

Wolff, W. J. 2000. Causes of extirpation in the Wadden Sea, an estuarine area in the Netherlands. *Conservation Biology*, 14(3): 876–885.

World Wildlife Fund. 1991. *The Common Future of the Wadden Sea.* Husun, Germany: World Wildlife Fund for Nature, WWF-Wattenmeerstelle, 64 pp.

van Zutphen, J. P. 1989. Nature management and sustainable development in the Dutch Wadden Sea. In W. D. Verney (ed.) *Nature Resources and Sustainable Development: Proceedings of an International Congress* [Groningen, The Netherlands, 6–9 Dec. 1988], pp. 302–311. Amsterdam: IOS.

van der Zwiep, K. 1990. An integrated system for conservation of marine environments: Pilot Study: Wadden Sea. In M. Marchand and H. A. Udo ed Haes (eds.) *The Peoples Role in Wetland Management, Proceedings of an International Conference* [Leiden, The Netherlands, June 5–8 1989], pp. 716–722. Centre for Environmental Studies, Leiden University.

Van der Zwiep, K. and C. Backes. 1994. *Integrated System for the Conservation of Marine Environments, Pilot Study, Wadden Sea.* Baden-Baden: Nomes Verlagsgellschaft, 244 pp.

Chapter 3
The Axios River Delta – Mediterranean Wetland Under Siege

Introduction

This chapter takes a look at wetland conservation in the Mediterranean and more specifically at a Ramsar wetland site in Greece – the Axios River Delta complex. The delta complex actually includes three rivers, the Axios, Loudias, and Aliakmon rivers, but for the purposes of this chapter is referred to as the "Axios Delta". The questions here to be examined is what has been the recent impact of European wetland conservation policy expressed by the Grado declaration, MedWet, and Greece's wetland policy on a specific wetland area? The other issue is what specific roles did the WWF Greece-sponsored project (the NGO in this case) play in wetland management policy affecting the Axios River Delta area?

Context of Greek Wetland Conservation

The main problems facing Greek wetlands are from development of agriculture, livestock, fishing, and lumbering (Maragou and Montziou 2000):

- Agriculture: There is steady pressure for the expansion of cultivation areas, the development of irrigation schemes, and the use of fertilizers and pesticides.
- Livestock: Grazing is poorly controlled with long-term consequences for the farmers and the ecosystem.
- Fisheries: The over-exploitation of fisheries has resulted in the reduction, and in some cases the disappearance of some species of fish. Many problems have developed because proper fishing methods were not used.
- Lumbering: Until recently, unprogrammed lumbering and forest fires led to the destruction of enormous forest areas and resulted in heavy sedimentation in lakes.

In addition, in the 2002 Ramsar report, main problems causing wetland degradation in Greece included dam construction and river alteration causing alteration in hydrologic regime, over pumping, clearing of natural vegetation,

R.C. Smardon, *Sustaining the World's Wetlands*,
DOI 10.1007/978-0-387-49429-6_3, © Springer Science+Business Media, LLC 2009

illegal hunting, water pollution from industry and agriculture, expansion of farms, and housing development in wetland areas (Bassoukea and Markopoula 2002, Zalidis 1993).

Legislation

In recent years decentralized services with appropriate technical staff have been established in order to apply legislation to each area. This legislation concerns the conservation of nature and exploitation of natural resources. It now applies to all Greek wetlands whether they are covered by international law or are covered only by Greek legislation. Indeed some wetlands covered by national legislation are more important than some included in the lists of the International Conventions.

The following wetlands have been listed under the Ramsar Convention:

1. Evros Delta
2. Amvrakikos Gulf
3. Vistonis Lake plus the Porto Lagos Lagoon
4. Delta of the Rivers Axios, Aliakmon, and Alyki Kitros
5. Mesologi Lagoon
6. Nestos Delta
7. Lake Mikri Prespa
8. Mitrikou Lake
9. Artificial Lake of Kerkini
10. Lakes Volvi and Lagada
11. Kotichi Lagoon and Strofilia Forest

For each of these wetlands, studies have been elaborated by the Ministry of the Environment, Physical Planning and Public Works, with the participation of scientists of different university faculties in order to assess the wetlands' ecological characteristics, to register the dangers that threaten them, and finally to make a proposal about their delineation. The author worked on such a study for the Delta of the Rivers Axios, Aliakmon, and Alyki Kitros to determine whether remote sensing signatures could be used to delineate wetlands in the area.

Developing a Conservation and Protection Strategy

The Ministry of the Environment has started plans leading hopefully to management wetlands and their immediate periphery in order to reduce man's negative impact. Priority is given to areas like the Delta of the Rivers Axios, Aliakmon, and Alyki Kitros, which are directly threatened. Zoning is based upon the following principles:

• The delineation of the core as an area under complete protection and the definition of minimum activities are in balance with the wetlands function.

- The control of land uses in the immediate periphery.
- Management measures, which will allow use of the natural resources in a manner, which is compatible with conservation of the ecosystem.

A national strategy for wetland conservation was the topic for the meeting of April 18, 1989, in Thessaloniki, Greece. The following is a summary of the major findings and conclusions of that meeting (Gerakis 1992). During the discussion of "Present status, trends and forces of change" in Greek wetlands all speakers (Psilovikos, Papaylannis, Kourteli, and Economou) agreed that there had been great losses of wetland resources in Greece, due to both natural causes and human action.

The human activities, which had caused these losses, were identified as follows:

  (i) Diversion of fresh water away from wetlands (both within Greece and in shared water catchments beyond Greece's frontiers)
 (ii) Agricultural interventions (land reclamation and water control)
(iii) Urbanization
(iv) Development of tourism

The speakers also agreed during the 1989 meeting that the remaining wetlands are under threat at present from continuation of earlier threats and

  (i) overuse of agricultural fertilizers and pollution by chemicals;
 (ii) intensive aquaculture;
(iii) sewage, urban, and industrial pollution;
(iv) urbanization and particularly resort housing.

The speakers identified the following human activities, which in addition to the continuation of current threats may affect Greek wetlands in the future:

  (i) Unmanaged visitor pressure on wetlands, including influx of foreign tourists, hunters, fisherman.
 (ii) Changes in water management practices provoked by changes in agricultural crop subsidies.
(iii) Intensification of productive activities in and around wetlands and consequent over-exploitation of biotic resources.
(iv) Further development of tourism (damage of wetlands for construction of tourist installations such as airports, roads, and mosquito control measures).

It was noted that the basic reason for these changes was the import of unsustainable development models, inappropriate local conditions, which provided profits for individuals rather than the wider population. Among examples of these models were postwar aid programs, EC Integrated Mediterranean Programs, and subsidies for aquaculture and agriculture, all of which led to uses of land and resources competitive with wetland conservation.

Speakers, discussants, and participants also noted a number of positive points in relation to wetland conservation in Greece (Gerakis 1992):

(i) Growing concern for conservation matters among the Greek public, thanks to promotion by schools and NGOs especially as EC subsidies decrease.
(ii) Consequent increased sensitivity to environmental issues, both at the governmental level and in political parties.
(iii) Changes in EC policies, and in particular introduction of an obligation to draw up Environmental Impact Assessments at the inception of all projects.
(iv) Improvement of the management of public affairs in Greece, which will hopefully improve as time goes on.
(v) Increasing governmental action to achieve wetland conservation in the framework of the Ramsar Convention, the EC Birds Directive, and the new Habitat Directive (e.g., rehabilitation of the Drana Lagoon in the Ervos Delta). The meeting also expressed its appreciation of the work of dedicated officials in the administration concerned, suggested they needed greater support, and welcomed their work in the interministerial committee and their collaboration with national and international NGOs. The latter is not a trivial matter for the interministerial committee as it was the first time that agencies such as the Ministry of Environment and Agriculture had worked together on such a matter.

The session noted the following points for future action and consideration:

(i) Because of the lack of clarity over precise responsibilities in the field, it is often difficult to implement laws; better training of regional and field staff would help overcome the problem.
(ii) There is an urgent need for mapping of wetlands, to show current area, and historical extent. Such mapping can better be done on the basis of studies of current and old maps (scale preferably 1:50,000) and ground truth studies.
(iii) Sea fisheries, which are economically important in Greece, depend on inflow to the sea of freshwater from rivers through deltas. Wetland reclamation will affect these fisheries.
(iv) Conservation of wetlands involves conservation of soil and vegetation throughout the catchment, but particularly in its upper part, to avoid erosion. A national wetland policy therefore needs to be linked to an overall natural resources planning operation.
(v) The need to strengthen the wetland conservation administration and to increase its budget could be further promoted by the establishment of a national wetland committee – as established under Ramsar auspices in other countries – bringing together natural and regional authorities, NGOs, and experts.

The desirability of establishing a matrix of information on Greek wetlands, covering all their many functions and values was discussed and initial matrix of such was made.

Institutional Policy Constraints and Obstacles

Political will for protecting multiple values of wetlands varies from positive to extremely negative depending on the case and the people who happen to make the decisions. An expression of positive will is not the result of knowledge and ecological (environmental) sensitiveness due to pressure from Greek public opinion, but rather from pressure by the EC and other international organizations.

A known example of weak political will is neglecting an obligation, according to the Ramsar Convention, to set boundaries for the 11 wetlands mentioned in the list. Related studies by the Ministry of Environment were made during 1984 (and published in 1986). Naturally there was to follow the implementation of proposals included in those studies and the compensation of managerial plans for each wetland. The process continued for only two wetlands. A preliminary plan was completed for the National forest of Prespas, where a Ministerial Decision was issued for Ambrakikos, according to Paragraph 21 of Law 1650/86 (which according to many experts was inadequate). For the other nine areas, procedures did not progress in spite of continuous appeals from responsible public officials of the Ministry of Environment, EC's recommendations, and the pressure from environmental associations and even local residents.

Mediterranean Context – the Grado Declaration

In 1991, IWRB organized a major symposium at Grado in northern Italy entitled "Managing Mediterranean wetlands and their birds for the year 2000 and beyond" (Hollis et al. 1992). Some 300 experts attended from 28 countries. The Mediterranean wetlands are now so degraded and destroyed and pressure on those remaining are so severe that they are among the most threatened ecosystems on earth. A simple goal was enunciated at the end of the symposium – "To stop and reverse the loss and degradation of Mediterranean wetlands". To achieve this strategy along the following lines:

(1) That supra-national and international organizations, governments and financial institutions recognize Mediterranean wetlands as a common natural heritage of the region and assume individual and joint efforts for their conservation; that they ensure coherence between all their policies and actions concerning wetlands; and further, that the European Community undertakes much greater financial commitment to the conservation, enhancement, and restoration of wetlands of the whole Mediterranean region.

(2) That policy-making bodies at all levels submit present and future policies, programs and projects that may have an impact on wetlands to a strict economic and environmental appraisal in order to guarantee the

sustainable use of natural resources and achieve the maximum long-term benefits from wetlands for the people of the Mediterranean.

(3) That a free flow of information and an open consultation procedure be adapted in managing wetlands.

(4) That the Contracting Parties to the Ramsar Convention and the states yet to join develop a regional approach to the conservation of wetlands through greater international cooperation and the effective implementation of wise use as relevant to Mediterranean wetlands and related river basins.

(5) That non-government organizations develop a more substantial member-ship base and act in a more coordinated fashion to increase awareness of the value of Mediterranean wetlands, to ensure that any use of wetland resources is sustainable and to monitor the status of wetlands and activities affecting them; that in addition, they strive to play a crucial role in retaining close collaboration between the people's of the Mediterranean for the conservation of wetlands.

(6) That research directly relevant to achieving the goal is undertaken, includ-ing the evaluation of existing and proposed policies; that institutional capacity to conserve and manage wetlands effectively be increased by vigorous education and training programs.

(7) That priority sites for wetland restoration be identified and techniques be developed and tested for their complete rehabilitation.

(8) That integrated management of all activities concerning wetlands, their support systems, and the wider area surrounding them be carried out by properly funded and well-staffed multi-disciplinary bodies, with the active participation of representatives of the government, the local inhabitants, and the scientific and non-government community.

(9) That government of all Mediterranean countries adopts and in particular enforces national and international legislation for a better management of the hunting activity.

A Mediterranean Wetlands Forum (MedWet) has been set up between international bodies, governments, and non-governmental organizations to develop the strategy and draft the action plan. The Italian Government is hosting a small secretariat in Rome. The European Community has been asked to provide 6 million eco$ in support of MedWet activities. A feeling of optimism now prevails instead of the environmental despair that had charac-terized the region previously. MedWet has been extended to non-EU countries.

Role of International NGOs

The number of Greek associations concerned with the protection of the natural environment has increased during the last decade. The importance of this fact, as an indication of increasing awareness of the public in environmental issues as well as the achievements and difficulties of these associations deserve closer attention and study. Some of the most well-known associations have a long history of activities, especially for wetlands. Two of them, namely the Greek Society for Protection of Nature (oldest Greek NGO) and the Greek Society for the Protection of the Environment and Cultural Heritage are known widely as the "Greek Society". Their activities vary from the publication of flyers and posters, elaboration of research on wetlands, organization of seminars and lectures to the establishment of two biology research stations. The research station of the first company is located in the Deltas of the Evros River and the second in Mikri (little) Prespa. These stations have hosted many researchers and naturalists, both Greek and foreign. The Greek Company for the Protection of Nature carries one of the largest reputations in Greece, for information on protection of nature in general and wetlands specifically.

The Greek (Hellenic) Ornithological Company, with its numerous local chapters in several wetland areas, in spite of its recent founding, has been addressing wetland protection issues. It is not limited to scientific activity such as the midwinter waterfowl census, but has taken on strong partisan initiatives. The Society is systematically monitoring populations of Pygmy Cormorant and the Lesser White-footed Goose in the Axios Delta and other wetlands in Greece. There are other national (Pan-Hellenic) associations, especially scientific, which among other interests include wetlands' protection (e.g., Greek Botanical Company, Greek Forestry Company, Greek Zoological Company, Greek Hydrotechnical Company, Association of Greek Ecologists, Association of Law on the Environment).

Also important is the formation of small groups, interested especially in protecting certain wetlands. Some of them are functioning under a regular constitution where others are citizen groups. There are such associations of citizen groups for Prespes (actually two of them, one local and one based out of Athens), for Nestos Delta, for Loudias, for Trichonis, for Vegoritida, and possibly others. The numerous ecological movements (meaning the green movement which managed to capture a temporary seat in parliament) appeared in almost every district because they all included protection of wetlands on their agendas (e.g., Komotini, Lamia, Xanthi, Patra [Thessaloniki]).

All these organizations have problems with resources and the Greek government rarely invites representatives of ecological associations to participate in decision making. Legal protection (or environmental law in general) has rarely been used by the ecological associations during their campaigns. This is even though existing special legislation on wetlands and the environment, in combination with more general laws, present opportunities for legal action.

The support from foreign national and international, non-governmental associations and organizations for the protection of nature, offered to Greece for many years, is very significant. Their activities are taking place in combination with related Greek associations, universities, or independently. They include research programs, training for groups of experts, general information for the public, publishing written material, mailing of newsletters, and even appeals to the Greek government for forthcoming ecological disasters. The cooperation of WWF and the Greek Company for Environmental Information and Education has produced much monitoring material for environmental education in general as well as for special districts such as Evros, Rodopi, Drama, and Xanthi.

There are several reasons why the efforts to conserve Greek wetlands have not been very effective. One is insufficient knowledge among politicians, administrative officials and the public of the functions and multiple values of wetlands. Another is the fragmentation of forces supporting wetland conservation, which results from the lack of a national strategy and absence of a widely accepted action plan. In order to help increase understanding of the importance of Greek wetlands, the problems they face, and design of an action plan to address these, a project entitled *Conservation and Management of Greek Wetlands; Strategies and Action Plan* was organized in April 1988 by the World Wide Fund for Nature (WWF), the Laboratory of Ecology of the Faculty of Agriculture of the University of Thessaloniki, and the World Conservation Union (IUCN).

The aims of this project were

(a) to review functions of wetlands and their values to society;
(b) to identify the present status of Greek wetlands, the forces and trends of change, and research priorities;
(c) to develop a strategy and action plan for Greek wetlands and of management proposals for specific sites;
(d) to make the general public, decision makers, and resource managers aware of the needs for conservation and the benefits from effective management of Greek wetlands.

The project had three phases. The first phase consisted of the preparation of studies on the functions and values of wetlands, on the changes witnessed in Greek wetlands during the present century and on factors of change. Also, case study reports on three representative wetland sites were prepared. This material was brought together for the second phase which was a workshop held in Thessaloniki from April 16 to 21, 1989. One of the products of that workshop was an Action Plan for the Conservation and Management of Greek Wetlands. The third phase is the promotion of the Action Plan, which includes the publication of the Workshop Proceedings and of shorter illustrated editions for non-specialists, the organization of open discussions in various cities, etc. The Workshop Proceedings were published in Greek in July 1990 and the English version was published by IUCN in 1992 (Gerakis 1992).

Greek Plan for Action

In the beginning of 1989, a small panel of Greek scientists composed the preliminary plan which was the subject of extensive conversations during the International Workshop on the Protection of the Greek Wetlands, which took place in Thessaloniki from April 17 to 21, 1989, including participation of 30 Greek and foreign scientists from several disciplines. These discussions drove to a semi-final plan, which was presented for discussion to the participants of the open daily meetings in Thessaloniki (April 21, 1989), Athens (June 6, 1989), Xanthi (January 31, 1990), and Patra (March 8, 1990). The results of those meetings drove to the creation of the final Plan for Action. This plan was adopted by the government (Politia here, meaning more than the government alone) and be used as a base of a national policy for our wetlands.

The remainder of this chapter will be an in-depth look at one of the Greece's Ramsar wetlands that has been under great stress, Delta of the Rivers Axios, Aliakmon, and Alyki Kitros. We will review the history of this particular wetland, analyze its functions and problems, and see what affect if any the new Greek Wetland Conservation Strategy has had on the management of this wetland.

Introduction and History of Case Study

The Axios River is one of the major rivers of the Balkan Peninsula, with a total length of 380 km. It starts in the Serbo-Albanian Scardos Mountains, and most of its length lies within the boundaries of the former Republic of Yugoslavia. Upon entering Greece, the river flows for approximately 80 km across the prefectures of Kilkis and Thessaloniki before ending its journey in the Thermaikos Gulf (Gulf of Thessaloniki) (see Figs. 3.1 and 3.2). Its watershed area is 23,747 km^2 90% of which lies within the former Yugoslavian territory and the rest in Greece – in the prefecture of Florina and the valley of the Axios itself. Key sources for this section include Athanasiou (1990), Athanasiou et al. (1994), Gerakis et al. (1988), Konstandinidis (1989), Newly (1995), Psilovikos (1988), and Zalidis (1993).

The Axios Delta is part of a larger complex of wetlands, which also includes the mouth of the Gallikos River to the east, and the mouth of the Loudias and the delta of the Aliakmon River to the west.

In the fifth century BC the area which is now the plain of Thessaloniki (also known as Kampania) was covered by the sea; and Pella, at that time the capitol of the Kingdom of Macedonia, was virtually a coastal city, lying on the Thermaikos Gulf. The three rivers – the Gallikos, or Echedorus, the Axios and the Aliakmon – and a number of smaller steams fed the gulf. By the first century BC, the alluvial material deposited by the two biggest rivers (Axios and Aliakmon) had encircled an expanse of sea off the port of Pella, creating the Loudias lagoon. By the fifth century AD, the lagoon had been completely cut off from the sea and had become a lake, known as Lake Yannitsa.

At the beginning of the twentieth century this shallow lake, with its extensive marshlands and its divisive vegetation, covered the entire central portion of the plain. The Axios flowed to the east of the Halastia, discharging into the

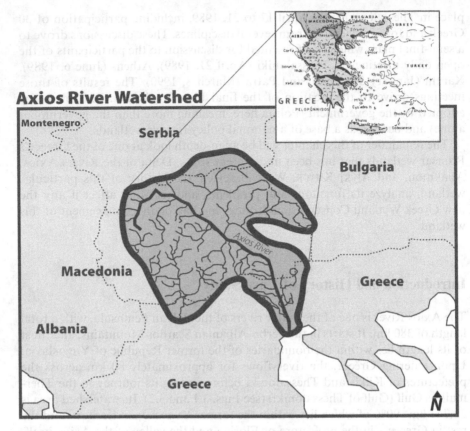

Fig. 3.1 Location of Axios Delta in Greece and Axios River watershed and drainage to the Mediterranean Drawn by Samuel Gordon. Sources: Alphamentor 2005. *Small Hydro Projects in Greece: The Case of Axios River.* PowerPoint presentation made in Krakow, Poland, on September 19–20, 2005, http://www.alphamentor.gr; And European Commission, undated, Eurocrat: European Catchments and Coastal Zone–Axios River Catchment–Axios: The Axios River Catchment http://www.cs.iia.cnr.it/EUROCRAT/Axios%20ingles.htm

Thermaikos Gulf, south of Kalohori, where it formed a delta. During the periods of heavy rainfall the river would overflow its banks, inundating large areas of the plain, where its alluvial deposits threatened to block the entrance to the port of Thessaloniki. The Loudias River, which drained the lake, created extensive marshes on its way to the Gulf.

Loss of the Interior Wetlands on the Thessaloniki Plain

The population of the area was 41,607 people in 1920, almost doubled in 1928 when 70,477 people were recorded, and tripled in 1940 when 107,590 people

Fig. 3.2 Low oblique aerial photo of Axios River Delta. Source: Alphamentor 2005. *Small Hydro Projects in Greece: The Case of Axios River.* PowerPoint presentation made in Krakow Poland on September 19–20, 2005, http://www.alphamentor.gr

were recorded (National Statistical Service of Greece). The great influx of Greek refuges from Asia Minor that arrived and settled in the area after 1922 created high demand for more agricultural land as well as land for housing. The wetlands of the plain were considered as sites suitable for expansion and the state introduced the reclamation scheme of Thessaloniki plain in 1925. The population continued increasing after 1951, but at much lower rates.

The construction of the peripheral canal that cut off the small rivers flowing from the surrounding mountains into Aliakmon River deprived Lake Giannitsa from its water supply. The conversion of the Loudias River into a drainage canal led to almost complete drainage of Giannitsa Lake (13,313 ha) and the inland marshes of the plain surrounding the lake (7,787 ha) and Loudias marshes (4,525 ha) by 1935. Only a small fraction of inland marshes survived after the completion of the first stage of drainage works. Part of the drained lake and marsh area was replaced by meadow and scrub area at the same period. The rest was converted to agricultural land.

The diversion of Axios River into its present course and the dikes built along its bed deprived the marsh area along the coast from its water and sediment supplies and initiated the phase of the coastal marsh reclamation that was to take place during the second phase of the reclamation works. At this stage only a small fraction of these marshes were lost (1,600 ha), a small area became meadow or scrubland while a small area of marsh (600 ha) was gained until 1935, by the diversion of the sediment carried by the Axios River to its present location, and the sediment deposited by the Aliakmon River. The distribution of sand deposits changed too after the diversion of the same river. Those deposited by its former marsh eroded away while new ones appeared along its course right after the construction of the new river bed.

Major reductions of the floodplain and riverine forests took place at the same time. The increase of the population of the plain led to direct forest clearance to satisfy increased needs for fuel wood, timber as well as agricultural land. Part of the forest was replaced by meadows and part by scrubland as a result of the combined effect of forest clearance and major alterations of hydrology of the area that followed the diversion of the Axios River, the diking of both Axios and Aliakmon Rivers, and the drainage of Giannitsa Lake and Loudias Marshes, but the lagoon area remained unaltered.

During the following period (1935–1970) further reclamation works that took place led to complete drainage of all marshes left in the central part of the plain (3,175 ha) and to major reductions of the coastal marsh area. Dikes were built along the coast, and filling of the marsh area with land was the marsh reclamation practice of the period. About 4,281 ha of coastal marsh were lost due to the conversion to agricultural land, another 525 ha were lost due to erosion, and sediment deposits created subsidence that took place at the old Axios River Delta, while 725 ha of new marsh were created by the two river mouths.

A further reduction of forest area for the same reasons as was mentioned previously was the other main feature of the period. Another 4,450 ha of forest were lost while scrubland area declined as well (by 4,181 ha). Meadow area was converted to agriculture while lagoon area remained almost the same. An increase of lagoon area is due to further extension of the newly formed Axios Delta. No sand deposits were recorded at the end of this period as the building of dams along the upper course of the two main rivers as well as the diversion of Axios sediment into the deeper parts of the gulf caused the loss.

The final account of the delta during 1970–1990 represents further reductions of coastal marsh area due to further reclamation. The other major feature of the period was the reclamation of 469 ha of lagoon area between Loudias and Aliakmon Rivers. Remnants of riverine forest are now limited only along the riverbeds of the two major rivers. The slight increase of scrubland area was due to the declaration of a game reserve along the Axios River bed that excluded farming from the designated area.

The natural part of the wetland is covered by salt marshes, which are the predominant feature of the coastal area (see Fig. 3.2). The rich material carried down by the river has created within the delta a series of shallow lagoons and sandy islets, frequently colonized by a wealth of dense vegetation. Bushes and tall trees line the banks of the islets and the riverbed (see Figs. 3.3 and 3.4). The wilderness areas of the wetlands are criss-crossed by the drainage ditches delineating the arable land. In the spring and summer, when the extensive rice paddies are flooded, they enhance the landscape of the natural wetland and create a unique landscape.

Fig. 3.3 Axios River Floodplain from the air. Source: Alphamentor 2005. *Small Hydro Projects in Greece: The Case of Axios River*. PowerPoint presentation made in Krakow Poland on September 19–20, 2005, http://www.alphamentor.gr

Fig. 3.4 Axios River Floodplain from the riverbank. Source: Alphamentor 2005. *Small Hydro Projects in Greece: The Case of Axios River*. PowerPoint presentation made in Krakow, Poland, on September 19–20, 2005, http://www.alphamentor.gr

Despite the changes that took place over the past century when the wetland lost almost a third of its original size, the delta continues to impress scientists and visitors with its diversity, especially for fisheries and bird habitat.

Wetland Description

The following sections describe current wetland vegetation, fauna, and land use. Major sources for the following section include Athanasiou (1987, 1990), Athanasiou et al. (1994), Gerakis (1988), Jerrentrup et al. (1988), Kazantzidis (1996), Nazirides et al. (1992), Newly (1995), Psilovikos (1992), Tsiouris and Gerakis (1991), and Valaoras (1992).

Wetland Vegetation

In coastal wetlands, vegetation varies according to the humidity and salinity of the soil. The Axios Delta comprises six distinct vegetation zones:

– Halophytic vegetation: Halophytes are plants, which thrive, in a saline environment like salicornia, which dominates one of the plant communities. Sapphire (*Salicornia europaea*) is an indicator species for one community that includes *Aeluropus littoralis, Halimione portulacoides, Sperrula* sp., and asters (*Aster tripolium*). Another salt marsh community has rushes (*Juncus maritimus*) as an indicator species with other members being black grass (*Juncus gerardii*), *A. littoralis,* sapphire (*S. europaea*), and asters (*A. tripolium*). The third salt marsh community has two indicator species: alkali grass (*Puccinellia festuciformis*) and *H. portulacoides* with the other member being *A. tripolium*. The fourth salt marsh community has two species of *Arthrocnemum* as indicators: *Arthrocnemum fruticosum* and *A. glaucum* with other members being *H. portulacoides*, sea lavender (*Limonium gmelinii* and *L. bellidifolium*), asters (*A. tripolium*), alkali grass (*P. festuciformis*), sapphire (*S. europaea*), and foxtail (*Hordeum maritimum*). The fifth community consists of one species *Arthrocnemum perennis* and the sixth community also has one dominant species, *Halocnemetum strobilaceum*.
– The tamarisk scrubland community (see Fig. 3.5) is found mostly flanking the river, but also further inland in the delta. Nearer to the sea the tamarisk ceases to thrive and is gradually replaced by halophytes, except right along the riverbed where the increasing salinity is tempered by the freshwater of the river. The indicator species is tamarisk (*Tamarix hampaenan*) and the halophytic under story includes European alkali grass (*Puccinellia distans*), *A. littoralis*, and *H. portulacoides*. Non-halophytic under story vegetation includes Bermuda grass (*Cynodon dactylon*) and foxtail (*Hordeum murinum*).
– Rush meadows (*Juncus* spp.) are chiefly found in areas protected from the effects of saltwater. Large expanses once covered with rushes have been reclaimed for farmland. Indicator species is rush (*Juncus acutus*) and other species include peas such as Fabaceae Leguminosae and Fabaceae Gramineae.

Fig. 3.5 Axios River Floodplain – *Tamarix* plant community. Source: photo by Dylan Lloyd that appeared in Newly (1995, p. 4)

- Reed beds are found at the mouth of the river and along the riverbanks and drainage canals. Three phytosociological units were recognized: *Bolboschoenetum martimi* is an indicator species for one community with other species being common reed (*Phragmites australis*) and narrow-leafed cattail (*Typha angustifolia*). The second community is dominated by common reed (*P. australis*) with other species being common cattail (*Typha latifolia*), narrow-leafed cattail (*T. angustifolia*), water fennel (*Oenanthe aquatica*), flowering rush (*Butomus umbellatus*), *Bolboschoenus maritimus*, spike rush (*Eleocharis palustris*), purple loosestrife (*Lythrum salicaria*), water veronica (*Veronica anagallis-aquatica*), and water mint (*Mentha aquatica*).
- Hydrophytic species, like duckweed and hornwort, flourish whenever there are shallow expanses of freshwater such as irrigation canals, drainage ditches, and rice paddies. Specific species include pondweed (*Potamogeton nodosus*), sago pondweed (*Potamogeton pectinatus*), and *Potamogeton perfoliatus*, water milfoil (*Myriophyllum* sp.), hornwort (*Ceratophyllum* sp.), European frog's bit (*Hydrocharis morsus-ranae*), duckweed (*Lemna minor* and *Lemna trisulca*), and mosquito fern (*Azolla filiculoides*).
- Riparian forest (see Fig. 3.6) can be found along the banks of the river and on the many islets formed in the riverbed. The principal species are white poplar (*Populus alba*), black poplar (*Populus nigra*), black alder (*Alnus glutinosa*), white willow (*Salix alba*), *Salix triandra*, silky-osier willow (*Salix viminalis*), and purple-osier willow (*Salix purpurea*). Under story vegetation includes bramble (*Rubus* sp.), hops (*Humulus lupulus*), ivy (*Hedera helix*), birthwort (*Aristolochia clematitis*), *Cynanchum acutum*, silk vine (*Periploca graeca*), mint (*Mentha* sp.), sedge (*Carex* sp.), and European water horehound (*Lycopus europaeus*).

Fig. 3.6 Axios River Floodplain – grassland community. Source: photo by Dylan Lloyd that appeared in (Newly 1995, p. 5)

Wetland Fauna

The delta's extensive undeveloped areas are more isolated from human disturbance, and provide an ideal habitat for a whole variety of wildlife. Both saltwater and freshwater fish live in the Axios Delta. Thirty-six different species have been identified, of which 33 are indigenous and 3 have been introduced. They include perch, carp, eels, mullet, needlefish, and one endemic species of roach (*Rutilus macedonicus*).

Although the amphibian and reptilian populations of the delta have not been studied in detail, six species of reptiles have been observed. Frogs, terrapins, and water snakes are found in the canals and drainage ditches, while turtles, snakes, and lizards thrive in drier areas. The Alyki wetland adjacent to the Axios Delta is a site of herpetological importance. A large population of tortoise (*Testudo hermonni*) began a slow recovery between 1990 and 1999 following catastrophic habitat destruction in 1980 (Hailey and Goutner 2002).

By far the most impressive inhabitants of the delta are its birds. Some 215 different species have been identified; of these, 109 are waterfowl or shorebirds, which are dependent on water presence. These species come to the delta to nest, to winter, or to rest during their long migratory journeys. Waterbirds that utilize the Axios, Loudias, and Aliakmon estuaries include Eurasian spoonbill (*Platalea leucorodia*), little bittern (*Ixobrychus minutus*), black-crowned night heron (*Nycticorax nycticorax*), squacco heron (*Ardeola ralloides*), little egret (*Egretta garzetta*), Dalmatian pelican (*Pelecanus cripus*), Pygmy cormorant (*Phalacrocorax pygmeus*), Eurasian oystercatcher (*Haematopus ostralegus*), black-winged silt (*Himantopus himantopus*), Pied avocet (*Recurvirostra avosetta*), Kentish plover (*Charadrius alexandrinus*), black-tailed godwit (*Limosa limosa*), collared pratincole (*Glareola pratincola*), yellow-legged gull (*Larus cachinnans*), Mediterranean gull (*Larus melancephalus*), little tern (*Sterna albifrons*), and greater short-toed lark (*Calandrella bracydactyla*) (Bird life International 2006).

In the spring and summer months the most conspicuous occupants of the delta are the herons, whose nesting colonies are among the largest in Europe (Erwin 1996, Kazantizidas and Goutner 1996, Kazantzidis et al. 1997, Papakostas et al. 2005). From April to September little egrets, night herons, squacco herons, and purple herons can be seen feeding in the extensive rice paddies, the coastal marshes, the canals, and the riverbanks. During these months there are also spoonbills, glossy ibis, and cormorant as well as waders such as redshank, black-winged stilt, avocet, and terns.

Thousands of ducks winter in the delta. Most have flown south from their northern nesting grounds to spend the winter months in a milder climate, but some species, including shelduck and mallard, are full-time residents. A total of 112 different species of ducks have been observed. In the winter there are also many birds of prey, including buzzard, long-legged buzzard, falcons, other species of herons, such as the great white egret (which formerly nested in the delta) and the grey heron, as well as a variety of shore birds, including stints, curlews, and turnsones. Most shore birds, however, are seen during the spring and autumn migration, for the delta is a major nesting place for migratory birds on their long journeys.

Foxes, jackals, badgers, martens, weasels, hares, and wildcats – even wolves occasionally – and at least 10 other species all make their homes among the dense vegetation along the shores of the river. The observant visitor may well catch a glimpse of a European suslik on the embankments, or a coypu (which is not a native species but has been introduced from South America) in a canal or a drainage ditch. According to a study by Newly (1995) evidence of occurrence of a number of mammals within the Axios River Delta is provided in Table 3.1.

Table 3.1 List of mammals found in the Axios River Delta area

Species found	Evidence
Eastern hedgehog (*Erinaceous concolor*)	Dead animals found
Bi-colored white shrew (*Crocidura leucodon*)	Species captured
Lesser white-tailed shrew (*Crocidura suaveolens*)	Species captured
Mole (*Talpa* spp.)	Signs
Red fox (*Vulpes vulpes*)	Sightings
Wolf (*Canis lupis*)	No direct sightings
Golden jackal (*Canis aurous*)	Reports
European badger (*Meles meles*)	Dead animal found
Marbled polecat (*Vormela peregusna*)	No direct sightings
Beech Martin (*Martes foina*)	No direct sightings
Stoat (*Mustela erminea*)	Sightings
Weasel (*Mustela nivalis vulgaris*)	Sightings

Table 3.1   (continued)

Europa Suslick (*Spermophilus citellus*)	Sightings
Steppe Mouse (*Mus spicilegus* var. *hortulans*)	Species captured
House mouse (*Mus musculus domesticus*)	Dead animals found
Pigmy field mouse (*Apodemus microps*)	Species captured
Wood mouse (*Apodemus sylvaticus*)	Species captured
Brown rat (*Rattus norvegicus*)	Sightings
Coypu (*Myocastor coypus*)	Sightings
Brown hare (*Lepus europaeus*)	Sightings

Source: Data from Newly (1995).

Human Activities and Land Use

The principal crops grown in the area are rice, maize, and cotton. Rice cultivation, which requires that the fields remain flooded for several months of the year, is considered particularly valuable for the ecosystem, especially during summer when the freshwater supply drops substantially. The rice paddies provide a habitat for a wide variety of insects, amphibians, reptiles, and birds (Fasola et al. 1996). Extensive networks consisting of channels, ditches, and dikes modify the water flow and divert it from the natural systems for agricultural needs.

In former years livestock grazed over a much greater area than at present, when so much land has been given over to agriculture, leaving only a narrow strip along the coast and the river banks for grazing purposes (Maragou and Montziou 2000). Cattle and sheep, although perfectly adapted to the habitat, have largely replaced water buffalo although more recently water buffalo is being reintroduced in the Axios River Delta and elsewhere (Gatteniohner et al. 2004). There are overgrazing problems along the river floodplain tamarisk scrub area.

There is extensive fishing – for both fish and shellfish, and especially oyster, horse mussel, and warty venus – in the coastal zone. The particularly favorable considerations in the delta itself, with the blend of salt and fresh water and nutrients, have led to the development of shellfish farming, with an emphasis on mussel production (Maragou and Montziou 2000).

There is sand extraction throughout the riverbed, which provides sand in sufficient quantity to meet the needs of the construction industry in the entire surrounding area. Hunting attracts not only people from the neighboring villages but also large numbers of city dwellers. There has been some scientific projects, some complete and some ongoing, to study the geomorphology, flora, fauna, culture, and history of the area. There is potential for environmental education and nature-based recreation, but the author saw little ongoing activity in his visit to the delta.

Threats to the Delta

Despite the protective status of the area, the value of the wetland is steadily being eroded by a whole series of threats.

Urban waste and industrial effluents from the entire area drained by the river constitute a water quality problem. Most of the waste and effluent originates within the former Republic of Yugoslavia to the north plus many towns and villages along the Greek portion of the river, which ends up in the delta. Unabsorbed fertilizers and agro-chemicals from adjacent farmlands also end up in the delta with negative effects on the quality of the water and all that lives in it.

Studies have indicated that organochlorides, metals, POPs, and other chemicals have ended up in birds and benthic organisms with the Axios Delta. For instance, Goutner et al. (1997) have examined organochloride insecticide residues in eggs of little tern (*S. albifrons*) in the Axios River Delta. Goutner et al. (2005) have also examined colonial waterbirds (Aves, Charadriiforms) for PCBs and organochloride pesticide residues in eggs. All pollutants were detected in all species in all areas expect Deldrin in the Mediterranean gull. Percent levels of higher chlorinated PCB congeners (IUPAC 118, 138 and 180) were greater than other compounds in all species and all areas, probably due to their bioaccumulation properties. Significant differences between Mediterranean gulls and avocets (at Ervos) were found with regard to PCB 138 and PCB 180, whereas differences between Mediterranean gulls and common terns (at Axios) were found for all PCBs except PCB 8 and PCB 20. Maximum pesticide concentrations in all samples were below 50 ppb, except for B-HCH and 2.4'DDD for all areas and species. In summary, agro-chemical sources are dominant over industrial pollution, but their levels were too low to have adverse biological effect over the species studied (Goutner et al. 2005).

Albaige (2005) also looked at levels of persistent organic pollutants (POPs) in different biotic (bivalves, fish, marine mammals, and sea birds) and abiotic compartments (air, seawater, and sediments) of the Mediterranean Sea. No conclusions were drawn due to scarcity of emission data and shortage of measurements of good quality.

Janssens et al. (2002) and Goutner and Furness (1997) looked at heavy metals, especially mercury in the Axios River Delta for water birds. Janssens et al. (2002) analyzed heavy metals (sliver, arsenic, cadmium, copper, and mercury. Lead and zinc concentrations have been found in feathers of nesting great tits (*Parus major*). There was a gradient of higher concentrations of silver, arsenic, mercury, and lead for those specimens closer to the pollution source and no significance in cadmium between sites. Goutner and Furness (1997) measured mercury concentrations in feathers of little egret and black-crowned night heron chicks and their prey in the Axios River Delta. Significantly major concentrations occurred more in night herons than in little egrets in 1993. Diets differed considerably between the two species due to different foraging habitats. Mercury concentrations in the pumpkinseed sunfish (*Lepomis gibbosus*), goldfish (*Cariassius auratos*), and dragonfly (*Odonata larvae*) were highest among the prey. Frogs and water beetles (Dytiscids) had moderate concentrations, whereas saltwater fish and terrestrial prey had very low mercury concentrations. The implication is that deltaic marshes are the habitat most polluted with mercury.

Ghatzinkolaou et al. (2006) recently assessed change in lotic benthic macro invertebrate assemblages along the Axios–Varde River in Greece and Macedonia. Macrozoobenthos and water samples were collected during summer 2000 and autumn of 2001. Total dissolved solids and total suspended solids were found to be the primary factors affecting the structure of benthic communities. Additionally, species composition was impacted by untreated sewage effluent, industrial discharges, agricultural runoff, intense water usage, and impoundments.

Studies by Karageorgis et al. (2005, 2006) have documented long-term impacts of anthropocentric pressures on the Axios River Delta and the Inner Thermaikos Gulf coast. Presently, more than 11,800 tons of nitrogen and 3,400 tons of phosphorus are released annually into the marine system. During the last 20 years, freshwater discharges have decreased and riverine nutrients have increased, whereas inputs from domestic and industrial effluents have a decreasing trend. Nutrient over-enrichment impacts such as eutrophication, harmful algae blooms, and hypoxia still have to be addressed according to Karageorgis et al. (2005). Van Gils and Argiropoulos (1991) call for a consistent methodology to analyze water quality of the Axios River basin and similar systems.

Damming, principally in the former Republic of Yugoslavia, plus heavy pumping for irrigation purposes, has resulted in lower freshwater levels and a drop in the water table, which in turn has resulted in increasing salinization or saltwater intrusion further inland. Decreased water flows also make it more difficult for the river and the delta to flush itself of domestic and industrial effluents. In the Delta of the Axios–Loudias–Aliakmon there are plans for the diversion of part of the Aliakmon River for irrigation of the Thessaloniki plain for the supply of drinking water to the City of Thessaloniki (Maragou and Montziou 2000).

Unauthorized construction and other illegal activities impact some of the most sensitive areas of the delta impacting large areas of prime habitat for fish spawning and bird reproduction. Abandoned refuse and open burning mar the beauty of the delta and the floodplain landscape.

Sand extraction operations, creeping expansion of farmlands toward the river, uncontrolled grazing plus illegal timber harvesting all contribute to a reduction of the floodplain riparian vegetation which is important to erosion protection and wildlife habitat. In 1987 sand extraction operation destroyed an entire islet in the river, which was the nesting ground of hundreds of herons and egrets.

Grazing is not controlled, resulting in overgrazing and deterioration of soils and vegetation. Another problem with livestock is that in spring the animals frequently tread on the nests of ground nesting birds like terns and stilts. Also burning of the tamarisk scrubland in hope of establishment of richer grazing grounds has taken place in more recent years.

Hunting is heavily practiced and hunters do not always respect the status of protected areas. Wardening is lacking, which means that preservation of the rare birdlife in the wetland depends largely on the ecological awareness of the hunters.

Illegal reclamation of salt marsh for secondary home development or establishment of facilities either for shellfish cultivation or establishment of stables for animals becomes the most serious threat for further reduction of coastal marsh areas.

Affect of Wetland Loss on Major Functions

The following section outlines some major impacts of past wetland loss on the delta's functions including sediment trapping, flood control, shoreline erosion reduction, groundwater recharge and discharge, retention/removal of nutrients, food chain support, fish habitat, wildlife habitat, and socio-economic utilization.

Sediment trapping: Walling and Webb (1987) have examined the data on the transport of material by the Worlds Rivers and they have included Greece in the highest category of sediment yield with more than 500 tons/km per year. One of the most important functions of the Axios Delta was sediment trapping. Archeological evidence together with pollen analysis and carbon 14 analysis (NEDECO 1970) indicates that a considerable increase of land has taken place since the fifth century BC in the Thessaloniki Plain. The plain represents the latest stage in the infilling of a subsiding basin.

The principal agents in transporting this material have been the Axios, Aliakmon, and Galikos Rivers. The three rivers historically have been areas of dynamic change as a result of fluvial processes advancing their deltas seaward into the Gulf of Thessaloniki. Alterations of their routes to the sea have often happened every few hundred years or during exceptional floods, when the rivers abandoned their old routes in favor of shorter adjacent ones. Various estimates have been done to estimate the amount of sediment transported by the three rivers and deposited into the plain. These estimates range from 16×10 m of sediment per year to 16.3×10 m/year, which would seem capable of filling the plain with its area and original depth 2,500 years ago to its present altitude.

Fears of infill of the Thessaloniki Gulf with sediment that would prevent the Thessaloniki port from being navigable by large ships by 1925 initiated the diversion of the Axios River that was completed by 1934. Most of the sediments and nutrients carried by the river are now being deposited in deep parts of the Thessaloniki Gulf and driven away by sea currents and therefore do not contribute to the construction or maintenance of the coastal wetlands at the same rate as before the diversion.

The diversion of the Axios has created the modern bird foot delta of Axios, while the abandoned delta is in a stage of decay. Almost 144 ha of sand deposited by the former mouth of Axios and Galikos Rivers eroded away by 1945, followed by 525 ha of coastal marsh due to erosion and subsidence by 1970. As rates of subsidence and erosion of an abandoned delta follow a decelerating pattern, slight additional erosion is expected to take place.

The flood control measures taken along the routes of the two rivers of the plain do not allow the river water to spread on the marsh areas where marsh plants could trap the sediment. In addition, a number of impoundments built along the courses of the Axios and Aliakmon Rivers for irrigation and energy production purposes also retain part of the sediment. So in essence, sediment is reduced from upstream sources, has little chance to deposit sediment along the river courses, and once it reaches the delta edge moves directly to the deeper portions of the gulf. The natural deposition period is closed now by construction of dikes along the rivers and the seashore. Only catastrophic floods and irrigation water can bring fresh sediments to the area.

Flood storage and desynchronization: Mediterranean wetlands are likely to provide localized flood control function during winter when the annual rainfall may cause seasonal flooding (Maltby et al. 1988). What little opportunity for flood storage and desynchronization was provided by the former floodplain of the Axios River. Because of the dikes and control structures on either side of the river, this function is greatly reduced.

Shoreline erosion reduction: The tide is at a minimum in the Mediterranean but storm and wave action can contribute considerably to coastal erosion. Marsh-protected coasts suffer comparatively little damage from storms and wave action. The only evidence of coastal erosion in the Axios Delta was provided by the retreat of the coastal marsh in the vicinity of the old Axios River Delta. About 525 ha of coastal marsh were lost during 1945–1970 due to erosion and subsidence of the abandoned delta (Athanasiou 1990).

Groundwater recharge and discharge: A water balance groundwater study was done by NEDECO from 1968 to 1969. The study indicated that the amount of downward and upward groundwater flows was negligible. A vertical hydraulic gradient exists between the phreatic zone and the top of the artesian aquifers over much of the plain. The gradient in the study area was generally downward and small while the sediments over the bulk of the plain are particularly clayey and therefore highly impermeable. The same study concludes that most of the surface water bodies Galikos, Axios, Loudias, and Aliak Rivers gain from groundwater. However, all rivers flow within bounded areas, protected by cutoff drains over a large part of their length so that their influence as an input to water bodies is negligible.

Exploitation of aquifers for irrigation and water supply may result in a fall in the water table, saline penetration, and reduced standing water leading to significant reductions in waterfowl and fisheries habitat. Possible seawater seepage in both phreatic and artesian systems in the area was considered by NEDECO (1970) to be negligible. Groundwater quality, even in the vicinity of the sea, for both phreatic and artesian waters was found to be significantly different in composition from seawater. Nevertheless, high evaporation rates during periods of low precipitation combined with the shortage of irrigation water derived from the rivers (Axios River has been dry during most of the summer the last few years) increase the demand for irrigation and place additional pressures on reduced freshwater supplies.

Retention/removal of nutrients: Nutrient retention involves the fixation and storage of nutrients, in particular nitrogen and phosphorus within the substrate or vegetation. Storage may occur on a short- or long-term basis associated with the substrate. These processes contribute to the maintenance or even improvement of water quality for the rivers and the delta.

Pollution of coastal water in the Mediterranean is high in areas that have high effluent discharges as a consequence of intensifying inorganic fertilizer applications in agriculture and the discharge of increasing volumes of sewage effluent. There is a large amount of phosphorus from detergents and synthetic fertilizers, and the river edge wetlands could play a role in reducing the pollution load before entering the Mediterranean Sea. Such phenomena are quite common in the Thermaikos Gulf. It is certain that the extensive shallow marshes of the plain had a high nutrient removal capacity that have been greatly reduced by direct loss of the majority of marsh and swamp areas and the reduction of the wetlands capacity to trap sediment. As a result the Thermaikos Gulf suffers high eutrophication and pollution rates that have directly affected the fish stock and recreation potential of the area.

Food chain support: Little information is available on the food chain support functions of Greek wetlands. Nevertheless, the dikes being built along the seashore of the delta area have minimized the tidal or wave action inundation of coastal marshes. It can be seen that there is minimal plant diversity within the coastal marshes. The flood control works rose along the main rivers of the plain and the canalization of others have minimized the active mixing of fresh and saline water as well as nutrient availability through sedimentation and increased salinity. The food chain support function is linked both to fish and wildlife habitat qualities.

Fish habitat: There has been no attempt to make analyses of fisheries dependence on wetlands in Greece. The drainage of extensive freshwater marshes of the plain as well as drainage of Giannitsa Lake has had a direct effect on fish catch. The only pre-drainage delta on fish catches in Giannitsa Lake refers to 640,000 kg of fish caught in 1930 (Konstandinidis 1989). This has been lost with the drainage of the lake. The major hydrologic alterations that took place in the plain and the reduction of freshwater influx to Thermaikos in combination with the pollution of the gulf must have affected both inland and coastal fisheries strongly. The extensive reclamation of the coastal zone, the infill of the lagoon between Loudias and Aliakmon Rivers in the early 1970s, and the rise of dikes along the coast line have decreased the available nursery and wintering areas for coastal species.

Wildlife habitat: Pre-drainage information concerning the plain's wetland fauna is scarce and is based on description of the area given by aged inhabitants of the area or novels inspired by the area (Delta 1937). They mention an abundant wildlife that includes wolves, foxes, wild boar, martens, reptiles, fish, and thousands of waterfowl that lived in the shallow Giannitsa Lake and the extensive marshes of the plain. Sources refer to the imperial eagle and the little bustard as well as very large numbers of nesting waders that have now been reduced significantly (Jerrentrup et al. 1988).

More recent information is closely related to the destruction of the remaining habitats of the study area. The 1970 IWRB mission stated, "particularly all Anatidae (water birds) were on the large lake between Loudias and Aliakmon, should this be drained results would be catastrophic for the wildfowl wintering in the area". Unfortunately plans went ahead and the area was drained in the early 1930s (Hafner and Hoffman 1974).

Midwinter counts underestimate the populations wintering in Axios–Loudias–Aliakmon Delta because of the difficult access of the coastal habitats in the mouths of the rivers (Jerrentrup et al. 1988). However conclusions driven from comparison between the 5-year means of midwinter counts conducted in the period 1968–1974 with those conducted in 1982–1986 show that the site is the strongest example of wetland destruction in Greece.

The delta used to qualify as a wetland of international importance both for its total number of wintering wildfowl (50,800 birds on average during the first period of the study with a maximum count of 141,800 birds in 1971) and for its concentrations in widgeon (20,600 on average during 1968–1974, maximum record was 70,000 in 1971), coot, and shelduck (13,600 and 700 on average, respectively, during 1968–1974). The latter figures mean that the wetland no longer qualifies as a Ramsar site under quantitative criteria. The recent counts (1982–1986) give an average of 4,300 birds in total wintering on the site (maximum count in 1982 was 10,100 birds) and an average of 1,000 widgeon, 200 coot, and 400 shelduck.

The site still qualifies as a wetland of international importance under qualitative criteria (Athanasiou 1987). Similar conclusions are drawn in the study of the overall results of the International Waterfowl Census in 1967–1986 (Monval and Pirot 1989). The direct decrease of inland wetland freshwater marshes and floodplain forests of the plain combined with the hydrologic alteration and degradation of coastal marshes deprived wildlife from the abundance of food offered by the natural marsh and the nesting and roosting areas found on forested islands in the river bed or the riparian forest.

A heronry is located on an island in the Axios Delta, which is among the most important features of the wildlife of the area. Almost 935 pairs of little egret, night heron, squacco heron, glossy ibis, spoonbill, pigmy cormorant, and cormorant breed on the riparian forest (Kazantzidis 1996). The heronry was previously located on an island outside the delta area, in the lower course of the river. Sand extraction operators in the riverbed knocked down the trees formerly used as roosting and nesting sites. The isolation of the Axios River Delta is difficult which permitted the survival of the heronry. The breeding populations of the little egret, night heron, spoonbill, and glossy ibis are of international importance. The glossy ibis concentration (50–70 pairs) is the most important in Greece (Jerrentrup et al. 1988). Significant numbers of black-winged stilt (Jerrentrup et al. 1988), avocet, redshank, collard pratnecks, and kential plover breed along the coastal strip of the salt marshes. The area also supports many pairs of white stork breeding in the surrounding villages

(Jerrentrup et al. 1988), while it is regularly used by Dalmatian pelicans (a world endangered species) as a feeding ground.

Socio-economic utilization: The Thessaloniki plain reclamation scheme has been considered one of the most successful schemes of its kind in Greece (Konstandinidis 1989). All economic evaluations or rentability studies (Altigos et al. 1962, NEDECO 1970, Konstaninidis 1989) have been done under the perspective of improvement of an apparent wasteland that has been put to good uses. No study took into account what was being lost in terms of productivity, life-support systems, water cycle control, or nutrient retention as explained above.

As pointed out by Athanasiou (1990) there is a significant missed opportunity cost. One of the important problems of the City of Thessaloniki faces at the moment is the disposal of sewage of more than 600,000 inhabitants as well as industrial effluents. Domestic sewage has been estimated as 21,000 m per day and there is no estimate concerning the volume of local effluents from local industries in the area. At the moment the effluents are deposited in the Thermaikos Gulf without any treatment, causing serious odor and aesthetic problems to the city, depriving adjacent coastal areas from their recreation potential, and having caused irreparable damage to the fisheries of the gulf. As of 1996, one-third of the sewage receives primary and secondary treatments and is then discharged to the Thermaikos Gulf. Smaller demonstration projects are being used to treat sewage with (a) lagoon systems for about 2,500 inhabitants and (b) artificial treatment wetlands for about 500 inhabitants. Treated effluent is then used for irrigation of nearby fields.

No studies have been done to estimate the cost of the recreational potential of the coastal areas or the coastal fisheries. Coastal fisheries in the Thermaikos Gulf at present suffer due to both high pollution of the gulf and loss of coastal marsh area.

Remaining Wetlands and Current Management Issues

Axios and Aliakmon Deltas: The two deltas certainly are of outstanding importance mainly because they remain relatively free from human intervention. They still perform functions such as sediment trapping, nutrient removal and retention of the load brought by rivers, shoreline anchoring, food chain support and are important habitats for wildlife (internationally important numbers of little egret, night heron, glossy ibis, and spoon bill breed in the Axios Delta).

The riparian forest: This habitat has undergone considerable decline through the years. They used to cover large areas of the floodplain while nowadays only patches are found in the deltas and between the agricultural land and the rivers. Research on the role of riparian forests in nutrient dynamics in agricultural watersheds indicates that sediment trapping and nutrient removal in riparian forests are of ecological significance to receiving waters and that they reduce or defuse pollution. Riparian forests also play a significant role in erosion prevention as well as wildlife habitat. In the river corridor they serve as a significant nesting and roosting habitat for waterfowl.

The marshes: This type of habitat has undergone a huge decrease of area due to conversion to agricultural land as was previously discussed. There are hardly any fresh marshes (380 ha) nowadays and those remaining are the most vulnerable areas in danger of further reclamation.

A waterbird survey on the accessible marshes of the area was conducted in spring 1989 and summer 1989 and 1990 (Athanasiou 1990). The results of the long-term monitoring of birds were presented in a delineation study in 1996. The marsh areas were classified for their vegetation and flooding regime and evaluation for their relative importance for waterfowl population. Athanasiou observed waterfowl during seven field visits over a total area of 1,814 ha consisting of tidal salt marsh (409 ha), permanently flooded non-tidal marsh (125 ha), lagoons (169 ha), and fresh water marsh (380 ha). The results demonstrate the impacts of the flooded areas for waterfowl populations during spring and summer. Seasonally flooded salt marsh (425 ha) (dry during survey) and dry salt marsh (306 ha) were also surveyed but no birds were observed. The same areas, besides being of significant importance for waterfowl during spring and summer also perform the functions of shoreline protection (tidal marsh), nutrient retention and removal (fresh water marsh, tidal marsh), and important habitat for fish and shellfish (tidal marsh and subtidal marsh).

The delineation of the area as a Ramsar site did not include a very important part of the marsh habitat. The Kalochori–Galikos area happens to include almost all permanently flooded non-tidal marsh habitats found in the whole wetland area as well as significant fresh water, tidal marsh, and lagoon areas. The permanently flooded non-tidal salt marsh, although just 7% of the surveyed area, concentrated on average of 32.6% of the observed waterfowl. It is also a very important roosting site of the little egrets breeding in the area, mainly used in late summer. If the Ramsar Convention is going to have any beneficial effect, this part of the wetland should enjoy full protection and should be involved in the relevant delineation.

Other management concerns for marsh areas include the following:

- Protection of all marsh areas (seasonally flooded salt marsh, dry salt marsh) as a buffer zone between the agricultural land and wet habitats.
- Stop illegal building of cottages on the salt marsh.
- Removal of rubbish disposal sites from the salt marsh as well as disposal sites for shellfish shells.
- Establishing the carrying capacity of marsh areas for grazing and/or control overgrazing and extensive destruction of nests of breeding waterfowl during spring.

The scrubland: The tamarisk scrubland area consists of an important habitat for wildlife. Further removal will not only reduce habitat diversity but will also eliminate vegetative screens, which shield wildlife. Any degradation of the scrubland will place new stress on the adjacent marsh and forest areas. Management concerns include carrying capacity of the scrubland for grazing animals as well as importance for wildlife habitat. Controlled burning should be considered vs. uncontrolled burning.

Rice Fields: Although this is an agricultural ecosystem, it is considered very important for the survival of the heronry of the area. Rice fields in 1987 covered 80% (8,785 ha) of the area in the vicinity of the remaining natural habitats (Gerakis et al. 1988). Many investigators have looked at factors influencing distribution of nesting colonies of herons and proximity of food sources in Europe. In general they support the fact that prey intake by herons in rice fields during their peak-breeding season is higher than it is in non-agricultural habitats. The abundance of prey (that consists mainly of fish, frogs, and aquatic insects), which reproduces in rice fields during May and June, is suitable for the tending of young herons that grow and leave the nest during the same period and seems to be a very important factor for the population of the Axios heronry as well (Kazantzidis 1996).

Current Institutional Framework and Role of NGOs

So what has happened by way of wetland management in Greece and specifically for the Axios River Delta? The follow-up data are based on a letter supplied by Athanasiou (1996) in response to the author's questions. The first part is a review of new institutional developments, the second part outlines the role of the WWF Red Alert Project in Greece, and the last part is a summary of what has been achieved to date.

Joint Ministerial Decision

The draft Joint Ministerial Decision (JMD), which defines the boundaries of the wetland and the permitted human activities within two different zones, has been prepared by the Greek Ministry of Environment, Physical Planning and Public Works. The Ministry of Environment, the Ministry of Agriculture, and the Ministry of Industry, Energy and Technology signed the JMD. The JMD, besides defining the limits and zoning of the wetland, describes the permitted activities in each zone and are valid for 2 years. Within these 2 years the state must issue a permanent Presidential Decree for the protection of the site. The JMD basically foresees three degrees of protection in zones: (A) high protection, (B) peripheral protection zone, and (C) protection of watershed. Among other regulations, control measures will be exercised upon polluting, habitat fragmenting/degrading, and species molesting development activities, such as installation of light industry plants, housing, crop farming, road construction, grazing, obstruction of riparian vegetation, and disposal of sewerage wastes.

The JMD that concerns the Axios Delta together with adjacent Galikos and Loudias estuaries, Aliakmon Delta, and Alyki Kitros has been circulated to the associated authorities whose views have been collected and processed. The official JMD has been issued and is still in force (WWF-Greece 2000).

Management Scheme

Until today, there has been no single wetland management scheme for any wetland or any protected area in Greece. What this means is that all activities affecting wetlands (crop and animal farming, irrigation, fisheries, housing and industry development, hunting, etc.) are managed, but what is missing is an integrated approach to wetland management.

In 1996 the Ministry of Environment put into operation a scheme that will utilize the information center already constructed in the vicinity of Axios Delta (town of Chalastra) as well as the warden hats (three at the Axios Delta). This scheme consisting of two scientists and three guards–guides will operate for 2 years. Among the duties that will be undertaken by the scheme will be (1) establishment within the local society, (2) getting acquainted with local problems and demands, and (3) facilitation of communication of associated parties, the coordination of positive actions, and the collaboration with local NGOs. At the end of the 2 years, there will possibly be an opportunity for the development of a wetland management scheme for the site. WWF thought that its Red Alert Project had also been invited to participate in the advisory committee for the wetland management scheme (Athanasiou 1996).

Habitat Directive

The Axios Delta (and Axios River) has been included in the proposed list of sites eligible to be included in the NATURA 2000 network of Directive 92/43/EEC also known as the Habitat Directive. The implementation of this directive attempts to put forward a network of sites ("Sites of Community Importance") and also to designate sites as Special Areas of Conservation, in order to contribute to nature conservation within the territory of the European Union.

The Red Alert Project of WWF Greece

The project has been operating from 1990 to 1997 at the Axios Delta and is the only active national/international NGO involvement (Maragou and Montziou 2000). The work program has three major components and corresponding sub-elements. These are given below:

- *Detection and monitoring of threats* to the site, through regular site visitation and the development of a network of contacts and constant information gathering (local authorities, public services, local NGOs, unions of farmers, fishermen, hunters, etc.).

- *Take action needed in order to avert threats through*

  - alerting decision makers and wetland-managing authorities both at the local and national levels;
  - supporting wetland-managing authorities in their efforts to promote and implement sound management measures;
  - documentation of threats;
  - evaluation of possible effects of threats on wetland functions and values;
  - investigation of possible alternatives that are not detrimental to the site or proposing conservation measures;
  - dissemination of information to all possible allies, scientific/conservation bodies, the Greek Biotype Wetland Center, Greek and International NGOs, the relevant ministries (Environment and Agriculture), and the public services, the European Commission and the Ramsar Bureau.

- *Raise public awareness on wetland functions and values through*

  - publications addressed to the public;
  - articles to the press and participation in TV and radio broadcasts;
  - provision of material for environmental education teachers;
  - presentations and guided tours for school children.

Progress to Date with the Red Alert Project

This section relies upon details provided by Athanasiou (1996):

Monitoring of threats and actions taken – The project's constant presence at the site allowed WWF Greece to acquire good knowledge of both existing and emerging threats to the site and to take action to avert such threats. It also allowed WWF Greece to develop good relationships with public services, local authorities, and users of the site. As a result human activities in and around the wetland are now much better known, not only by the project staff but also by each group of users and the government services for which the project acts as an information source. Specific activities for detection and resultant actions taken by the project include the pollution of the Axios River, the illegal constructions at the Axios Delta, industrial waste disposal in the Galikos area, and vandalism of the cormorant colony in the Axios Delta.

Additionally the WWF Greece Red Alert Project has been invited to participate in various local initiatives for the protection of the site including

- The "Committee on Wetlands" of Thessaloniki Prefecture, which consisted of about 10 members representing local civil services, involved with wetland management as well as local NGOs. The committee was created in order to provide recommendations in the actions needed to protect wetlands of Thessaloniki (two Ramsar and several non-Ramsar sites), which include

- "Exchange Experience and Knowledge between Mediterranean Countries on Wetland Conservation" a local community's union project funded by EEC.
- The "Committee for the Protection of Axios", which consists of representatives of all communities located along the Axios River, academics, and NGOs. The committee has been created in order to address problems, alert local authorities, and seek cooperation with local authorities in former Yugoslavia.
- The "Committee on the Natural Environment of Northern Greece" an initiative of the Ministry of Macedonia and Thrace.

Action for the promotion of legal protection status of the site has focused on the following issues:

- The delineation of the Axios Delta and the approval of its protection status.
- The protection of the Galikos estuary (Kalochori–Galikos marshes). The site has been included in the draft IMD that addresses the Axios Delta.

WWF Greece reports on the threats of the site that were used as support material in the Ramsar Conference of the contracting parties at both Kushiro (Japan) and Brisbane (Australia). These same report results have also received a lot of publicity in Greek newspapers. The contribution of the WWF Greece project to the conservation of wetlands has been acknowledged in the official report of the Greek State (prepared by the Ministry of Environment) to the contracting parties of the Ramsar Convention on the occasion of the Brisbane meeting.

The Red Alert Project staff participated actively in the delineation study of the Axios Delta site and has been asked to comment on the delineation proposals prepared by the ministry for the protection of the site in collaboration with the Greek Biotype Center. As a result of the WWF Greece work documenting the value of the Galikos River estuary, the Ministry of Environment has developed a plan for the protection of Galikos estuary including the area in the Axios Delta delineation proposal.

Dissemination of results from the WWF Greece project: The project is communicating their reports to the Ministry of Environment and the Ministry of Agriculture. Both agencies have displayed a positive interest in the project. WWF Greece's reports concerning threats and their activities were communicated to the European Union, to the Ramsar Bureau, to the IUCN, and to a selected number of local authorities and NGOs. The amount of information circulated by the project has made local authorities more aware of threats to wetlands and supposedly more careful in their decision making.

Support of wetland management authorities: In several cases WWF Greece provided specific technical support to local authorities in the preparation of plans or proposals for the designation of protected areas, the promotion of public awareness about wetlands, and the application for funding of projects integrating development with wetland conservation.

Then there was implementation of the WWF Project "Partnerships for sustainable development in the regions' funded by the European Union". The project was aimed at providing sustainable development through EU-funded development activities. The Red Alert Project joined with the Greek Biotype Center to prepare the "Training and Needs Assessment for Wetland Conservation in Greece" (MedWet).

Raise public awareness on wetland values and functions: WWF Greece's awareness efforts included production of posters and stickers highlighting the values of the site as well as production of information leaflets in cooperation with various local authorities. The production of these information leaflets focused on the qualities of wetlands in general and was aimed at a general audience. Titles included are as follows:

- What are wetlands and what is their importance?
- Threats to wetlands and wise use
- The Ramsar Convention and wetland protection
- The EC Birds Directive, the Bern Convention, and the Greek Frame law of the environment with respect to wetlands
- How local NGO's' citizen groups and schools can be involved in wetland protection
- Wetland vegetation
- Wetland wildlife
- Fisheries and wetlands
- Annual farming in wetlands

The leaflets were distributed to civil servants, municipalities, politicians, clubs, teachers, and local NGOs. They were often reproduced in newspapers. Total distribution was estimated at 1,500 copies and according to WWF Greece the demand is still growing. Preparation of an educational package on the delta was done for use by the local schools as well as in the region. Presentations at conferences, school lectures, guided tours, and articles in the local and national press were a common occurrence.

New Axios River Delta Management outcomes – post-2000 include the following: On a pilot scale, a number of experimental restoration projects have been implemented within the framework of the LIFE Nature II project "Conservation of the Pygmy Cormorant and Lesser White-footed Goose in Greece" through the creation of a series of three ponds in the flood plain bed of the Axios River; the deepening and clearing of a drainage ditch in the same area in Evros Delta during summer months, in order to increase the fresh water habitat; and the planting of trees in the Axios River Delta, Evros Delta, Lake Kerkini, and Porto Lagos in order to restore and increase riverine formations and riparian forests.

A private firm from Thessaloniki has undertaken management of a 136 ha area located in the flood bed of the Axios River, within the protected zone, with funds from the Agro Environment Regulation 2078/92. Management measures applied included removal of animal grazing from the area, wardening and fencing of the area, management measures for the vegetation of the area, as well as development of educational and recreational activities.

Environmental education programs: In the Delta of Axios it is estimated that about 500 people per month participate (WWF-Greece 2000). During the years 1999–2002 the Hellenic Ornithological Society organized boat tours in the Thermaikos Gulf and Axios River to inform citizens about the birds living in the gulf and the need of protection. The Hellenic Ornithological Society and the World Wildlife Fund for Nature (WWF) have developed, since 1997, a project for the protection of the *Ancer erythropus* and *Phalacrocorax pygmaeus* in 10 wetlands (including the Axios Delta). The WWF has prepared three educational packages, two of which are dedicated to the Axios Delta and Evros Delta, respectively (Bassoukea and Markopoulou 2002).

Institutional Development, Innovation, and Evaluation

One of most important achievements of the Red Alert Project has been the development of a "methodology" and a "standard of work" that other NGOs can use in order to develop similar activities. The kind of work carried out by the Red Alert Project, which in the early 1990s was considered innovative activity, in its course came to be considered necessary and was adopted first by the Greek Biotype Wetland Center and now the state itself (Athanasiou 1996).

On the basis of the Red Alert Project, the Greek Biotype Wetland Center developed the "Wetlands Monitoring Project" which has been running for 2 years in approximately 50 sites, while today the Ministry of Environment is also adopting the idea and wants to implement it in eight other Ramsar wetlands. The Red Alert Project has been invited to participate in the Advisory Committee of the above-mentioned scheme.

The above cited development of local organization and monitoring capability is very important for Greece at this time. The country had no precedent of environmental NGOs that would undertake such activity or recognize such values. Second, the government ministries were under pressure from the EC, Ramsar Bureau, IUCN, and others because the wetlands were in such bad shape and so little progress made toward integrated management. In fact one can see the official criticism in the Ramsar Conference meeting in Japan of the lack of progress toward management of Greek wetlands. So the Red Alert's progress was welcome indeed.

The project management team needs to get more actively involved in the management of the Axios Delta while at the same time maintaining close cooperation with the JMD scheme created by the Ministry of Environment. The management issues that the Red Alert Project plans to become involved with have been chosen on the basis of serving as pioneer projects in the field of wetland management in Greece while placing much emphasis on the active involvement of local people. Two of such projects are as follows:

- Rice Farming at the Axios Delta – how can it become more compatible with natural wetland functions?
- Protection restoration of the riverine forests and *Tamarix* scrub woodland in the Axios Delta related to needs of colonial nesting birds.

These will be important management activities as we have seen in the current management issues section of this case study. It remains to be seen how the JMD, the management scheme, and the role of the Red Alert project will work out. In 1999 there was an "Expression of opinion" was filed with the Ramsar Convention (1999) for possible removal of the Delta of the Axios, Loudias, and Aliakmon rivers. The Axios Delta and many Greek wetlands have some tough management problems to solve, a history of wetland degradation, and little environmental advocacy. At least with the help of WWF Greece, some in roads have been made toward basic recognition of wetland functions, values, and health. The role of WWF Greece Red Alert project as an environmental conscious and monitoring presence should be especially noted.

Acronyms

EC: European Commission
EEC: European Economic Community
IUCN: International Union for the Conservation of Nature
IWRB: International Wetlands Research Board
JMD: Joint Ministerial Decision
NEDECO: Netherlands Engineering Consultants
PCB: polychlorinated biphenyls
POP: persistent organic pollutants
WWF: Worldwide Fund for Nature – World Wildlife Fund

References

Albaige, J. 2005. Persistent organic pollutants in the Mediterranean Sea. In *Handbook of Environmental Chemistry*, pp. 89–149. Berlin/Heidelberg: Springer.
Altigos, N., K. Kyraikos, A. Maheras, and N. Nikolaidis. 1962. *Project d' Irrigation de la Plaine de Salonique, sur Rentabilite*. Athens: Royaume de Greece, Ministere de la Corrdination.
Athanasiou, H. 1987. *Past and present Importance of Greek Wetlands for Wintering Waterfowl*. Slimbridge UK: IWRB Pub., 63 pp.
Athanasiou, H. 1990. *Wetland Habitat Loss in Thessaloniki Plain, Greece*. M.Sc. Dissertation, London: University College, 45 pp., plus appendices.

Athanasiou, H. 1996. FAX/Letter in response to author's questions, 4 pp.

Athanasiou, H., A. Dimitricu, and S. Kazantzidis. 1994. *The Axios Delta*, Athens: WWF-Greece.

Bassoukea, E. and S. Markopoulou. 2002. *National Report for COP7; National Report for the 7th Meeting of the Contracting Parties to the Convention on Wetlands and Hellenic Ministry of Environment*, Athens, Greece, 34 pp.

Birdlife International. 2006. GRO28 Axios, Loudias and Aliakmon estuaries, http://www.birdlife.org.

Delta, P. S. 1937. The secrets of the marsh. *Estia* (in Greek).

Erwin, R. M. 1996. The relevance of the Mediterranean region to colonial Waterbird conservation. *Colonial Waterbirds*, 19(1): 1–11.

Fasola, M., L. Canova, and N. Saino. 1996. Rice fields support a large portion of herons breeding in the Mediterranean region. *Colonial Waterbirds*, 19(1): 129–134.

Gatteniohner, U., M. Hammeri-Resch, and S. Jantschke (eds.). 2004. *Reviving Wetlands-Sustainable Management of Wetlands and Shallow Lakes*, pp. 82–83 and 91–93. Radolfzell, Germany: Global Nature Fund.

Gerakis, P. A. (ed.). 1988. *Conservation and Management of Greek Wetlands*. Gland Switzerland: IUCN, 439 pp.

Gerakis, P. A. (ed.). 1992. *Conservation and Management of Greek Wetlands*. Proceedings of a Workshop on Greek Wetlands [Thessaloniki, Greece 17–21 April 1989], Gland Switzerland: IUCN, pp. xii, +493.

Gerakis, P. A., D. S. Veeresoglou, and K. Kalambourzti. 1988. *Agricultural Activities in Axios Delta and Evaluation of their Potential Hazardous Effects on the Wetland Habitat*. Thessaloniki: Aristotelian University of Thessaloniki, Department of Agriculture, Laboratory of Ecology.

Ghatzinkolaou, Y., V. Dakos, and M. Lazaridou. 2006. Longitudinal impacts of anthropologic pressures on benthic macro invertebrate assemblages in a large transboundary Mediterranean river during the low flow period. *Acta Hydochimica et Hydrobilogica*, 34(5): 453–463.

Goutner, V., T. Albanis, and I. Konstantinou. 2005. PCB's and organochloride pesticide residue in eggs of threatened colonial Charadriiforms species (Aves, Charadriiformes) from wetlands of international importance in northeastern Greece. *Belgium Journal of Zoology*, 135(2): 157–163.

Goutner, V., I. Charalambidou, T. A. Albanis. 1997. Organochloride insecticide residues in eggs of the Little tern (*Sterna albifrons*) in the Axios Delta, Greece. *Bulletin of Environmental Contamination & Toxicology*, 58(1): 61–66.

Goutner, V. and R.W. Furness. 1997. Mercury in Feathers of the Little Egret (*Egretta garzetta*) and Night heron (*Nycticorax nycticorax*) chicks and their prey in the Axios Delta, Greece. *Archives of Environmental Contamination and Toxicology*, 32(2): 211–216.

Hafner, H. and L. Hoffmann. 1974. *The 1970 IWRB Mission to Greece*. IWRB Bull. 37.

Hailey, A. and V. Goutner. 2002. Changes in the Alyki Kitrous wetland in northern Greece: 1990–1999, and future prospects. *Biodiversity and Conservation* 11(3): 357–377.

Janssens, E., T. Dauwe, L. Bervoets, and M. Eens. 2002. Inter- and intraclutch variability in heavy metals in feathers of Great Tit nestlings (*Parus major*) along a pollution gradient. *Archives of Environmental Contamination and Toxicology* 43(3): 323–329.

Jerrentrup, H., M. Gaetlich, A. H. Joensen, H. Nohn, and S. Brogger-Jensen. 1988. *Urgent Action Plan to Safeguard Three Endangered Species in Greece and EC: Pygmy Cormorant, Great White Egret, White Tailed Eagle*. Report to EC-DGXI, 153 pp.

Karageorgis, A. P., V. Kapsimalis, A. Kontogianni, M. Skourtos, K. T. Turner, and W. Salomons. 2006. Impact of 100-year Human Interventions on the Deltaic Coastal Zone of the Inner Thermaikos Gulf (Greece): A DPSIR framework analysis. *Environmental Management*, 38(2): 304–315.

Karageorgis, A. P., M. S. Skourtos, V. Kapsimalis, A. D. Kontogianni, N. Th. Skoulikidis, K. Pagou, N. P. Nikolaides, P. Drakopoulou, B. Zanou, H. Karamous, Z. Levkov, and Ch. Anagnostou. 2005. An integrated approach to watershed management within the DPSIR framework: Axios River catchment and Thermaikos Gulf. *Regional Environmental Change* 5(2–3): 138–160.

Kazantzidis, S. 1996. Breeding ecology of the Egret garzetta L. (Little Egret) and the *Nycticorax nycticorax*, L. (Black-Crowned Night Heron), (Ardeidae, Aves) in Axios River Delta, Greece.

Kazantizidas, S. and V. Goutner. 1996. Foraging ecology and the conservation of feeding habitats of Little Egrets (Egretta garzetta) in the Axios river delta. Macedonia, Greece. *Colonial Waterbirds*, 19(1): 115–121.

Kazantizidas, S., V. Goutner, M. Pyrovetsi, and A. Sinis. 1997. Comparative nest site selection and breeding success in 2 sympatric ardeids, Black-Crowned Night herons (*Nycticorax nyticorax*) and Little Egret (Egretta garzetta) in the Axios delta, Macedonia, Greece. *Colonial Waterbirds*, 20(3): 505–517.

Konstandinidis, K. A. 1989. *The Reclamation Works in Thessaloniki Plain.* Thessaloniki: Greek Geotechnical Union, 217 pp (in Greek).

Maltby, E., R. Hughes, and C. Newbold. 1988. *The Dynamics and Functions of Coastal Wetlands of the Mediterranean Type.* DGXI, 112 pp.

Maragou, P. and D. Montziou. 2000. *Assessment of the Greek Ramsar Wetlands*, Athens: WWF-Greece, 59 pp.

Monval, J. Y. and J. Y. Pirot. 1989. *Results of the IWRB International Waterfowl Census 1967–1986.* Slimbridge, UK: IWRB Special Publication No. 8, 129 pp.

National Statistical Service of Greece. *Population Censuses 1920, 1928, 1940, 1951, 1971, 1981.*

Nazirides, T., H. Jerrentrup, and A. J. Crivelli. 1992. Wintering Herons in Greece (1964–1990). In T. Hollis et al. (ed.). *Managing Mediterranean Wetlands and their Birds.* Slimbridge, UK: IWRB Special Publication No. 20, IWRB and Insituto Naziorde d. Biologia della Selvaggina, Ozzano Emila, Italy, pp. 73–75.

NEDECO. 1970. *Regional Development Project of the Salonika Plain.* Vols. A, B. C. and D.

Newly, S. 1995. *The Occurrence and Distribution of Mammals at the Axios River Delta, Northern Greece.* Thessaloniki, Greece: COMETT Programme, Greek Biotype and Wetland Centre. http://www.bangor.ac.uk/~bss035/projects/comett/mammals.html

Papakostas, G., S. Kazantzidis, V. Goutner, and I. Charalambidou. 2005. Factors affecting the foraging behavior of the Squacco Heron. *Colonial Waterbirds*, 28(1): 28–34.

Psilovikos, A. 1988. Changes in Greek wetlands during the twentieth century: The cases of the Macedonian inland waters and of the river deltas of the Aegean and Ionian coasts. In P. A. Gerakis (ed.) *Conservation and Management of Greek Wetlands*, pp. 175–196. Gland, Switzerland: IUCN.

Psilovikos, A. 1992. Prospects for wetlands and waterfowl in Greece. In T. Hollis et al. (ed.) *Managing Mediterranean Wetlands and their Birds*, pp. 53–55. Slimbridge, UK: IWRB and Insituto Naziorde d. Biologia Della Selvaggina, Ozzano Emila, Italy.

Ramsar Bureau, 1999. *Expression of Opinion on Greek Ramsar Wetlands and Possible Removal from the Montreux Record.* Gland, Switzerland: Ramsar Convention Bureau, and at http://www.ramsar.org/key_montreux_greece1_anx6.htm

Tsiouris, S. E. and P. A. Gerakis. 1991. *Wetlands of Greece: Values, Alteration Protection.* Thessaloniki: Aristotelian University of Thessaloniki Department of Agriculture, Laboratory of Ecology and Environmental Protection (in Greek), 96 pp.

Valaoras, G. 1992. Greek wetlands: Present status and proposed solutions. In T. Hollis et al. (ed.) *Managing Mediterranean Wetlands and their Birds*, pp. 262–266. Slimbridge, UK: IWRB Special Publication No. 20, IWRB and Insituto Nazionale d. Biologia della Selvaggina, Ozzano Emila, Italy.

Van Gils, J. A. G. and P. Aigiropoulos. 1991. Axios River basin water quality management. *Water Resources Management* 5(3–4): 271–280.

Walling, D. E. and B. W. Webb. 1987. Material transport by the world's rivers: evolving perspectives. In *Water for the Future: Hydrology in Perspective*, pp. 313–329. International Association Scientific Hydrology Publication 164.

Zalidis, G. 1993. International wetlands inventory for Greece: prospects and Progress. In M. Moser et al. (eds.) *Waterfowl and Wetland Conservation in the 1990's; A Global Perspective*, pp. 178–184. Slimbridge, UK: IWRB Special Publication No. 26, IWRB.

Chapter 4
The Kafue Flats in Zambia, Africa: A Lost Floodplain?

Introduction

In parts of sub-Saharan Africa, as well as South America, India, and southeast Asia, interior riverine wetlands are stressed or altered by dams and reservoir projects for hydroelectric, irrigation, and flood control "benefits" (Dugan 1988, Nelson et al. 1989, Scudder 1989). Resultant impacts from hydroelectric alteration and lack of flooding downstream affect both biodiversity and human use of floodplain wetlands for agriculture, fiber, and medicinal plant usage (Mathooko and Kariuki 2000, Tockner and Stanford 2002).

We know relatively little about many of the interior wetland systems of Africa as few wetland inventories have been done to document existing African wetlands distribution, value, and function prior to the 1980s. Wetland inventories were brokered by the IUCN and/or the Ramsar Bureau from the 1980s on. In Zambia a comprehensive wetland inventory was completed in 2002.

Because of the impacts on biodiversity and floodplain-dependent agriculture, scientists around the world are re-examining the possibility of emulating flood flows to re-establish lost floodplain functions (Acreman 1994, Bayley 2006, 1995, Giller 2005, Horowitz 1994, Junk et al. 1989, Standford et al. 1996, Ward and Stanford 2006, Welcomme 1995). In fact, specific reintroduction of flood flows have been partially implemented for riverine wetland systems in South Africa (Brock and Rodgers 1998, Le Maitre et al. 2002), Cameroon (Evans et al. 2003, Mouafo et al. 2002, Scholte et al. 2000, Wesseling et al. 1996), for the Phongolo floodplain (Bruwer et al. 1996), northern Nigeria (Thomas 1999), as well as work proposed for the Zambezi River (Beilfuss and Davies 1999, Gammelsrod 1996, Scudder and Acreman 1996).

Kafue Flats, as we will see, is one of the most studied and unique floodplain riverine systems in Africa. This case study is important as it races both the biodiversity and human livelihood changes within a wetland system with little dependence on government or NGO intervention and management until recent times.

The Kafue River is a major north bank tributary of the Zambezi, which joins downstream of the Chirundu, and approximately 75 km below the current Kariba Dam (Fig. 4.1). Its basin lies wholly within the Republic of Zambia,

R.C. Smardon, *Sustaining the World's Wetlands*,
DOI 10.1007/978-0-387-49429-6_4, © Springer Science+Business Media, LLC 2009

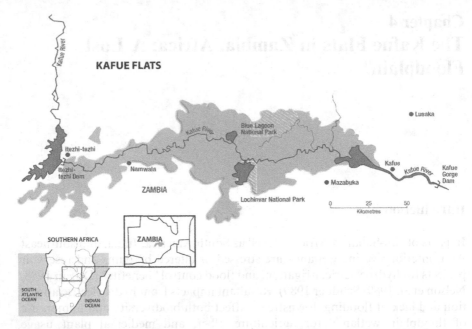

Fig. 4.1 Location map of Kafue Flats on the Zambezi River. Drawn by Samuel Gordon and adapted from WWF-Partners for Wetlands Zambia WWF, Kafue Flats, http://www.pan da.org/about_wwf/where_we_work/africa/where/zambia/kafue

and covers an area of 154,000 km, which is approximately one-fifth of the total area of the country. The river rises near the Zaire border and initially flows southeastward through the copper belt. Thereafter it adopts a generally southwesterly direction until it enters the Kafue National Park, through which it strikes southward to Itezhitezhi. At Itezhitezhi a ridge of resistant rock, which crosses it at that point, constricts what has been a fairly open valley. Below Itezhitezhi the river leaves the Kafue National Park and swings eastward through the Kafue Flats (Fig. 4.1). The Kafue takes a meandering and indeterminate course through the flats, the main stream splitting and joining in places. Many blind-ended "lagoons" are linked to the river, and these and other oxbow lakes indicate former river courses. The gradient of the main channel is notably low, the river falling only 10 m in the 450 km of channel length between Itezhitezhi and the Kafue Rail Bridge at the eastern end of the flats. This low gradient and the width of the riverine plain, combined with the constriction of the valley in the Kafue Gorge downstream of the flats, is the reason for the seasonal flooding which occurs when the rainy season discharge of the Kafue enters this section of its valley. At the gorge the river abruptly descends into the rifted trough of the Zambezi, falling approximately 600 m in just 25 km. The Kafue is confluent with the Zambezi 40 km below the gorge.

The Kafue Flats formed a large annually flooded plain (see Figs. 4.2a and b) approximately 255 km long and up to 56 km in width, along the borders of the Kafue River (see Fig. 4.1), a major tributary of the Zambezi. About 3,000–5,000 km of the total area of 7,000 km is inundated for a period ranging from 1 to about 7 months. When dry, most of the grasslands on the Kafue Flats are grazed by cattle, owned by local herdsman of the Ila and Tonga tribes. The

Fig. 4.2a Nyimba looking northwest from east of Nampewgue in April 1970. Photo credit: University of Michigan Fisheries Research Team

Fig. 4.2b Chunga–Namp Gag in April 1970. Photo credit: University of Michigan Fisheries Research Team

annual flooding largely prevents any other type of traditional land use on the flats proper, but in the dryer parts of the transition zone some corn is grown locally. Fishing villages are found on the higher levees along the main river and its tributaries. The traditional inhabitants of the flats, the Twa, live in mostly permanent villages, while migrants from other areas of Zambia and from neighboring countries usually occupy semi-permanent villages that have to be abandoned during high floods.

On the Kafue Flats only a few large-scale agricultural projects have been implemented. In the southeastern part of the flats near Mazabuka (Fig. 4.2) some 10,000 ha of sugarcane are grown under irrigation. Apart from this, use of the Kafue River water for farming on a commercial basis is confined to some small-scale private estates in the same area.

General Floodplain Ecology

The following references have provided source material for the following section: Chapman et al. (1971), Chabwela (1998), Chooye and Drijver (1995), Ellenbroek (1987), Handlos (1997), Howard and Williams (1982), Perera (1982), Sheppe and Osborne (1971), Rees (1978a, b), and Williams and Howard (1977). Until the beginning of this century, the Kafue Flats presented, like other Central African floodplain systems, a very rich wildlife area. The high primary productivity of the floodplain grasslands allowed a rich fish and birdlife. During the course of the dry season, when the plains are dry, large herds of wild ungulates being expelled from the burned upland savannas invade the area. Today, this natural situation persists in two wildlife sanctuaries on the Kafue Flats, the Blue Lagoon National Park and the Lochinvar National Park, respectively, on the north and south banks of the river (Fig. 4.1). Both parks are former cattle ranches in which large carnivores were systematically exterminated. Wild herbivores, however, were protected against poaching and this is certainly the main reason for the fact that these areas still harbor large concentrations of wild ungulates today.

As regards birdlife, the parks are listed among the 10 best-stocked sanctuaries in the world. The Kafue Flats are renown for their enormous numbers of waterfowl, ducks, geese, pelicans, stilts, storks, and egrets (Douthwaite 1974b, Douthwaite 1982, Howard and Aspinual 1984, Osborne 1973). Also, the quite rare wattled crane (*Grus carunculatus*) occurs in large numbers there (Douthwaite 1974a). In a normal year fewer than 1,000 cranes are present at high flood; but as the water level subsides the population increases, and in the latter half of the dry season it numbers some 3,000 birds. Following widespread flooding in 1972 at least 300 pairs nested as the water fell. Many full-grown birds molt their remiges between January and April, and are then flightless. The diet is largely of rhizomes dug from soft mud and suitable feeding grounds in the dry season are created by a falling water level.

Among the larger herbivores, by far the most important inhabitant of the Kafue Flats in terms of numbers of individuals, however, is the Kafue lechwe (*Kobus leche kafuensis*) (see Figs. 4.3a and b). This antelope, which is endemic to the Kafue Flats, lives a semi-aquatic way of life. Its special hoof structure enables it to walk on very soft and sticky clay soils during flooding and allows it to graze the emergent vegetation in the shallows, up to 50 cm deep water.

Fig. 4.3a Red Lechwe, pelican (pink back), wood ibis and spoonbills along Kafue Flats in 1970. Photo credit: Donald Stewart

Fig. 4.3b Red Lechwe buck in April 1970 Photo credit: University of Michigan Fisheries Research Team

Lechwe spend most of their time on the floodplain (see Bell et al. 1973, Handlos et al. 1976, Howard et al. 1988, Howard and Sidorowicz 1976, Robinette and Child 1964, and Sayer and van Lavieren 1975). During high flood, however, the deep water forces the lechwe to leave the floodplain and almost completely strip this area of plant cover.

Apart from lechwe, a number of other animals still present today at Lochinvar and Blue lagoon, must be mentioned. Zebra (*Equus burchelli*), the second most abundant species at Lochinvar, may be found on the floodplain only when the soils are dried out and hard. During the rainy season and the early dry season these animals are largely confined to the termitaria grasslands. Wildebeest (*Connochaetes taurinus*), though much less numerous than zebra, shows more or less the same seasonal migration pattern. Other large ungulates on the flats and adjacent woodland areas at Lochinvar include buffalo (*Syncerus caffer*), Oribi (*Ourebia ourebi*), reedbuck (*Redunca arundinum*), and kudu (*Tragelaphus strepsiceros*). See Ellenbroek (1987), Perera (1982), Sayer and van Lavieren (1975), Sheppe and Osborne (1971), and Williams and Howard (1977) for more background on mammals.

Larger carnivores such as lion (*Panthera leo*) and wild dog (*Lycaon pictus*) occasionally visit the area. Spotted hyena (*Crocuta crocuta*), serval (*Felis serval*), and side-striped jackal (*Canis adustis*) are still permanent inhabitants. Two interesting nocturnal mammals, still common at Lochinvar but rarely seen, are the peculiar termite-feeding aardvark (*Orycteropus afer*) and the vegetarian-created porcupine (*Hystrix* spp.). Hippopotamus (*Hippopotamus amphibius*) occurs in small herds in most parts of the Kafue Flats (Fig. 4.4) while sitatunga (*Tragelaphus spekei*), a true aquatic ungulate, is confined to papyrus and reed marshes, has been observed locally. At Lochinvar the activities of hippos hardly produce any visible signs of utilization of the floodplain grasslands.

Fig. 4.4 Hippos in the Zambezi in April 1970. Photo credit: Donald Stewart

Every year the Kafue River floods the Kafue Flats to a depth of up to 5 m for several months. The flats are 235 km long and up to 40 km wide. The life of the flats is conditioned primarily by the alternating rainy and dry seasons and by the floods. About 80 cm of rain falls from November to April. The Kafue rises slowly during the rains, is highest in May, and falls during the latter part of the dry season. Vegetation is composed primarily of grasses, especially *Oryza barthii* (wild rice). The main vegetation zones include (1) the Main river; (2) levees; (3) lagoons and depressions; (4) floodplain grassland; (5) water meadow; and (6) the littoral zone. A Vetivera belt includes (7) lower termitaria zone; (8) upper termitaria zone; and (9) transition zone (see Figs. 4.5a and b). Finally there is the upland consisting of (10) Munga woodland and (11) Miombo woodland (Ellenbroek 1987). Because of the abundant water the primary productivity of the flats is much greater than surrounding woodlands as is secondary productivity.

Every year there is an alternation of aquatic and terrestrial faunas. During the floods fish move onto the flats from the Kafue River, and most spawning takes place there. Terrestrial species are driven off, but as the floods recede they reoccupy the floodplain and use what is by far the best grazing in the region. Large mammals find shelter in tall stands of grass on the floodplain, small mammals in the thick mat of vegetation that covers much of the ground or in the deeply cracked soil.

There is a gradient to use of the floodplain; some species (hippopotamus, otter) always stay near the water at low water, others (lechwe, zebra, wildebeest) go for varying distances onto it, and more than half the animals (squirrel, vervet, aardvark) go onto it little if at all. Failure to use the floodplain seems to be due to the absence of suitable habitats or food, rather than exclusion by

Fig. 4.5a Lechwe at a distance. Photo credit: Donald Stewart

Fig. 4.5b Flooded termitaria grassland near Chunga in April 1970. Photo credit: Donald Stewart

the floods. The most abundant large mammal on the floodplain is the lechwe. Several shrews and mice, especially *Mastomys natalensis*, are common on the floodplain and breed there during the rains. During the floods they leave the flood plain or take refuge on natural levees along the Kafue. Crocodile (*Crocodylus nilotica*) and monitor live near the water's edge and move in and out with the floods like the hippopotamus. Some snakes are common on the flats, but turtles and frogs are not. Terrapins are present in some areas. Although ants and termites are abundant in the surrounding region, they are largely excluded from the flats by the floods. An excellent source on the vegetation of the Kafue Flats is Ellenbroek (1987) and for fauna, see Sheppe and Osborne (1971).

Of all the birds that use Kafue Flats, and are wetland dependent, the Wattled Crane is one of the most significant and threatened species. The large river basins like Kafue Flats are their preferred habitat in shallow wetlands with minimal human disturbance. Their diet consists primarily of aquatic vegetation such as *Cyperus* and *Eleocharis* spp. and water lilies (*Nymphaea*), but also includes seeds, insects, and waste grain in drier habitats (Douthwaite 1974a). The cranes are non-migratory but do irregular local movements in response to water availability (Burke 1992).

Human Use and Land Use History

Background for the following section came mainly from Chabwela (1998), FAO (1968), Jeffrey (1990), Jeffrey and Chooye (1990), Lehmann (1977), and Seyam et al. (2001). Various Bantu Tribes inhabit the region of the Kafue Flats, primarily Tonga in the southeast and Ila in the northeast and west. They practice a largely subsistence agriculture, with corn and other staple crops. Much of the economic and social life centers on cattle. These tribes originally practiced shifting agriculture, but increasing population and habitat deterioration have caused more people to be crowded on less land resulting in constant use of the same land and subsequent deterioration of the soil.

The flats themselves are inhabited only by fishermen, who live in scattered villages on the natural levees along the Kafue River (see Figs.. 4.6a, b and c). The original inhabitants were Batwa, but when commercial fishing became important in the 1950s people from the other fishing tribes moved onto the flats and now seem to be replacing the Batwa. Fisherman on the Flats was estimated at about 1,000 in 1964 (Pike and Corey 1965), many of them present only at low water, the main fishing season. This is one of the most important fisheries in Zambia (Chapman et al. 1971, Dudley and Sculley 1980, Lagler et al. 1971, Muyanga and Chipungu 1982). Fish are taken by gill net or seine, set from dugout canoes and fiberglass boats. Most of the catch is sundried, bundled, and

Fig. 4.6a Mound with village at Nyimba in April 1969. Photo credit: KFL

Fig. 4.6b Village mound at Nyimba. Photo credit: Donald Stewart

Fig. 4.6c Flooded village at
Luwanta in April 1969.
Photo credit KFL

sold to itinerant African traders who come out to the river during the dry season
in trucks or on bicycles.

In addition to fishing, the permanent residents grow small patches of corn,
vegetables, and tobacco for their own use, and many of them have cattle. High
floods cover even the highest levees and the residents may live in water for
several weeks.

After the flood recedes, tribesmen from wooded areas around the flats bring
tens of thousands of cattle to the river. Grazing on the uplands is poor at best
and almost worthless during the dry season, and most of the growth of the cattle
takes place during the few months when they are in the flats.

Most of the flats area was native reserve, but early in this century several
large areas were given to European settlers for ranching and some are still used
this way. There is a large agricultural research station at Mazabuka. East of
Mazabuka a small area of the floodplain has been diked and farmed, but the
heavy clay soil makes farming impractical.

Except for the extermination of game herds, use by man seems to have done
little to disturb the ecology of the flats according to Sheppe and Osborne (1971)
up until the hydroelectric dam development. Cattle may remove all of the
exposed vegetation in some areas, but it is replaced by an equally heavy growth
during the next rain and flood season. It has been suggested that sorghum
thickets may form where there were cattle pens during the years of commercial
ranching. Numbers of winter-thorn and perhaps other trees on the levees have
been cut to make dugouts and used for firewood, and the villages and their
gardens have changes in small areas of the levees. No new plants or animals
are known to have colonized the flats as a result of man's activities in 1971,
except perhaps the pied wagtail (*Motacilla aquimp* Dumont), which is said to
occur on the flats only around human habitation. Recently man has contributed to the invasion of Kariba weed (*Salvinia molesta*), which becomes
entrapped in fishing gear, and then the seed is transported to non-infested
areas (B. Kamweneshe).

The National Parks

The Kafue Basin encompasses three national parks with a total area of 23,440 km and most of the nine game management areas (GMA) with an area of 34,750 km (Fig. 4.4) (Douthwaite and van Lavieren 1977, Howard 1977). The Kafue National Park (KNP) is famous as one of the largest and best-stocked reserves in the world. This park includes most of the western watershed of the Kafue River from the Busanga Plain to the start of the Kafue Flats below the Itezhitezhi Dam. The Kafue Flats area contains two small national parks (Lochinvar and Blue Lagoon) with a total area of 840 km while a further 9,000 km comprises Namwola, Kafue Flats, and Mazabaka GMAs, which is most of the area on the flats.

The value of the three national parks as preserves of wildlife, natural heritage, and as attraction for tourists is undisputed. All three parks also act as stocking reserves for wild animals, which may be hunted under license in the GMAs of the Kafue Flats and surrounding areas. Without these parks the GMAs would become depleted of large mammals.

In the wet season from January to May only two centers are open to tourists at KNP (at Ngoma and Chunga) as most roads are impassable and much of the wildlife is spread throughout that vast reserve. However, the KNP has a unique frontage onto the Kafue and Lufupa Rivers with roads (passable throughout the dry season) for hundreds of kilometers near the water. It is here that most of the wildlife congregate in the drier months making game viewing, bird watching, and scenic touring possible between the eight tourist lodges and camps scattered down the length of the park. KNP is accessible by road from all directions and has many international visitors who fly into Ngoma from Lusaka.

Lochinvar National Park (LNP) is one of the few national parks that is open to tourists throughout the year and its value is enhanced because it is close to Lusaka. Good roads are available for game viewing and touring while a visitor's lodge caters to overnight stays or longer tours. Blue Lagoon National Park is even closer to Lusaka and its causeway provides a unique opportunity to drive several kilometers out into the flooded Kafue Flats to observe the proliferation of birds, wildlife, and aquatic vegetation in a way that could only, otherwise, be done by boat.

The special value of Lochinvar and Blue Lagoon NPs is their uniqueness in preserving and providing the last remaining natural refuges of the Kafue lechwe, mentioned earlier, which was previously found throughout the entire length and breadth of Kafue Flats (Howard and Sidorowicz 1976). These parks are also regarded as being "one of the most important freshwater wetlands for waterfowl in east, central, and southern Africa" (Douthwaite 1974a). It has even been suggested that these two areas be combined with an area of the GMA between them to form a "Lechwe National Park" (FAO 1968, vol. V). Together the two established reserves of the Kafue Flats contain numerous other mammals beside the especially adapted antelope including zebra, wildebeest, buffalo, bushbuck, Oribi, impala, reedbuck, sitatunga, hippo, warthog, bush pig as well as vervet monkeys, baboons, side-striped jackals, wild dogs, hyena, common duiker, greysbok, and smaller cats, mongoose, squirrels, pangolin, aardvark, porcupines, etc.

Some of the management issues with these national parks include cropping the antelope to maintain a staple population and/or providing other forms of wildlife from the Kafue Flats as source of food for local population, e.g., spur wing geese, ducks, guinea fowl, and francolin. The other issue is the compatibility of other land uses/activities with park activities. For instance, the Mindeco Small Mines Gypsum Plant inside the LNP seems to have little direct impact and have been disbanded. Similarly, other activities such as fishing under license and passage of fish traders through the parks are fine as long as soils, vegetation, and wildlife are not disturbed. However, there have been instances where the presence of fish traders, fish transport trucks, and other people have affected the breeding of the Lechwe by interfering with the animals forming breeding groups. There have been instances of destruction of vegetation (for firewood, house-building, unbogging of vehicles) and the erosion of tracks into watercourses by increasing numbers of people associated with legal activities within the parks. Most people are within the parks without the required permits for entry, so it is difficult to control their movements and their presence encourages and disguises the activities of poachers.

Even given all these issues the overall tone until 1972 is compatible in usage patterns both within and outside the parks and the Kafue Flats. The situation sketched until this time changes with the advent of hydroelectric development on the Kafue River. So sustainable usage patterns underwent a wrenching change after 1972. The following sections outline the hydroelectric development schemes and their impact on the Kafue Flats. This is followed by the role of NGOs and other groups in reaction to these changes and the search for more sustainable regulation of the river and endangered floodplain resources.

Hydroelectric Development on the Kafue River

Key sources for the following section include Howard and Williams (1982), Rees (1978c), Schuster (1980), Scudder (1989), Sheppe (1985), SWECO (1967, 1968, 1969, 1971), Tiffen and Mulele (1993), Williams (1977), and Williams and Howard (1977). Eleven months (October 1966) after the unilateral declaration of independence (UDI) President Kaunda announced that the Kafue hydroelectric scheme would be undertaken and in 1967 work began on the Kafue Gorge Dam. The UN was approached, and in May 1961 a project was approved for a multipurpose survey of the entire Kafue Basin to determine the optimum use of its land and water resources in accordance with the needs of the country with FAO as the executive agency. Work began in 1962 and was completed in

1966. Its seven-volume report was published 2 years later (FAO 1968). As expected the report viewed favorably the development of hydroelectric potential, but its recommendations led to modifications to the scheme on ecological grounds (particularly in response to changes in the flooding regime). In the meantime work had gone ahead on the engineering aspects and in April 1967 the Swedish engineering consultants engaged on the work SWECO presented their report on the Stage I power station, followed by that on the gorge storage reservoir in 1968 (SWECO 1967, 1968). The first stage of the Kafue Gorge scheme commenced in 1967 and was completed in 1972.

At the Kafue Gorge the river descends approximately 600 m over a distance of 25 km. To develop the hydroelectric potential of the site a three-phase scheme was proposed:

Stage I: Construction of the first power station within the gorge utilizing the upper 400 m head of water, and having an initial capacity of 600 MW. Regulated flow for this stage was to be provided by the construction of the shallow Kafue Gorge Reservoir, inundating the eastern edge of the flats.

Stage II: The addition of two additional generating units at the Kafue Gorge (upper) Power Station to provide a total output of 900 MW. This required improved water regulation, which would be possible with the construction of the Itezhitezhi Dam, 250 km upstream, and above the flats section of the river.

Stage III: Construction of the Kafue Gorge Lower Power Station is to harness the remaining 200 m head of water to have a 450 MW capacity, making a total installed generating capacity for the gorge of 1,350 MW. A possible fourth stage involves the raising of the height of the Itezhitezhi Dam and the installation of a small generator unit there.

Preliminary work started at the Kafue Gorge Station in 1967, and in October 1971 the first generating unit was commissioned. Completion had been planned for 1970, but not until April 1972 were all four 150 KW units in operation. The power station is located in a chamber cut out of solid rock some 500 m below the ground to which water is conveyed from the dam in a 10 km tunnel. The dam itself is an earth rock fill structure 50 km high and 375 m long. As can be seen from these dimensions, it was contained in a very narrow valley and could be quite easily increased in height.

The impoundment area, behind the dam, which covers an extensive area of the eastern part of Kafue Flats, is shallow and so a slight increase in water depth would result in a massive extension of the flooded area, and corresponding enormous increase in evaporation. For this reason, the normal maximum design level for the dam is 976 m. This will flood an area of 800 km, although the maximum design level for the dam is 979.0 m, which would give a flood area over three times as large; approximately 3,200 km, which is almost half the total area of the flats. Such extensive flooding was viewed as unacceptable in view of its effects on other activities in the flats but during the 3-year period 1973–1975 the reservoir level was raised to a temporary high level of 977.8 m to ensure adequate water regulation during the period when the Itezhitezhi Dam was being constructed.

Work on Stage II of the project began in 1972, for planned completion in 1978. At the Kafue Gorge the original four generating units were augmented by two additional units, giving a station generating capacity of 900 MW. At Itezhitezhi construction began in June 1973 and the dam was completed in 1977. Itezhitezhi Lake created by the dam covers an area of 370 km extending 30 km along the Kafue and its tributary the Musa River. Ninety per cent of the area of the lake sits within Kafue National Park, and the natural hydrologic regime was substantially modified by this development.

Water Regulation

The real problem or issue is not so much the construction of dams, but that of water level regulation and its effect behind the dam at Kafue Gorge. The seasonal pattern of rainfall over the Kafue Basin results in wide seasonal discharge variations in the river from a low flow of 50 m/s to a high of 700 m/s at the Kafue Gorge. In order to produce a more constant flow, both between seasons and between wet and dry years, a reservoir is required to retain the peak discharge, which can be released progressively when natural drainage levels decline. It is most efficient, in an engineering sense, to have the reservoir at, or a short distance upstream of, the power station. In the case of the Kafue, the gorge reservoir immediately upstream of the power station is inadequate for water regulation, for its volume is limited by unfavorable basin shape. It has a maximum capacity of only 800 million m at the normal maximum water level which would only ensure a "firm" power output of only 207 MW in dry years which is below station capacity.

It was therefore necessary to locate the main storage reservoir at Itezhitezhi, above the flats. This has a storage capacity of 4,950 million m. The location of the main storage reservoir upstream of the flats, and some 250 km distant from the power station that it serves, gives rise to the problems, which are causing wide-ranging ecological concern.

The flats have marked annual flooding regimes, and vegetation, wildlife, fish, and man are all adapted to this. The widespread flooding of the flats is not desirable from the viewpoint of regulation for hydroelectric purpose as large water losses occur by evaporation, and regulation at Itezhitezhi involves holding back part of the flood. Changes in the natural flooding regime are inevitable and drastic. These changes involve the eastern and western parts of the flats differently.

In the eastern part of the flats the effect of the Kafue Gorge dam, in general, will be an increase in the amount of flooding (SWECO 1971). The maximum reservoir level, especially in the entrance of the gorge, will be significantly higher than natural flood levels. As the major function of the regulation is to ensure adequate water supply at the gorge, during the drier parts of the year there will be significantly more water in this part of the flats than under normal conditions. Water storage at the Kafue Gorge Reservoir will be much less efficient than at Itezhitezhi on account of high evaporation and evapotranspiration losses from aquatic vegetation, which abounds in shallow waters. Perennial shallow water conditions would also encourage the growth of aquatic plants such as papyrus, increasing further loss by transpiration, and causing trouble with the penstock intakes of the dam. Water releases from Itezhitezhi will thus not be increased until the level at Kafue Gorge Reservoir has been dropped below its maximum, and as there is a substantial lag time in the passage of water through the flats, considerable variation of water level in the eastern part of the flats happens.

In the western part of the flats the situation is reversed (SWECO 1971) with a reduction in either the amount or duration of flooding or both. In wet or even normal years the amount of water spilled from the Itezhitezhi after filling the dam is likely to be no less than the peak river flow under natural conditions. However, the duration of this maximum discharge will be considerably reduced, so that even normal levels of flooding are reached, they will be reached only briefly. The main problem will be in dry years, when the flood peak does not fill the reservoir and so there will be no spilling of peak discharge. Thus no flooding will take place. This could have disastrous effects on the ecology and economy of this part of the flats, and so to simulate natural conditions in these dry years a "freshet" of 300 m/s is to be released over a 5-year period in March (the normal period of peak discharge) to produce a partial flooding. Even with a freshet, it is clear that in dry years there will be a very significant reduction in the amount of flooding in the western part of the flats.

Ecological and Other Impacts from a Modified Hydrologic Regime

Given the development of the first two phases of the hydroelectric development on the Kafue River the following section outlines ecological and other impacts from the modified hydrologic regime to date.

Vegetation

The floodplain vegetation, mostly consisting of grasses, depends on the annual flooding cycle and has died out in the new bodies of permanent water, which in places now support large submerged mats of aquatic vegetation. *Lagarosiphon ilicifolius* and *Potamogeton thumbergi* have been identified from Chunga Lake.

Thickets of plants that require permanently moist soil are becoming established along the shore of new bodies of water. These plants include papyrus

(*Cyperus papyrus*), Kariba weed, and cattail (*Typha domingensis*), which were formerly excluded by the dry soil that prevails during the low-water season (Mumba 2003).

As had been expected, the elimination of floods on much of the flats has reduced the productivity of the grasses there, though in 1983 it was not possible to distinguish this affect from the affect of the drought. Food elimination has also permitted the invasion of the floodplain by woody plants that formerly were killed by the floods. The most common is *Mimosa pigra*, a tropical American shrub that is now a pest in many tropical areas around the world. *Hibiscus diversifolius* var. *rivularis* also occurs in place.

Affects on Fauna

Unidentified ants are now widespread in places where they formerly did not occur (Sheppe and Osborne 1971). Surprisingly, termite mounds were not seen on the former floodplain, although, before the dams were built, colonies were sometimes temporarily established in sites that were not flooded during the years of low floods.

It had been expected that the dams would benefit fisheries, but initially this has not happened (Dudley and Scully 1980). Experimental sampling in the mid-1970s showed reduced populations of several major fish species, although it is not clear whether this is a long-term trend and, if so, whether it is caused by the dams, overfishing, years of low rainfall, or other factors.

T. O. Osborne (in lit.), former Park Biologist at Lochinvar National Park, believes that the altered flooding regime has adversely affected ungulates, cattle, birds, and fish. Elimination from large areas of the former floodplain grasses has reduced the populations of herbivores and their predators. The floodplain fish were primarily herbivores, and populations of both fish and fish-eating birds have been reduced. There are now many fewer of the formerly abundant herbivorous snails, and the openbill storks (*Anastomus lamelligerus*) that fed on them. A reduction in herbivorous insects has also led to there being now smaller numbers of Jacanas and insect-eating birds.

Parts of the floodplain that once had only transient populations of rodents and shrews now have presumably permanent populations in the thickets of papyrus and other plants that have developed on perennially marsh grown around new bodies of water. Specimens of the shrews *Crocidura marquensis* and *C. occidentalis* and the rodents *Praomys (mastomys) natalensis* complex and *Dasymys incomtus* were trapped in such habitats in 1983. One specimen of the black or roof rat, *Rattus rattus*, which had not been previously recorded from that part of Zambia, was trapped on the riverbank opposite a fishing village.

The only animal that has been carefully monitored is the lechwe, which was described before. Rees (1978a,b) believes that the altered flooding-regime threatens the lechwe by reducing its food supply, while Schuster (1980) suggests that the altered regime may threaten the lechwe directly by interfering with its reproductive behavior and hence lowering the birth rate.

Each species will be apt to respond to the changing flood regime differently, depending on how the precise details of the regime relate to its own needs. Concern has been raised on effects of nesting of species such as the yellow-billed stork (*Mycteria ibis*) and other water birds of the Kafue Flats. Change in flooding hydrology regime is a principal threat to the wattled cranes habitat and reproduction. Douthwaite (1974) noted that the number of pairs attempting to nest on the Kafue Flats depended on the degree of flooding. After an average flood (6.4 m), 40% of the pairs attempted to breed. After minimal flood (5.0 m) only 3% of all pairs of all pairs bred. From 1971 to 1973 aerial surveys of wattled cranes were conducted on the Kafue Flats, Busanga Plain, and Lakanga Swamp (Douthwaite 1974). In 1987, 369 wattled cranes were counted in an aerial survey of the Kafue Flats which projects to 2,500 birds for the entire area (Burke 1992).

Effects on Consumptive and Non-consumptive Uses

The Gorge Dam has affected the fishermen directly by eliminating large areas of the emergent floodplain vegetation that formerly protected the floodwaters from the wind. The open water that has replaced it is exposed to the wind and at times becomes quite rough. The fishermen are not accustomed to this and being unable to swim, some of them have drowned when the boats capsized. In 1978 Itezhitezhi Dam threatened to leak, and so the water level was rapidly lowered, thus creating sudden and unexpected flooding on the flats that endangered fisherman, cattle, and wildlife.

Itezhitezhi Reservoir now covers what was formerly one of the most productive parts of Kafue National Park, including an extensive riverine grassland area that was known as Puku Flats. The puku (*Kobus vardoni*) that lived there have disappeared altogether from this part of the park.

In the gorge, the river formerly fell over a long series of rapids, cascades, and falls, forming distinctive habitats and some of the most attractive scenery in Zambia. Now the water is bypassed through the headrace tunnel and the river in the upper gorge is dry except occasionally when excess water is released from the dam directly into the river.

So the modified hydrologic regime has had drastic effects on the Kafue River and Flats landscape. More drastic effects may be forthcoming on species dependent on the vegetation and former hydrologic regime. This includes the lechwe and other water-dependent mammals, water birds such as the Yellow-billed stork and fish species. These effects, in turn, directly affect fisherman, cattle herders, and people associated with the national parks and their visitors. So the question arrives what role did the local population and local and international NGOs play as events unfolded.

The other stress on wetland health for the region is the impact of copper mining effluent containing heavy metals and resultant degradation of the Kafue River ecosystem and in this case the sediment and biota of the Kafue Flats wetlands (Mwase et al. 1998, Norrgren et al. 2000). Mining effluent has entered the waterways of the copperbelt for the past 70 years (van der Heyden and New 2003), resulting in extensive environmental impacts detected as far downstream as the Kafue Hook bridge, 700 km from the mining area (Backstrom and Jonsson 1996). Several geo- and hydro-chemical studies have quantified the impact of the mining industry on Kafue River chemistry. Kasonde (1990) and Pattersson and Ingri (2001) have documented the increased concentration of dissolved and suspended heavy metals in the Kafue River and the marked accumulation of cobalt, copper, iron, and manganese within the river sediment.

Metal accumulation within the Kafue River ecosystem has been associated with various toxicological impacts. The disappearance of hippopotamus (*H. amphibius*) from the Kafue River in Chingola (van der Heyden and New 2003), the proliferation of water hyacinth, and the bioaccumulation of heavy metals within wildlife tissue have been associated with pollutants in the Kafue River ecosystem (Sinkala et al. 1977; Syakalima et al. 2001).

Mwase et al. (1998) found elevated levels of copper in river sediments and associated this with increased pathology of fish in Kite, Itezhitezhi, and Kafue Town. Mwase et al. (1998) and Norrgren et al. (1998, 2000) demonstrated increased fish mortality and decreased aquatic productivity following exposure of caged fish eggs and fry to Kafue River water and sediment from the mining activity.

The other impact from mining effluent is metal concentrations in wetland plants, which affect plant function and productivity (van der Heyden and New 2003). Some aquatic plants are less tolerant, but both *Typha* spp. and Cyperus spp. are more tolerant and this leads to homogeneity in wetland vegetation composition. Sources report that over 99% of the wetland vegetation at New Dam, Zambia, is composed of *Typha* spp. and *Cyperus* spp.

Role of CBOs and NGOs

As far as the author can determine local CBOs and NGOs had no impact on most of the events described to 1985. The Government of Zambia before and after the unilateral declaration of independence (UDI) did not encourage public participation processes and shared decision making. Thus decisions were centrally made based on political motives and the influence of international development agencies and large technical consultant companies. However, there have been two types of involvements of NGOs plus a recent shift in government policy on some issues affecting the Kafue Flats and its people.

One type of NGO involvement has been local and international academics studying the Kafue Flats. The Kafue Flats ecosystems have been extensively studied both before and after construction of the dams. In the 1960s an FAO team produced a multiple volume report on the resources of the area (FAO 1968). The all-important hydrology of the flats was studied by the FAO team and more recently by a Dutch team (DHV Consulting Engineers 1980). Soils and agriculture were studied by FAO.

FAO (1968), Douthwaite and Lavieren (1977), and Ellenbroek (1987) have described the vegetation. The fish and fisheries have been studied by Chapman et al. (1971), Lagler et al. (1971), Dudley and Scully (1980), and Muyanga and Chipungu (1982), among others. Several workers have studied bird populations (Osborne 1973, Douthwaite 1974a,b, 1982).

There have been numerous studies of lechwe, including those by Robinette and Child (1964), Bell et al. (1973), Sayer and Lavieren (1975), Handlos et al. (1976), Schuster (1980), Rees (1978a,b), and Howard and Jeffrey (1981, 1983). Other mammals have received less attention (Sheppe and Osborne 1971, Sheppe 1972, 1973).

The University of Zambia has had an active interest in the area since the 1960s and its Kafue Basin Research Project (KBRP) continues to study some aspects of the ecology and human use of the flats, including an annual aerial census of lechwe populations. KBRP has also sponsored several conferences and publications on the area (Williams and Howard 1977; Howard and Williams 1982).

Despite all this academic activity, there has been very little activist or NGO activity until just recently. As Sheppe (1985) puts it "Our understanding of the basin is still inadequate for satisfactory protection and management of its resource and second, what we do know has little effect on policy decisions". The construction of the power project was approved without regard to its probable environmental effects, and its design and operation have been based almost entirely on a desire to produce the greatest possible amount of power – without regard to other interests.

In 1983, during regional discussions concerning a wetlands program for southern African states belonging to the Southern African Development Coordination Conference (SADCC) it was suggested that wetlands management be integrated with community development. This was the first formal recognition of the strategic significance of Zambia's wetlands and the dependence of their conservation on the socio-economic well-being of resident communities. This initiative was further developed by the government of Zambia as part of a joint WWF/IUCN Wetlands Program. This program culminated in a consultative workshop, which was held in 1986 for representatives of local communities from two of Zambia's largest wetlands, government and party political officials, technical experts, and other concerned organizations and individuals.

It was against this background that the WWF-Zambia Wetlands project was established with WWF-I support in 1986. The project's aims are appropriate to government policy as specified in the National Conservation Strategy for Zambia adopted in 1985. The project grew out of concern that conventional management of Zambia's wetlands was failing to coordinate development and regulate natural resource utilization. This was manifested in the environmental impacts of hydroelectric power development in the Kafue Flats, as just previously described, and declining fisheries and wildlife population in the Kafue Flats and Bangweulu Swamps. Current thinking at the time was that the demise in wetland management ability is due both to neglect by the central government and not linking to local community development needs.

On the premise that no development can succeed without the security of resources being assured, the fundamental aims of the WWF-Zambia Wetlands Projects are to conserve wetlands' natural resources and enhance their natural productivity. Commensurate with these aims is the objective of improving the standards of living of the wetland's local communities through the sustainable utilization of natural resources.

It became clear that the role of local communities is an influential factor in wetlands management and had been underestimated. The development aspirations of local communities had long been neglected and combined with progressive alienation from their traditional resources of a centralized government yielded a negative relationship. Furthermore, local communities perceived that outside interests (developers and consumers of hydro-electric power) had been given preferential access to resources with little or no return to local people. Simultaneously, the capacity of local government to manage and control the wetland declined. All these factors together, exacerbated by Third World economic depression and population growth, led to progressive abuse of both the Kafue Flats and Bangweulu Swamps.

A program was thus required which would retain more control of, and benefits from, wetlands utilization to the traditional communities by incorporating them in wetlands management and community development processes. It was also realized that wetlands management and community development interests would have to be integrated, local communities would have to participate willingly, and national and international interests in the wetlands would have to be accommodated.

Fortunately, the project's evolution closely followed two fundamental shifts in government policy aimed at promoting self-sufficiency. A decentralized system of government was introduced through the Local Administration Act of 1980. Then in 1983, government departments were given the legal means to develop their own revolving funds to support their functions. Building on these opportunities, a transitional phase of project implementation was elaborated in 1989, beginning with integration with the Department of National Parks (hereafter called the department) and Wildlife Service.

While the project's scope is broader than one department, two factors favored integration. First, the wetlands project areas consist predominately of game management areas and national parks, which fall under the jurisdiction of the department. Secondly, through the department's Administrative Management Design Policy for game management areas and its Wildlife Conservation Revolving Fund, existing facilities are available for integrating natural resource management and community development.

The project areas are divided administratively into wetlands management units incorporating groups of chiefdoms on a geographically manageable basis (communications are often on foot or by canoe). After briefing the district councils (i.e., local government) concerned with these areas, the project is introduced to the chiefdoms by the project team operating within the framework of the National Parks and Wildlife Service. The opportunities for restoring local vested interests in wetlands management and the potentials for earning revenues for community development from the sustainable utilization of resources are carefully explained to chiefs and community members at public meetings.

Under the traditional leadership of the chiefs, it is then up to the communities themselves to elect and run committees, named Community Development Units after the chiefdoms they serve, to take advantage of the services and facilities provided by the project. The interface between customary society and conventional government is provided by Community Development Units and the Wetland Management Authorities established for each Wetland Management Unit. The district governors of the principle districts concerned are normally appointed chairman of these authorities. The chiefs and the Community Development Units on their respective wetlands management authorities represent the interests of the participating chiefdoms. The authorities are also strongly supported by representatives of the project, local leaders, district political and government officials, extension officers, and representatives of associated departments and organizations. Project implementation in the field is thus not only supported by central government through the National Parks and Wildlife Service and associated departments but also reinforced by linkages with central government through district and provincial councils.

The project has had an encouraging reception among its participating communities and local governments. All district councils and chiefdoms have been briefed on the project's objectives, and nearly half the anticipated total number of Community Development Units has been formed. In the meantime, community development activities are proving catalytic in building support for the project. Notable examples are manpower training and employment, construction and rehabilitation of two rural health clinics, and rehabilitation of rural schools, wells, bore holes, roads, and canals, The latest initiatives emphasis community self-sufficiency through the development of economic activities such as tourist enterprises, cottage industries, wildlife cropping, hunting, fishing, agriculture, and livestock improvement and marketing.

Management infrastructure and capabilities are constantly being improved as the project grows. A significant shift in attitude favoring the project's statutory responsibilities to manage the project areas sustainability has been recorded among the communities. Two wetland management unit leaders and five village scouts selected by their own communities have already been trained and deployed in their home areas. Two more unit leaders and 15 village scouts are required for training during the latter half of 1989.

Financial Arrangements and Revenue Streams

The WWF-Zambia Wetlands project is funded By WWF-International, with an external aid budget of approximately half a million Swiss Francs per annum for 4 years initially. The Government of Zambia contributes financially and materially to the project by secondment of civil servants, provision of working facilities and infrastructure, and logistical support in the field from regular staff of the national parks and wildlife service and associated departments. At present, therefore, wetlands management and community development activities in the core project areas are supported almost entirely by the project, the National Parks and Wildlife Service, and a few associated departments such as fisheries. It is due to their isolation that the core project areas receive minimal input from district councils and other aid agencies. It should be noted that the core project areas and their communities may benefit from district facilities and developments at large such as feeder roads, schools, clinics, communication, and agricultural and marketing activities.

In Zambia, statutory government revenues from license and permit fees, levies and taxes, etc., are usually paid to the central government. The treasury allocates capital and recurrent votes annually to finance government operations following submission of estimates. Total central government current revenue for 1987 was the equivalent of US $280 million at current exchange rates. Total statutory revenue collected in 1988 by the department on behalf of central government was the equivalent of US $3.3 million, of which at least 97% was related to hunting activities.

However, the department's Wildlife Conservation Revolving Fund has also been accruing non-statutory revenues, which it can retain to support its own operations and community development programs in important wildlife areas. The revolving funds total revenue for 1988 was the equivalent of US $1.4 million of which nearly 70% was from sales of ivory and 30% from payment of hunting rights by commercial safari hunting operators. Of this, the equivalent of $30,000 is held in the revolving fund on behalf of the wetlands management and community development programs, pending the formation of the wetlands management authorities.

To date, this revenue is derived exclusively from hunting rights. Considering the strategic national and international significance of the wetlands water rights and diverse economic activities such as hydro-electric power generation, it is unlikely that the project areas will ever be able to retain statutory government revenues in total. It is thus believed that a more realistic approach is to negotiate for greater and more diverse return to the project areas from both statutory and non-statutory revenues, and to stimulate indigenous economic activity.

It is anticipated therefore that central government funding through the civil service and district government will continue to play a vital and expanding role in wetlands management, supplemented by revenues generated and retained in the project areas. On the other hand, community development should eventually become self-sufficient from local revenues and expansion of indigenous economic activity.

Project Results to 1990s

It is still too early to determine whether the project will result in sustainable use of wetland resources. Some issues such as management of water rights will take many years to negotiate with the powerful interests concerned.

Some early encouraging indicators should be noted. The Kafue lechwe population has increased by 10% to 65,000 since 1983 (Howard et al. 1988). While the black lechwe population in the Bangweulu swamps has declined by 18% overall to 34,000. Since 1983, the greater part of this is attributable to poaching in one area in the vicinity of a new trunk road through Kalasa-Mukoso Game Management Area (Howard et al. 1984, Howard et al. 1988). A population increase was recorded in the more inaccessible interior of the swamps where a hunting-free zone was in force for several years.

The fisheries of the Kafue Flats and Bangweulu are showing some signs of recovery, and the recently enforced control of fishing pressure during the breeding season enhances this trend (Subramanian 1986). The project areas include sites of major international significance to the conservation of wattled cranes and shoebills (Howard and Aspinal 1974).

Legal offtakes of wildlife are usually not significant (less than 1%). The indications are that sub-optimal population growth or decline in fisheries and wildlife are functions of illegal offtakes, although encroachment, competition, and environmental constraints may also play a part. It is hoped that by bringing illegal fishing and hunting under more effective control, the legal industries may be expanded on a sustained yield basis to encourage the return of more consistent and valuable benefits to local communities.

Raising sufficient finance to cover the 4-year transitional phase of this project is crucial to the aim of achieving community development self-sufficiency by 1993. The project has had some internal difficulties. Progress has been "hampered" by villagers who misunderstand the project. Such natural suspicion can only be overcome by patient dialogue and tangible demonstration of the benefits of participation. Educational components of the project can assist, but should be practically oriented. Short-term benefit realization vs. long-term sustainable practices will be difficult to sell.

Between 1978 and 1990 large family groups moved from Mazabuka, Monze, and Choma districts to the region north of Lusaka and central provinces. These mass movements are caused by several factors according to Chabwela (1998):

- The vast growth of human and cattle populations required much land for settlement and grazing. By 1990, human population in the region had grown from 96,000 to 946,000 while the cattle population had expanded to more than 250,000. A sharp land use conflict resulted, which required the

establishment of a commission of inquiry in a land matters dispute in southern province in 1982 (GRZ 1982).
- The location and expansion of the Nakamba sugar estate affected cattle
  movements into the Kafue Flats in the Mwanachingwala and Sionjalika
  communities.
- The construction of the hydroelectric dam (previously covered) Itezhitezhi
  caused serious concern as people in the area relied on regular flood patterns
  for improving the quality of the ranges used by their cattle.
- These years had very little rainfall. Areas such as Choma and Kafue recorded
  mean rainfall of less than 800 mm (Tiffen and Mulele 1993).

Current population movements into the area follow the increase in fishing
and the improved market for fisheries products. While only Twa people fish in
the southern province, a large population of migrant fishermen have moved
into the area from western Luapula and northern provinces of the country.
They have established semi-permanent villages in the flood plane in the Luwato,
Nyimba, Wanki, and Namalyo areas.

All the above factors, especially the hydrologic alteration due to the dam
operation, have reduced flood levels, changed timing, and reduced duration of
water levels, in tern causing

- significant decline of fish production;
- threats to wildlife from poaching, poor grazing range, and loss of breeding
  grounds;
- reduction of livestock grazing;
- risk to human settlements from uncertainty of flooding.

Summary and Evaluation of NGO Roles Post-2000

With this background WWF, who had previously established the two national
parks in the area started a new initiative in 1998–1999 called the "Partners for
Wetlands – Kafue Flats, Zambia" (see Schelle and Pittock 2005) and http://
www.panda.org/about_wwf/where_we_work/africa/where/zambia/kafue. The
following is a timeline of recent events coordinated by WWF.

In 2000, the Zambian Wildlife Authority and the Tourism Company Real
African Safaris sign an agreement to work together to rehabilitate facilities and
develop ecotourism in 50,000 ha of the Blue Lagoon National park.

In 2001 WWF, a local community chief and representative of five commercial sugar farms sign an agreement to work together on establishing the
50,000 ha Mwanachingwala Conservation Area.

In February 2002, the Kafue integrated water resources management project
is launched. In June 2002, there is development of an integrated water resource
management strategy and a memorandum of understanding is signed with the
government of the Republic of Zambia.

In 2003, a tripartite agreement was signed by the WWF, the Ministry of Energy and Water Development, and the Zambian Electricity Supply Company. WWF, the Zambian Wildlife Authority, and the tourism company Star of Africa sign an agreement to work together to rehabilitate facilities and develop ecotourism in 60,000 ha of Lochinvar National Park. In July 2003, there is implementation of the new water management system for Kafue Flats.

In general WWF's stated goal in Kafue Flats is to persuade traditionally non-conservation-oriented stakeholders to integrate the concept of "wise use" of wetlands, including nature conservation, into their own business/livelihood activities. According to WWF, this is achieved "through adopting an intermediary and catalytic role, creating partnerships, bringing in expertise and developing projects on the ground" (see Schelle and Pittock 2005, WWF at http://www.panda.org/about_wwf/where_we_work/africa/where/zambia/kafue).

In Zambia, formal partnerships have been established with stakeholders that are key to achieving integrated water management in Kafue Flats. These involve the sugar industry (Zambia Sugar, Manga and Ceres Farms), the Zambian Electricity Supply Company (ZESCO), the Ministry of Energy and Water Development (MEWD), the Zambian Wildlife Authority (ZAMA), Chiefdom of the Tonga people (Chief Mwanachingwala), and two private tourism companies (Star of Africa and Real African Safaris).

With the sugar industry, WWF is working to restore 50,000 ha of the Kafue Flats – the Mwanachingwala Conservation Area. This is being achieved through a combination of measures including raising awareness among local communities, the introduction of wise use practices, translocation of animals, and ecotourism. WWF is also encouraging sugar farms to pre-treat their effluent through bio-filters (small artificially created wetlands and reedbeds) to lower nutrient levels and therefore reduce the growth and spread of water hyacinth. The plants grown as biofilters can also be used to make a modest income such as basket making from reeds.

With ZESCO and MEWD, WWF is working to improve the management of water resources in the Flats by improving the operating procedures of the Kafue Gorge and Itezhitezhi Dams. The objective is to mimic natural water flows as closely as possible in order to restore wetland functions and values. The first step of this partnership produced the Integrated Water Resource Management Strategy, which has been accepted by key stakeholders. Computer models were also developed to simulate potential water management scenarios and to study their likely impacts.

The second step began in July 2003, and over 9 months, focuses on implementation of the new water management system for Kafue Flats. Reestablishment of the hydro-meteorological monitoring networks, further refinement of computer models, dam operation, and legal and institutional frameworks are the main components of this phase. Testing of the new dam operating procedures was expected in 2004.

The Integrated Water Resource Management Project is part of the Kafue pilot project being implemented by the Ministry of Energy and Water Development through the Water Resources Action Program (WRAP). It is hoped that such a program will act as an example and catalyst for sustainable water resources management in the whole region, notably the wider Zambezi River Basin.

So, we have a dynamic situation of a natural flood-driven system with traditional fishing and cattle grazing that is suddenly transformed by two large hydroelectric facilities. In-migration and resource use pressures are further stressing both ecosystems and local populations' traditional uses. The current WWF partnership initiative seeks to

- provide both socio-economic and biodiversity benefits for seemingly conflicting stakeholders;
- establish partnerships, especially with non-conservation-oriented sectors such as electricity supply companies and the sugar industry;
- develop model sites – where ownership lies clearly with partners – allowing eventual phased withdrawal of WWF;
- promotion of ecotourism as means for diversifying economic opportunities within protected areas as well as infrastructure financing;
- magnification – using Kafue Flats as a model for integrated water management to extend throughout the Zambezi River Basin.

Acronyms

FAO: Forest and Agricultural
GMA: game management area
IUCN: International Union for the Conservation of Nature
KFP: Kafue National Park
NP: national parks
LNP: ochinver National Park
KBRP: Kafue Basin Research Project
MEWD: Ministry of Energy and Water Development
SADCC: Southern Africa Development Coordination Conference
SWECO: Swedish Engineering Company
WWF: World Wildlife Fund for Nature – World Wildlife Fund
ZAMA: Zambian Wildlife Authority
ZESCO: Zambian Electricity Supply Company

References

Acreman, M. 1994. The role of artificial flooding in the integrated development of river basins in Africa. In C. Kirby and W. R. White (eds.) *Integrating River Basin Development*, pp. 35–44. Chichester, UK: John Wiley and Sons.

Acreman, M. C. 1996. The IUCN Sahelian Floodplain Initiative, networking to build capacity to manage Sahelian floodplain resources sustainably. *International Journal of Water Resources Development*, 12(4): 429–436.

Backstrom, M. and B. Jonsson. 1996. *A Sediment Study in the Kafue River, Zambia*. Unpublished Masters Thesis, Lulea University of Technology, Lulea, 96 pp.

Bayley, D. B. 1995. Understanding large river systems: Floodplain ecosystems. *Bioscience*, 45(3): 153–158.

Bayley, D. B. 2006. The flood pulse advantage and the restoration of river floodplain systems. *Regulated Rivers; Research and Management*, 6(2): 75–86.

Beilfuss, R. D. and B. R. Davies. 1999. Prescribed flooding and the rehabilitation of the Zambezi Delta, Mozambique. In W. J. Streever, (ed.) *An International Perspective on Wetland Rehabilitation*, pp. 143–158. The Netherlands: Kluwer Academic Publishers.

Bell, R. H. V., J. J. Grimsdell, L. P. van Lavieren, and J. A. Sayer. 1973. Census of the Kafue Lechwe by a modified method of areal stratified sampling. *East African Wildlife Journal* 11: 55–75.

Brock, M. A. and K. H. Rodgers 1998. The regeneration potential of the seed bank of an ephemeral floodplain in South Africa. *Aquatic Ecology*, 61(2): 123–135.

Bruwer, C., C. Poultney, and Z. Nyathi. 1996. Community based hydrological management of the Phongolo floodplain. In M. Acreman and G.E. Hollis (eds.) *Water Management and Wetlands in Sub-Saharan Africa*, pp. 199–212. Gland, Switzerland: IUCN/The World Conservation Union.

Burke, A. 1992. *The Cranes Status Survey and Conservation Action Plan, Wattled Crane (Bugeranus carnunculatus)*. USGS, Northern Prairie Wildlife Research Ctr, 23 pp.

Chabwela, H. N. W. 1998. Case Study Zambia: Integrating Water Conservation and population Strategies on the Kafue Flats. In A. Sherbinin and V. Dompka (eds.) *Water and Population Dynamics: Case Studies and Policy Implications*, Washington, DC: AAAS, and http://www.aaas.org/international/ehn/waterpop/zambia.htm

Chapman, D. W., W. H. Miller, R. G. Dudley, and R. J. Scully. 1971. *Ecology of the Fishes in the Kafue River*. Rome: FAO FI: SF/Zam 11: Tech. Rep. 2, 66 pp.

Chooye, P. M. and C. A. Drijver. 1995. Changing views on the development of the Kafue Flats in Zambia. In H. Roggeri (ed.) *Tropical Freshwater Wetlands*, Dordrecht, Netherlands: Kluver Academic Publishers.

Douthwaite, R. J. 1974a. An endangered population of Wattled Cranes (*Grus carunculatus*). *Biological Conservation*, 6(2): 134–142.

Douthwaite, R. J. 1974b. *The Ecology of the Ducks (Anatidae) on the Kafue Flats, Zambia*. Lusaka: Final Report to the Kafue Basin Research Committee, Univ. of Zambia, 103 pp. (mimeograph.).

Douthwaite, R. J. 1982. Water birds: Their ecology and future on the Kafue Flats. In Howard and Williams, 1982, pp. 137–140.

Douthwaite, R. J., and L. P. van Lavieren. 1977. *A Description of the Vegetation of Lochinvar National Park Zambia*. Nat. Coun. Sci. Res. Tech. Rpt. (Lusaka), 34, 66 pp.

Dudley, R. G. and R. J. Sculley. 1980. Changes in experimental gill net catches from the Kafue Floodplain, Zambia, since construction of the Kafue Gorge Dam.

Dugan, P. J. 1988. The importance of rural communities in wetlands conservation and development. In D. D. D. Cook et al. (eds.) *The Ecology and Management of Wetlands, Volume 1: Management Use and Value of Wetlands*, pp. 3–11. Portland Oregon: Timber Press.

Ellenbroek, G. A. 1987. *Ecology and Productivity of an African Wetland System*. Dordrecht/Boston/Lancaster: Dr. W. Junk Pub. / Kluwer Acad. Pub. Group.

Evans, S. Y., K. Bradbrook, R. Braund, and G. Bergkamp. 2003. Assessment of the restoration potential of the Logone floodplain (Cameroon). *Water and Environment Journal*, 17(2): 123–128.

FAO. 1968. *Multipurpose Survey of the Kafue River Basin*, Zambia, Rome: FAO/SF; 35 ZAM, 7 vols.

Gammelsrod, T. 1996. Effect of Zambezi River management of the prawn fishery of the Sofala Bank. In M. C. Acreman and G. E. Hollis (eds.) *Water Management and Wetlands in Sub-Saharan Africa*, pp. 119–124. Gland, Switzerland: IUCN.

Giller, P. S. 2005. River restoration: seeking ecological standards. *Journal of Applied Ecology* 42(2): 201–207.

GRZ (Government of the Republic of Zambia). 1982. *A Report of the Commission of Inquiry into Land Maters of Southern Province.* Lusaka, Zambia.

Handlos, W. L. 1977. Aspects of Kafue Basin Ecology—. In Williams and Howard, 1977, pp. 29–39.

Handlos, D. M., W. L. Handlos, and G.W. Howard. 1976. A study of the diet of the Kafue Lechwe (*Kobus leche*) by analysis of rumen contents. In *Proc. Fourth Regional Wildlife Conf. East Central Africa* (Dept. NP &WS, Lusaka, Zambia), pp. 197–211.

Horowitz, M. M. 1994. The management of an African river basin: alternative scenarios for environmentally sustainable economic development and poverty alleviation. In Koblenz (ed.) *Proceedings of the International UNESCO Symposium: Water Resources Planning in a Changing World*, pp. 73–82. Karlsrue, Germany: International Hydrological Program of the UNESCO/OHP National Committee of Germany.

Howard, G. W. 1977. National Parks in the Kafue Basin. In Williams and Howard, 1977, pp. 47–56.

Howard, G. W. and D. R. Aspinal. 1974. Aerial census of Shoebills, Saddlebilled Storks and Wattled cranes at the Bangweulu Swamps and Kafue Flats, Zambia. *Ostrich*, 55: 207–212.

Howard, G. W. and D. R. Aspinual. 1984. Aerial census of shoebills, saddle billed storks and wattled cranes at the Bangweulu Swamps and Kafue Flats, Zambia. *Ostrich*, 53: 207–212.

Howard, G. W. and R. C. V. Jeffrey. 1981. Present distribution of Lechwe on the Kafue Flats. *Black Lechwe No. 1 NS*, pp. 17–20.

Howard, G. W. and R. C. V. Jeffrey. 1983. *Kafue Lechwe population status 1981–1983.* Chilangua: Report to the Director NPWS, mimeo.

Howard, G. W., R. C. V. Jeffrey, and J. J. R. Grimsdell. 1984. Census and population trends of Black Lechwe in Zambia. *African Journal Ecology* 22: 175–179.

Howard, G. W., R. C. V. Jeffrey, B. M. Kamweneshe, and C. M. Malambo. 1988. *Black Lechwe population Census, Bangweulu Swamps.* Chilanga: Report to the Director NPWS (mimeo.).

Howard, G. W. and J. A. Sidorowicz. 1976. Geographical variation of the Lechwe (*Kobus leche gray*) in Zambia. *Mammalia* 40: 69–77.

Howard, G. W. and G. J. Williams (eds.). 1982. *Proceedings of the National Seminar on Environment and Change: The Consequences of Hydroelectric Power Development on the Utilization of the Kafue Flats.* Lusaka: Kafue Basin Res. Com., Univ. of Zambia, Lusaka, 159 pp.

Jeffrey, R. C. V. 1990. A general appraisal of consumptive wildlife utilization in Zambia. In M. Marchand and H. A. Udo de Haes (eds.) *The People's Role in Wetland management: Proceedings of the International Conference* [Leiden, the Netherlands, June 5–8 1989], pp. 336–338. Centre for Environmental Studies, Leiden University.

Jeffrey, R. C. V. and P. M. Chooye. 1990. The people's role in wetland management; The Zambian initiative. In M. Marchand and H. A. Udo de Haes (eds.) *The People's Role in Wetland management: Proceedings of the International Conference* [Leiden, the Netherlands, June 5–8 1989], pp. 83–91. Centre for Environmental Studies, Leiden University.

Junk, W. J., P. B. Bayley, and R. Sparks. 1989. The flood pulse concept in river-floodplain systems. In D. P. Dodge (ed.) *Proceedings of the Large River Symposium*, pp. 110–127. Canadian Special publications in Fish and Aquatic Science, volume 106.

Kasonde, J. 1990. *Environmental Pollution Studies on the Copperbelt in the Kafue Basin.* Lusaka: National Council of Scientific Research WRRU/ERL/TR.

Lagler, K. F., J. M. Kaptezki, and D. J. Stewart. 1971. *The Fisheries of the Kafue River Flats, Zambia in Relation to the Kafue Gorge Dam.* Rome: FAO FI: SF/Zam 11: Tech. Rpt., 1, 161 pp.

Le Maitre, D. C., B. W. van Wilgen, C. M. Gelderblom, C. Bailey, R. A. Chapman, and J. A. Nel. 2002. Invasive alien trees and water resources in South Africa: case studies of the costs and benefits of management. *Forest Ecology and Management,* 160(1–3): 143–159.

Lehmann, D. A. 1977. The Twa: People of the Kafue Flats. In Williams and Howard, 1977 (qv), pp. 41–46

Mathooko, J. M. and S. T. Kariuki 2000. Disturbances and species distribution of the riparian vegetation of a Rift Valley stream. *African Journal of Ecology,* 38(2): 123–129.

Mouafo, D., E. Fotsing, D. Sighomnou, and L. Sigha. 2002. Dam, environment and regional development: Case study of the Logone floodplain in Northern Cameroon. *International Journal of Water Resources Development,* 18(1): 209–211.

Mumba. M. 2003. Vegetation and hydrological changes in the Kafue Flats Zambia associated with the construction of the Itezhitezhi and Kafue Gorge Dams. *Geophysical Research Abstracts,* 5: 14074.

Muyanga, E. D. and P. M. Chipungu. 1982. A short review of the Kafue Flats fishery from 1968 to 1978. In Howard and Williams, 1982, pp. 105–113.

Mwase, M., T. Viktor, and L. Norrgren. 1998. Effects on tropical fish of soil sediments from Kafue River, Zambia. *Bulletin of Environmental Contamination and Toxicology,* 61(1): 96–101.

Norrgren, L., B. Brunstrom, M. Engwall, and M. Mwase. 1998. Biological impact of lipophilic sediment extracts from the Kafue River Zambia, in microinjected rainbow trout yolk-sac fry and chick embryo livers exposed *in vitro. Aquatic Ecosystem Health and Management,* 1: 91–99.

Norrgren, L., U. Petersson, S. Orn, and P.-A. Bergqvist. 2000. Environmental monitoring of the Kafue River located in the Copperbelt, Zambia. *Archives of Environmental Contamination and Toxicology,* 38(3): 334–341.

Nelson, R. W., R. S. Ambasht, C. Ameros, G. W. Begg, A. A. Begg, A. A. Bonetto, I. R. Wais, E. Dister, E. Wenger, C. M. Finlayson, J. K. Handoo, A. K. Pandst, K. M. Mauuti, D. Parish, and D. Savey. 1989. River floodplain and delta management team: A project of the Worlds Wetland Project. In J. A. Kusler and S. Daley (eds.) *Wetlands and River Corridor Management,* pp. 75–82, Berne, NY: Association of Wetland Managers.

Osborne, T. O. 1973. Additional notes on the birds of the Kafue Flats. *Puku,* 7: 163–166.

Perera, N. P. 1982. Ecological considerations in the management of wetlands of Zambia. In B. Gophal, R. E. Turner, R. G. Wetzel, and D. F. Whigham (eds.) *Wetlands Ecology and Management,* pp. 21–30. New Delhi: National Institute of Ecology and International Scientific Publishers.

Pettersson, U. T. and J. Ingri. 2001. The geochemistry of Co and Cu in the Kafue River as it drains the Copperbelt mining area, Zambia. *Chemical Geology,* 177: 399–414.

Pike, E. G. R. and T. G. Corey. 1965. The Fisheries of Zambia, the Kafue floodplain. In M. A. E. Mortimer (ed.) *The Fish and Fisheries of Zambia,* pp. 76–84. Lusaka: Ministry of Lands and Natural Resources.

RAMSAR. 2002. A Directory of Wetlands of International Importance: Zambia. Wetlands International and RAMSAR, A Web document found at http://www.wetlands.org/RBB/Ramsar_Dir/Zambia/zm001D02.htm

Rees, W. A. 1978a. The ecology of the Kafue Lechwe: The food supply. *Journal of Applied Ecology,* 15: 177–191.

Rees, W. A. 1978b. The ecology of the Kafue Lechwe: Soils, water levels and vegetation. *Journal of Applied Ecology,* 15: 163–176.

Rees, W. A. 1978c. Do the dams spell disaster for the Kafue Lechwe? *Oryx,* 14: 231–235.

Robinette, W. L. and G. F. T. Child. 1964. Notes on the biology of the Lechwe (*Kobus leche*). *Puku,* 2: 84–117.

Sayer, J. A. and L. P. van Lavieren. 1975. The ecology of the Kafue Lechwe populations in Zambia before the operation of hydroelectric dams in the Kafue River. *East African Wildlife Journal*, 13: 9–38.

Schelle, P. and J. Pittock. 2005. *Restoring the Kafue Flats; A Partnership approach to environmental flows in Zambia*. Godalming, UK: Dams Initiative, Global Freshwater program, WWF International, 10 pp.

Scholte, P., P. Kirda, S. Adam, and B. Kadiri. 2000. Floodplain rehabilitation in North Cameroon: Impact on vegetation dynamics. *Applied Vegetation Science*, 3(1): 33–42.

Schuster, R. H. 1980. Will the Kafue Lechwe survive the Kafue dams? *Oryx*, 15: 476–489.

Scudder, T. 1989. River basin projects in Africa. *Environment* 31(2): 4–9, 27–32.

Scudder, T. and M. C. Acreman. 1996. Water management for the conservation of the Kafue Flats, Zambia and the practicalities of artificial releases. In M. C. Acreman and G. E. Hollis (eds.) *Water Management and Wetlands in Sub-Saharan Africa*, pp. 101–106. Gland, Switzerland: IUCN.

Seyam, I. M., A. Y. Hoekstra, G. S. Ngabirano, and H. H. G. Savenije. 2001. The Value of Freshwater Wetlands in the Zambezi Basin. *Globalization and Water Resources Management: The Changing Value of water*, AWRA/IWLRI – University of Dundee International Specialty Conference, 10 pp.

Sheppe, W. A. 1972. The annual cycle of small mammal populations in a Zambian floodplain. *Journal of Mammalia* 53(1): 445–460.

Sheppe, W. A. 1973. Notes on Zambian rodents and shrews. *Puku*, 7: 176–190.

Sheppe, W. A. 1985. Effects of human activities on Zambia's Kafue Flats ecosystem. *Environmental Conservation*, 12(1): 49–57.

Sheppe, W. A. and T. O. Osborne. 1971. Patterns of use of a floodplain by Zambian mammals. *Ecological Monograph*, 41(3): 179–205.

Sinkala, T. et al. 1977. *Control of Aquatic Weeds in the Kafue River – Phase 1: Environmental Impact Assessment of the Kafue River basin between Itezhitezhi Dam and the Kafue Gorge*. Lusaka: Ministry of the Environment and Natural Resources, Gov. of the Republic of Zambia.

Standford, J. A., J. V. Ward, W. L. Liss, C. A. Frisnell, R. N. Williams J. A. Lichatowich, and C. C. Cotant. 1996. A general protocol for restoration of regulated rivers. *Regulated Rivers*, 12: 391–413.

Subramanian, S. P. 1986. A brief review of the Bangweulu Basin and Kafue Flats. *Proceedings of the WWF-Zambia Wetlands Project Workshop*. Lusaka.

SWECO. 1967. *Kafue Gorge Hydroelectric Power Project*. Lusaka.

SWECO. 1968. *Kafue River Regulation: Project on the Main Storage Reservoir*. Lusaka, 2 vols.

SWECO. 1969. Kafue River Regulation: Water Management Stage I. Lusaka.

SWECO. 1971. *Kafue River Hydroelectric Power Development Stage III Itezhitezhi Reservoir and Extension of Kafue Gorge Upper Power Station Preliminary Study*. Lusaka, 3 vols.

Syakalima, M. S. et al. 2001. Bioaccumulation of lead in wildlife dependent on the contaminated environment of the Kafue Flats. *Bulletin Environmental Contamination and Toxicology*, 67(3): 438–445.

Thomas, D. H. L. 1999. Adapting to dams: agrarian change downstream of the Tiga Dam, Northern Nigeria. *World Development*, 27(6): 919–935.

Tiffen, M. and M. S. Mulele. 1993. *Environmental Impact of the 1991-2 Drought on Zambia*. Zambia Country Office, Lusaka: IUCN.

Tockner, K. and J. A. Stanford 2002. Riverine floodplains; present state and future trends. *Environmental Conservation*, 29: 308–330.

Van der Heyden, C. and M. New. 2003. *Natural wetland for mine effluent remediation? The case of the Copperbelt*. School of Geography and the Environment, Oxford University UK, 24 pp.

Ward, J. V. and J. A. Stanford. 2006. Ecological connectivity in alluvial river ecosystems and its disruption by flow regulation. *Regulated Rivers, Research and Management*, 11(1): 105–119.

Welcomme, R. L. 1995. Relationships between fisheries and the integrity of river systems. *Regulated Rivers*, 11(1): 121–136.

Wesseling, J. W., E. Naah, C. A. Drijver and D. Ngantou. 1996. Rehabilitation of Logone floodplain, Cameroon, through hydrological management. In M. C. Acreman and G. E. Hollis (eds.) *Water Management and Wetlands in Sub-Saharan Africa*, pp. 158–198. Gland, Switzerland: IUCN.

Williams, G. J. 1977. The Kafue hydroelectric scheme and its environmental setting. In Williams and Howard, 1977, pp. 13–26.

Williams, G. J. and G. W. Howard. 1977. *Development and Ecology in the Lower Kafue Basin in the Nineteen Seventies*; Papers from the Kafue Basin Research Committee of the University of Zambia, Lusaka.

WWF. 2001. *Investing in the Kafue Flats Wetlands to ensure clean water and wildlife for later years*. WWF document found at http://www.partnersforwetlands.org/report/report-zmbia-sept.2001.html

WWF, KafueFlats, http://www.panda.org/about_wwf/where_we_work/africa/where/zambia/kafue

Chapter 5
Community-Based Wetland Management: A Case Study of Brace Bridge Nature Park (BBNP), Kolkata, India

Introduction – Urban Wetlands Utilization

The East India–Bangladesh region has a rich history of coastal mangrove and swamp forest systems. But these same areas are under stress from land conversion and population explosion. This case study traces the development of an urbanized wetland in east Kolkata, India, and follows its multifunctional utilization for water quality treatment, aquaculture, garbage-fed agriculture, and urban wetland park. A crucial question for such urbanized wetland systems is whether they can be sustained in the face of mounting land conversion pressure plus other environmental stresses.

Urban or periurban wetlands whether they be coastal or freshwater, man-made or natural are under assault worldwide (Guntenspergen and Dunn 1998) and particularly in Asia because of land use conversion and urban growth pressures (Lee 2006, Smardon 2008, Zhao et al. 2000, ZongMing et al. 2004). There is an excellent overview of the state of Asian wetlands by Wong (2004) that includes China, Philippines, and Thailand. There are a number of studies by Indian researchers that include urban wetlands in India (Kumar and Reddy 2000, Patnaik and Srihari undated, Ramachandra 2001)) including the Kolkata wetlands within this case study.

Urban wetlands in Asia are being utilized for many environmental services including wastewater/storm water treatment in Phnom Penh, Cambodia (Irvine 2007), many communities in Australia (Hart and Ibarra 2006, Streeter 1998, Wong 2006), Luang Marsh in Laos PDR (Gerrad 2004), Ho Chi Min City in Vietnam (Costa–Pierce et al. 2005), China (Wong 2004), as well as the Indian Kolkata wetlands (Costa-Pierce 2005). Some urban wetland complexes provide water supply such as that in Western Australia (Tapsuwan et al. 2007).

Urban wetlands are also used for aquaculture such as the east Kolkata wetlands and in Ho Chi Min City in Vietnam (Costa-Pierce et al. 2005) as well as wildlife habitat. For both these functions there is concern over the ability

Case study written by A.K. Gosh and N.C. Nandi, Zoological Survey of India. M-Block, New Alipore, Calcutta 700053, with substantial editorial revision by the author of this book.

R.C. Smardon, *Sustaining the World's Wetlands*,
DOI 10.1007/978-0-387-49429-6_5, © Springer Science+Business Media, LLC 2009

of such wetlands to absorb heavy metals and other pollutants (Prain et al. 2006, Vymazel 2005) or pass them on to wildlife (Fasela 2002). There are also issues of hosting vectors for disease such as mosquitoes during wetland maintenance or reconstruction or even the perception of this threat (Smardon 1989, Zedler and Leach 1998). Other issues include suitability of altered urban wetlands to support migratory waterfowl populations and compatibility with human pollutions visitation of such populations (Antos et al. 2007, Zedler and Leach 1998). In some cases reconstructed urban wetlands have evolved to elaborate wetland parks complete with structures and interpretation facilities such as in Hong Kong and Taipei, Taiwan.

The overriding issue is examining the compatibility of multiple environmental services by urban wetlands (Emerton 2005, Zedler and Leach 1998) plus the connection to local livelihoods (FAO 2003, Ratner et al. 2004) in the face of overwhelming pressure for land use conversion.

Introduction for India–Bangladesh Region

The eastern India–Bangladesh region (see Fig. 5.1) has a rich history of traditional use of wetlands in terms of use of plants for food and fiber and for fisheries. Traditional commercial practices of West Bengal have been practiced over 300 years (Ghosh 2004) and prior to that people harvested wetland products for domestic consumption. Rural people in different states of India, particularly 24 Paragonas (south and north), Hugli, Haora, and Medinipar (east and west) were responsible for commercialization of major wetland

East Calcutta Wetlands

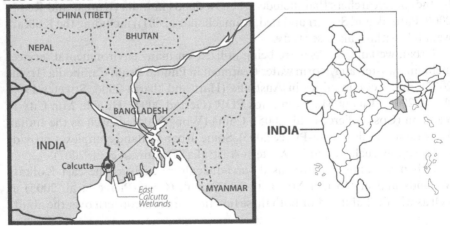

Fig. 5.1a Location of east Kolkata wetlands. Drawn by Samuel Gordon and adapted from wwfindia at http://www.wwfindia.org/calcutta_29php?fileid = 29

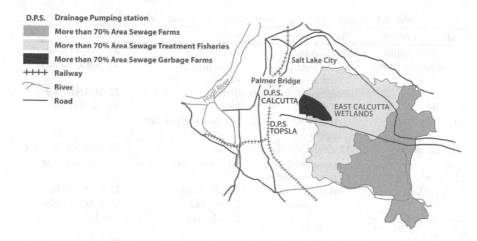

D.P.S. Drainage Pumping station

More than 70% Area Sewage Farms

More than 70% Area Sewage Treatment Fisheries

More than 70% Area Sewage Garbage Farms

++++ Railway

River

—— Road

Fig. 5.1b East Kolkata wetlands layout and use. Drawn by Samuel Gordon and adapted from D. Ghosh, 1998

Fig. 5.2 East Kolkata wetlands – view of treatment lagoons. Source: D. Ghosh, 1998, p. 1

products obtained from plant resources. These resources include *Typha elephantia* and *Typha domenginsis* (hugla or cattail), *Aeschynomene aspera* (shoal), *Cyperus pangorei* and *Cyperus corymbosus* (madurlathi or sedges), *Trapa natans* var. *bispinosa* (paniphal), and *Euryale ferox* (makhona) (see Table 5.1). In addition, several wetland plants have all been harvested by rural villagers as supplemental vegetables and medicinal plants. Kalmi (*Ipomoea aquatica*) and Kachu (*Celccasu esculenta*) are the most prominent (Ghosh 2004).

Table 5.1 Major traditional commercial practices using wetland plants[1]

Cultivation region	Plants used	Uses	Value
Paragonas (south and north) 3000 families	Hugla/cattail	Mats	Rs. 5,000/ha/year
	Typha elephantia	Thatching/roofs	Rs. 9,000/ha/year
	Typha domenginsis	Paper/decoration	
	Holga gunri	Sweets	25–30 INR[2]/kg
West Bengal	Shola	Hats	Rs. 40.000/ha
1.5 million	*Aeschynomene aspera*	Shola art/ornamental products	
West Bengal	Madurlathi/sedges	Mats	Rs. 100,000/ha/year
Sabang	*Cyperus pangorei*	+painting/printing	2.273 USD/year
2,000 ha	*Cyperus corymbosus*		
North Bihar	Makana/fox nut	Fruits/seeds	INR 16,000/ha
96,000 ha	*Euryale ferox*	Edible puff	INR 107.400/ha
West Bengal		Fried seeds	USD 2,330
900 ha			
West Bengal	Paniphal/water chestnut	Edible fruit	INR 26,000–
Ponds/pits	*Trapa natans var. bispinosa*		36,000/ha/season
Medinipar	Lotus	Flowers	Rs. 2,000– 3,600/
	Nelumbo nucifera		season

[1] Data obtained from Ghosh 2004.
[2] INR = Indian National Rupee – 1 USD = 45 Indian Rupees.

Other species include the following:

Hingche (*Enhydra fluctuans*)
Sushi (*Marsilea minuta*)
Dhenki shak (*Diplazium esculentum*)
Jalsahi or ban-hingche or alligator weed (*Atternthea philoxeroides*)
Shlak or water lilies (*Nymphaea nouchali* and *Nymphaea pubescens*)
Thankuni (*Centella asiatica*)
Kulekhana (*Hygrophilia schulli*)
Brahmi (*Bacopa monnieri*)
Shimralya or water cress (*Nasturtium officinale*)
Acorus calamus, an emergent medicinal herb that is also harvested in the
    wilderness

Against this history, we have the fish culturing activity within and adjacent
to the east Kolkata wetlands that dates from 1860 (Chattopadhyay 2001) and
the fisheries of the eastern periphery of Kolkata covers an area of about 2774 ha
which is by far the largest contiguous wetland fishery in the world according to
Mukherjee (1998).

Introduction for Case Study

Brace Bridge Nature Park (BBNP) (see Fig. 5.1) is a freshwater, sewage-fed
wetland located within the jurisdiction of Kolkata, the congested capital city of
the West Bengal state of India. The wetlands of BBNP comprise 10 ponds or

tanks of varying sizes lying between the latitudes 22, 31' 23''–22, 33' 30'' north and longitude 88, 27' 20''–88, 28' 54'' east, about 3 m above the sea level and covers a total area of 80 ha in area with 60 ha of water bodies and 20 ha of uplands and dike area. The landscape elements include approximately 70% fish ponds (less than 1–15 ha), 10% dike or embankment, 7% slum or squatter settlements, 5% sewage treatment ponds/channel, 4% deer park and garden, and the remaining 4% of land area is comprised of rail lines, solid waste fallows, and office establishment. According to Ramsar guidelines for classification of wetlands, BBNP is at present man-made, sewage-fed aquaculture fishponds.

Today the wetlands of east Kolkata encompass about 20,000 acres containing vegetable farms, rice paddies, and fish farms (see Figs. 5.3, 5.4, and 5.5).

Fig. 5.3 Nature park and treatment lagoons. Source: DFID, 2001, p. 2

Currently, only about one-third of the city's sewage water actually flows through the marshes. The Kolkata Municipal Corporation constructed two channels approximately 33 km long, one for the storm runoff during the monsoon season and the other for sewage outfall after primary treatment at Beutala in two sedimentation tanks. These tanks, however, have not worked in a decade so there is untreated sewage from the city being released directly to the dry flow channel. As sewage increased from the city, the storm flow channel has also been put to use for sewage outfall year-round (Patnik 1990). These channels take all but approximately one-third (utilized by the sewage farms and fisheries) of the sewage outfall to the Kulti Gong River. The water from the channel is released into the Kulti Gong through a lock-type system, which attempts to keep the backflow from the river out during that part of the day when the river water level is higher than the canals water level. The sewage for the fisheries and agriculture has been removed through a provision in the outfall

Fig. 5.4 Productive aquaculture and agriculture. Source: DFID, 2001. p. 1

Fig. 5.5a Early morning fishing in the east Kolkata wetlands. Source: D. Ghosh, 1998, p. 3

drainage scheme "to raise an adequate water head and to supply sewage to most of the fish ponds by gravity" (D. Ghosh 1990). So the east Kolkata marshes have gradually evolved into 12,000 ha for vegetable farms, wastewater-fed ponds, or Bheris and rice paddy cultivation.

Early morning fishing in the East Calcutta wetlands

Fig. 5.5b Early morning fishing in the east Kolkata wetlands: close up. Source: D. Ghosh, 1998, p. 3

Ecological History of the Brace Bridge Nature Park

In the early 1900s the wetlands of BBNP were part of a swamp covering more than 1235.5-ha area situated to the south of the Hugli River in the Garden Reach Area under the industrial southwestern part of Kolkata City. The swamp extended from Santoshpur in the south to the Majerhat in the north, a linear stretch of about 14 km of low-lying land of the pre-independence period. There was a regular inflow of water from the Hugli River at high tide and outflow during low tide in the past. But, due to siltation over the centuries, more than 70% of the original swamp was transformed into low flatland where a large number of industrial uses were established. The present park area represents the remnant of the original swamp.

History and Involvement of a Fisherman Community-Based Organization

BBNP belongs to the Calcutta Port Trust (CPT), a statutory body of government of India, but CPT, the owner of the wetlands and the adjoining area, granted fishing rights to a small group of fishermen in 1957–1958. These fishermen, immigrants of Amta Village in the Haora District of West Bengal, later formed a registered community-based organization (CBO) in November 1961. The CBO is known as the Mudialy Fisherman's Cooperative Society (MFCS) which was formed with the assistance and guidance of one land manager of CPT and one deputy director of the Department of Fisheries, Government of West Bengal.

The members of MFCS also gained the support of the CPT to launch conversion of the wetlands into an urban fishery-cum-recreational ecosystem. MFCS initiated eco-development activities under the guidance of the chief executive officer appointed by the Department of Fisheries, Government of West Bengal. In the process, in 1985, the MFCS established the water area and the surrounding embankment as an "ecological park" and subsequently developed the renamed area as a "nature park" in 1991 as a conservation measure.

The MFCS organization is the product of a long and determined struggle by a group of fishermen led by seven members who had immigrated to the wasteland near the Kolkata dock area (Metiaburj) in search of contract jobs around 1942 when the Damodar River had dried up (Dutta and Rapoor 1992). The foundation of this CBO was started with 53 fishermen depositing 25 paisa per member per day and at present there were 100 members with voting rights, 176 nominal members, and 150 casual laborers associated with the Mudialy Fisherman's Cooperative Society (Ghosh 1993a). So, a total of about 400 fishermen families are associated with the organization and the society. From 1961 onward the society grew tremendously in terms of membership, nature conservation activities, prevention of pollution, and supply of fresh fish to the Kolkata markets.

Past Use of the Locality

Historically, the colonial metropolis of Kolkata was developed as a major port city of the British Empire. The basic elements of the city's land use surrounding the wetlands were associated with port functions. In the past the role of Kolkata had been the role of impoverishing the countryside and fattening itself at the latter's expense. In the seventeenth and eighteenth centuries, the East India Company, and in the nineteenth century, its successor the British government exploited the whole of eastern India with Kolkata as its economic base. Settlement in this region had possibly contributed with its neglect of agricultural investment. During the post-independence period, the city grew precipitously in the south primarily with the huge influx of migrants from the then East Pakistan (now Bangladesh). The land survey of the Kolkata industrial region by the Calcutta Metropolitan Planning Organization (CMPO) in 1961 by photographic interpretation of an aerial survey indicates 39% built-up area and 61% vacant or agricultural area in the Behala–Garden Reach industrial region (Munshi 1991).

The original swamp was dominated by reeds (*T. elephantia*), sedges (*Cyperus* spp.), and pith plants (*Aeschynomene indica*) and was used by local people to catch fish in some portions and fish culture in other areas on a very small scale. During the middle of the present century, a prosperous businessman who had employed the immigrant fishermen of Amta village leased part of these wetlands. In 1952, these fishermen leased the wetland for 6 months and then again in 1958 on an annual lease (currently on a 3-year lease). After obtaining fishing rights for these marshy areas, the fishermen started a cleanup operation of the reeds and converted the marsh into man-made wetlands.

History of Wetland Plant Community Change

Plant Communities Prior to 1985

The wetlands of Brace Bridge Swamp witnessed the impact of human activities in the post-independence period, especially during the decades of the 1950s–1970s. The landscape profile till 1984–1985 includes (i) Jheel with island, (ii) swamp area, (iii) transitional mudflat area, (iv) dike area, and (v) low mudflat area. This flatland usually turns out to be low meadow during the monsoon months. The settlers made considerable change to the landscape as well as degradation of the vegetation on land and destruction of reeds within the swamp. The characteristic plant communities had five distinct habitat types prior to 1985 and are summarized in Table 5.2 and as recorded by Ghosh and Chattopadhyay (1990).

Major Changes in the Plant Community

Prior to 1985, the wetland area had a wide array of floral components comprised aquatic macrophytes, marginal amphibious forms, and terrestrial plant species. The major changes in the floristic communities are as follows:

- Systematic removal of reeds belonging to *Typha* and *Phragmites* species along with the filling up of the southern part of the swamp with solid waste.
- Cutting of native flora like *Aeschynomene indica, Sesbania cannabina*, gradually changing the ecological condition of the wetland.
- Introduction of exotic flora including *Acacia auricularis* and subabul, *Leucaena macrophylla*.
- Greening of the dike area with *Acacia indica, Aegle marmelos, Azadirachta indica, Carica papaya, Emblica officinalis, Zizyphus mauritiana*.
- Undertaking plantation program pertaining to leguminous, dust absorbing, bird attracting, and horticultural plants.
- Gradual predominance of phytoplankton communities belonging to Chlorophyceae and Cyanophyceae in the fishponds.

Table 5.2 Plant communities in Brace Bridge wetlands pre-1984

Habitat types/plant communities	Representative wetland species
Habitat Type I. Jheel area	
(i) Floating hydrophytes	*Eichhornia crassipes, Pistia stratiotes, Lemna polyrhiza/L. minor, Salvinia cucullata*
(ii) Suspended hydrophytes	*Ceratophyllum demersum*
(iii) Anchored submerged hydrophytes	*Hydrilla verticillata*
(iv) Anchored floating hydrophytes	*Nymphaea rubra*
(v) Emergent amphibious hydrophytes	*Panicum tripheron, Polygonum hyropiper*
(vi) Sedge	*Cyperus* spp.
Habitat Type II. Swamp area	
(i) Floating forms	*Eichhornia, Lemna,* etc.
(ii) Emergent forms	*Enhydra fluctuans, Aeschynomene aspera Marsilea quadrifolia, Ludwigia, Polygonum,* and *Rumex* spp.
(iii) Reeds	*Typha elephantia*
(iv) Sedge	*Cyperus* spp.
Habitat Type III. Mudflat area	
(i) Emergent amphibious forms	*Alternanthera sessilis, Hygrophila spinosus Ipomoea* spp., *Centenella asiatica*
Habitat Type IV. Dike area	
(i)Tree species	*Azadirachta indica, Borassus flabellifer, Phoenix sylvestris, Pithecolobium dulce*
(ii) Other plant species	Species belonging to the genera: *Capparis, Cardiospermum, Cassia, Cayratia, Cocculus, Cuscuta, Glycomis, Hibiscus, Passiflora, Sida, Tiliacora, Tinospora, Zizyphus*
Habitat Type V. Low flatland	
(i) Herbs	*Cyperus* spp., *Ludwigia* spp., *Croton* spp., *Cynodon* spp., *Solanum* spp., *Phyllanthus* spp., *Vernonia* spp.
(ii) Shrubs	*Lippia geminata, Desmodium gangeticum*

Note: Ghosh and Chattopadhyay (1990) reported as many as 143 plant species belonging to 55 plant families including submerged, semi-emergent, floating, and marginal vegetation in the wetlands as well as other herds and trees associated with low flatland, upland, and dike areas.

Present Wetland Plant Communities

BBNP includes wetlands and uplands representing both aquatic and terrestrial plant communities. In the aquatic environment of the fish ponds phytoplankton communities comprised microflora belonging to Cyanophyceae, Chlorophyceae, Euglenophyceae, Xanthophyceae, Chrysophyceae, and Bacillariophyceae. This algal community includes more than 15 species in which Cyanophyceae dominates in spite of a large number of species of Chlorophyceae (Deb and Santre 1995). Algal blooms occur due to species *Microcystis, Spirulina*, and *Oscillatoria* in summer and post-monsoon months. Among the Chlorophyceae, algal species of *Chlamydomonas, Volvox, Pediastrum*, and *Tetraspora* are predominant. The gross primary productivity (GPP) of fish growing ponds receiving sewage from the adjoining localities in Kolkata Port range from 1.48 to 1.96 $g^2/$ m^2/h. But the overall net primary productivity (NPP) shows conspicuously lower value due to higher community respiration (CR) in all seasons (Deb et al. 1994).

The major macrophytic plants are represented by *Hydrilla, Vallisneria, Eichhornia, Pistia, Trapa, Lemna, Wolffia, Spirodela, Azolla, Utricularia, Sagittaria, Elodea*, and *Potamogeton* (Deb et al. 1994). The free-floating community mostly comprised water hyacinth (*Eichhornia crassipes*), water lettuce (*Pistia stratiotes*), and duck weed (*Lemna* spp.), while the marginal vegetation has a great diversity in its floral composition. The emergent amphibious community is dominated by *Leptochole chinensis, Enhydra fluctuans, Alternanthera sessilis, Eclipta prostata, Ipomoea reptans, Jussiaea repens*, and *Marsilea quadrifolia* (Mukherjee 1991). This emergent amphibious community also abounds in the transitional mudflats and the marshy meadows. However, a small patch of remnant swamp bed still characteristically is dominated by a single species of reed (*T. elephantia*) with sporadic presence of sedge (*Cyperus* spp.) located in the northwestern part of BBNP.

The upland and the raised dikes in between the fish ponds have more than 90 species of herb, shrub, and tree communities belonging to about 40 families comprising grass and wild plants, garden and vegetable plants as well as decorating flower plants and fruit trees. Some of the families are as follows: Amaranthaceae, Anacardiaceae, Apocyanaceae, Aracaceae, Asclepiadaceae, Asteraceae, Boraginaceae, Brassicaceae, Cannaceae, Caricaceae, Casuarinaceae, Compositae, Cucurbitaceae, Euphorbiaceae, Labiatae, Leguminosae, Liliaceae, Malvaceae, Meliaceae, Moraceae, Musaceae, Myrtaceae, Nyctaginaceae, Oleaceae, Pandanaceae, Palmae, Pinaceae, Poaceae, Rhamnaceae, Rutaceae, Solanaceae, Sterculiaceae, and Verbenaceae (Mukherjee 1991). These plant species and their production functions are listed in Table 5.3:

Present Use of the Brace Bridge Nature Parks

At present, the principle land use of BBNP is fishery and co-development activities. The work of fishery and creating the "nature park" was taken up since 1985 through the Fish Farmers Development Agency (FFDA) project under World Bank-aided Inland Fishery Project investing over 9,500,000 rupees toward development activities. Major components of this development work (Ghosh 1993a) include

Table 5.3 List of BBNP upland plants by function

Functional categories	Representative genera/species
1. Garden plants	*Casuarina, Codiaeum, Pinus*
2. Vegetable plants	*Capsicum, Carica, Cucurbita, Momordica, Solanum*, etc.
3. Flower plants	*Aster, Anthocephalus, Bougainvillae, Celosia, Cestrum, Ervatamia, Gardenia, Hibiscus, Ixora, Impatiens, Murraya, Mussandra, Nerium, Tagates*, etc.
4. Fruit plants	*Borassus flabellifer, Carica papaya, Citrus aurantifolia, Mangifera indica, Musa paradisiacal, Phoenix sylvestris, Psidium guajava, Punica granatum, Syzygium* spp., *Zizyphus jujuba*, etc.
5. Wild plants	*Azadirachta indica, Calotropis procera, Croton bonplandianum, Ficus bengalensis, Ficus religiosa*, etc.
6. Exotic plants	*Acacia auriculiformis, Acacia nilotica, Eucalyptus* spp., and *Leucaena macrophylla*

- treating about 25 million liters of sewage daily and protecting the River Hugli from being polluted;
- providing a green patch in this industrial area through planting more than 100,000 saplings;
- improvement of the drainage condition of the dock area;
- producing fish by trapping wastewater nutrients;
- creating waterfront for recreation and water bird habitat.

However, it should be mentioned that, beside the fishery, eco-development, and pollution abatement activities, the MFCS hosts a "deer park", garden, pet animals for promoting ecotourism, and also conducts field-level training programs in fisheries and environmental management.

Present Activities

The MFCS primary activity is pisciculture (fish culture) (see Fig. 5.5a and 5.5b). The basic layout and landscape mapping were done by Ghosh and Sen (1992) in 1988. Tanks or ponds were serially organized to act as facultative, maturation, and polishing tanks. Fish are grown in maturation and polishing tanks only. The water sources are domestic wastewater and urban runoff. The fisheries function as multiple-pond wastewater aquaculture systems. Studies on physical–chemical characteristics of the wastewater entering and leaving the wastewater entering and leaving the wetland system and the performance of the wetlands treating wastewater were carried out by the National Environmental Engineering Research Institute (NEERI 1990) in Kolkata at the behest of the CPT.

Besides fish culture, the MFCS also has a plantation program under which a nursery has been set up for different varieties of plants. Species are generally selected on the basis of their utility. They are fruit-bearing trees, flowering plants, trees providing dense foliage, plants acting as bioindicators of air pollution and those varieties which survive in urban and industrial areas and also help in mitigating atmospheric pollution. From 1985 to 1989 about 96,000 saplings were planted, of which 59,000 (60%) have survived. The society aimed to grow 2,000,000 saplings by the year 1989/1990. The fishery area, when interspersed with trees attracts many birds. A total of 120 species of birds from 35 families have been identified (Chattopadhyay 1985).

Another objective of the MFCS is to build the fishery and the adjoining area as an amusement center or a waterfront recreation center, especially for children. A deer enclosure with 22 (in 1996) spotted deer (*Axis axis*) has already been built. Other recreational facilities and facilities for environmental education are also being contemplated including raising ducks that also fertilize the fish ponds.

During 1987/1988, the NFCS earned about 1,897,000 rupees (USD 99,842). Total investment was about 2,529,000 rupees (USD 133,105) comprising the cost of fish seed (42%), wages and incentives given to other workers (53%), and other input costs (5%). The total sales turnover was 4,426,000 rupees (USD 232,947) of which 3,500,000 rupees (USD 184,210) was from the sale of fish. From 1987/1988 to 1988/1989, production increased from 3.91 tones/ha to about 4 tons/ha. During 1989/1990 production was about 5.61 tons/ha. In 1978/1988 the costs of labor for routine fishery activities accounted for about 54% of the total investment. In 1989/1990, this went down to about 26% while renovation costs accounted for about 9% of the total investment.

In addition to the members, the society also provides occasional employment to the local people. The members of the MFCS obtain a daily wage ranging from 28 rupees (USD 1.47) to 55 rupees (USD 2.89) depending on the nature of the work. In 1987/1988 about 87,000 man-days were created. Besides the daily wages, the MFCS encourages and ensures savings by the members and also provides financial assistance in the form of aid, loans, and pensions to its members.

The entire activity of the MFCS depends on its own resources and no financial assistance is sought from the outside. The establishment of the MFCS originated from individual contributions of 0.25 rupees (USD 0.013) per day for 50 members and their personal labor. The society received effective leadership from the executive officer, appointed by the Department of Fisheries, who continues to act as a facilitator to the members of the MFCS in the present phase of its activities.

The following tables describe the three major activities of vegetable farming, wastewater-fed ponds, and rice paddy cultivation. It should be known that the wastewater-fed ponds and fishery activity started first and this was followed by the use of wastewater for vegetable farming and rice paddy cultivation.

So Tables 5.4 and 5.5 describe most of the ongoing production activity as well as the usage of sewage effluent as part of the production process. All is not well as there are constant threats to these very productive processes.

Table 5.4 Types of resource recovery in east Kolkata wetlands[1]

Use/activity	Purpose	Quantity
Vegetable farm Alternating land strips with channels of sewage water Vegetables grown on substrata of garbage + irrigated w/sewage	For irrigation Drawn off sewage channels twice/year	150 tons of vegetables per annum
Wastewater-fed ponds (Bheris) Wastewater pre-treated before adding test fish 57,000 fingerlings/ha released	For initial filling of ponds Secondary filling of ponds to stimulate plankton growth and maintain DO levels For ponds >40 ha maybe continuous inflow/outflow for 15–21 days	8,000 tons/ annum
Rice paddy cultivation Used to grow more than one crop	Post-sewage fishpond effluent Benefits= high in nutrients Purified through settling, biodegradation + heavy metal removal	

[1]Sources include D. Ghosh, 1990 and Patnik, 1990.

Table 5.5 Sequence of activities for sewage-fed aquaculture[1]

Pond preparation
- Pond draining
- Sun drying
- Desilting silt traps (sometimes done instead of complete pond draining, probably due to land tenure concerns)
- Tilling
- Repairing dikes

Primary fertilization
- Filling with sewage
- Facultative stabilization
- stirring

Fish stocking – primary species include major Indian carps, silver carp, common carp, and tilapia
- Test fish
- Fish stocking proper

Secondary fertilization
- Filling with sewage

Fish harvest
- Net selection
- Team management
- Haul disposal/sales/distribution

[1]Source: D. Ghosh 1990.

Disturbances and Threats to the BBNP

The Calcutta Port Trust (CPT), the owner of these wetlands, intends to convert them into "real estate" by reclaiming the area for extension of the dock, container park, container repair yard, truck terminals, warehouses, water basin facilities, and construction of a housing complex and a road (AWB and WWF 1993). The CPT started distributing portions of the wetland area back to the dock authorities and initiated disposal of solid waste and city garbage at the wetland site through Calcutta Municipal Corporation (CMC) up to 1989 (Ghosh and Sen 1992). Though solid waste activity has been stopped, the CPT has been increasing rent regularly. The rent paid by MFCS to CPT has undergone a 15% rise in 1988. In November 1990, the lease to MFCS was extended to 3 years with the condition of a 25% rent increase each year. The society (MFCS) requested consideration of these terms and asked for a long-term lease with more rational rent. The CPT did not accept the appeal and issued an order on July 15, 1992 to vacate the area by July 23, 1992. The Fisheries Department of the Government of West Bengal strongly reacted to this approach and requested the chairman of CPT to hand over the park area to the Government of West Bengal (Ghosh 1993a), This dispute has been referred to the courts since 1992. Now solid waste is utilized as part of the vegetable growing process as described in the previous section.

Mention should be made here that the existing laws are the (i) West Bengal Fisheries (Acquisition and Requisition Act and (ii) the Town and Country Planning Act which stipulates that no pond measuring five cottahs (0.03 ha) or more can be filled up. The West Bengal Inland Fisheries Act (1984) stipulates that the management of embankments is obligatory for the proper utilization of fishponds. But, to date, the existing laws are not strictly enforced. There is every reason to protect and preserve this unique ecosystem, which can serve as a model for low-cost options for municipal sanitation in the poorer parts of the world (Ghosh 1993a). Furthermore, the traditional rights of fishing have been observed since 1958, which is a prevalent practice in the region, and is an important regulatory factor in legislation and management options.

The water bodies of BBNP receive raw wastewater from the adjoining 152 industrial units (see Table 5.6), which are also imposing considerable threat to these wetlands. The conflict between the landowner and the leaseholder is the major constraint impeding the progress and productivity of park activities. The disposal of raw waste instead of treated waste is an

Table 5.6 Adjoining industrial units and waste flows to park

Sl. No.	Type of industry	Number of industries	Wastewater flow (approx. m^3/day)
1.	Engineering industries	65	1728.6
2.	Chemical industries	26	15,489.6
3.	Godowns/garages	42	1505.6
4.	Institutions	4	2015.6
5.	Miscellaneous industries	15	1050.6
	Totals	152	21790.0

Source: Deb et al. (1996).

important risk factor toward toxic pollutant hazard/health hazard for the
fish eaters of West Bengal. There are also other limitations and threats like
the flow of funds, grazing by pigs, washing, bathing, defecation, etc., as
shown in the utilization of BBNP (see Table 5.7).

Table 5.7 Utilization scenario of BBNP, Kolkata

Usage categories	Utilization status[1]
A. Wetland/water usage	
1. Fisheries including nursery pond	VH
2. Source of employment/economic support	VH
3. Recreational boating	H
4. Fresh water fish supply to Calcutta market	H
5. Reservoir of water for	
a. sewage water receptacle*	VH
b. waterfowl habitat biodiversity	H
c. bathing for slum dwellers*	M
d. washing for slum dwellers*	M
e. domestic water fro slum dwellers	M
f. fire fighting	L
g. irrigation for garden plants	L
6. Sewage disposal both industrial and domestic*	VH
7. Water purification/pollution abatement for Ganges	VH
8. Tourism/ecotourism	VH
9. Conservation/eco-development/microclimate	H
10. Defecation and afterwash*	H
11. Retention of floodwater/waterlogging prevention	H
12. Piggery (for slum dwellers)*	H
13. Duckery (domestic)	L
14. Research/training/environmental management	H
15. Solid waste disposal (up to 1989)*	H
16. Grazing (by pigs, goats, etc.)	H
B. Dry land/dike usage	
17. Gardening/greenery/nature park/aesthetics	VH
18. Air pollution amelioration by plants	H
19. Prevention of soil erosion	H
20. Cool greenshed for tired/retired people	M
21. Deer park and pet animals for children	M
22. Dating site for young people	M
23. Picnic spot	M
24. Railway line	M
25. Roosting/nesting site for birds	H
26. Supply of fodder for deer and domestic animals	M
27. Supply of fruits, flowers, and vegetables	L
28. Firewood collection site	L
29. Institution/office establishment*	L
30. Anti-social activities*	L

[1]VH = very high, H = high, M = medium and L = low.
Note: Asterisks (*) indicates threats to the wetland. Readers are referred to Ghosh and Nandi
(1996) and Mukherjee et al. (1996) for comparative utilization scenarios.

Economic and Social Values of Wetlands

The wetlands of BBNP have multifarious usage (see Table 5.5) and support an important urban fishery system in addition to improving water quality before release into the Hugli River. The production of fish in 1989–1990 was recorded as 286 tons with a gross profit of 2,945,992 rupees out of a total earning of 3,001,441 rupees (see Table 5.8).

Table 5.8 Fish yield and earning from fish sale by MFCS

Year of Production	Area under fish culture (ha)	Total production (MT)	Yield/ha (MT)	Total earning (Rs. 00000)
1980–1981	40	65	1.6	8
1981–1982	40	74	1.8	10
1982–1983	43	79	1.9	12
1983–1984	45	87	1.9	13
1984–1985	45	97	2.2	12
1985–1986	45	85	1.9	13
1986–1987	60	229	3.75	34
1987–1988	60	235	3.92	35
1988–1989	65	260	4.20	39
1989–1990	50	288	5.6	50

Source: MFCS records (Dutta and Kapoor 1992).

These wetlands have proved to be efficient in treating sewage water with industrial and domestic wastewaters comprising about 70 and 30%, respectively, of influent flow as well as removal of BOD by 80.52% and fecal coliform bacteria by 99.99% (NEERI 1990). These wetlands provide a livelihood to about 400 fishermen families and 80 retailers and about 5000 fish eaters have benefited from a fresh supply of fish through retailers to 37 markets. The society (MFCS) also initiated selling of processed fish in polyethylene packs to selected retail stalls in Kolkata.

The site has aesthetic value to the people living in Kolkata and its adjoining districts. About 400,000 people visited the nature park in 1993 (S. Mandal, 1996. Wastewater as a resource for Development, A case study of Mudialy Fisherman's Cooperative Society Ltd. In *Conservation and Management of Lakes/Reservoirs in India*, pp. 177–182, Japan, ILEC, personal communication). In the same year, 4000 farmers participated in the training on botany, aquaculture, and cooperative management. The society has established an Information Exchange Center and maintained birds and pet animals like deer, rabbit, guinea pig, monkey, goats, ducks, swans, turtles, etc., to attract school children in Kolkata as well as West Bengal.

Wetland Management Issues

The management of wetlands involves integration of both land and water management along with appreciation/assessment of the social, cultural, and economic issues (Kada 1991). The population residing in the catchment area and the local NGOs/CBOs have an important role to play in water resource utilization, planning, and management. The present CBO (MFCS) in this respect provides a unique case study of urban wetland management encompassing four areas of integrated wetland management, e.g., water quality, water level, vegetation/landscape, and aquatic species management.

Water Quality Management

The wetlands of BBNP receive an annual daily inflow of approximately 23 million liters of sewage water comprising industrial (70%) and domestic wastes (30%). The incoming untreated sewage is initially treated in two anaerobic tanks and six-segmented macrophyte channels using lime as the only chemical for treatment. Aquatic plants (water hyacinth, water lettuce, duckweed, reeds, etc.) are used as macrophyte filters to facilitate absorption of oil, grease, and micropollutants in the effluent water. The first anaerobic tank is dug out to reduce sludge deposition as needed. The water quality is monitored by trained staff and is also regularly checked for the occurrence of prematurely dead fish and engulfing/surface behavior of fish maintained in the macrophyte channel. These air-breathing fish are *Anabas testudineus, Clarias batrachus, Heteropneustes fossilis, Channa orientalis, Channa punctatus*, and *Channa striatus* which can endure enough toxic stress in their aquatic environment. During the process of purification, the sewage water is retained in the macrophyte channel for about a week for treatment. The semi-purified water is further treated in fishponds before release into the Hugli River through a canal known as Manikhal. If fish mortality occurs, the purified fishpond water is recirculated into the macrophyte channel through a system of sluices for dilution of the toxic elements in the system. The water quality is tested by the National Environmental Engineering Research Institute (NEERI 1990) for influent and effluent water from BBNP at the request of the wetland owner (CPT) and is presented in Table 5.9. The data indicate that this sewage-fed wetland (BBNP) is a self-sustaining system with significant BOD and fecal coliform removal capacity.

It should be noted that some species of fish *Chanda ranga, Chanda nama, Amblypharyngodon mola, Puntius sophore*, and *Puntius conchonius* are sensitive to toxic stress and have been found in some ponds in recent years indicating improvement of water quality taking place within the wetland system. It should be mentioned that natural purification of wastewater is accomplished and augmented in the presence of sunlight. The nutrients contained in the influent wastewater help with the nourishing of a healthy algae bloom in the fishponds. The algae remove the nutrients, which accumulate in the algae biomass. The

Table 5.9 Wastewater quality for MFCS wetland system

Parameters	Influent	Effluent
Flow m^3/h	99.3	947.0
Temperature °C	30/28	33/28
pH	7.95	7.50
Total solids	1152	788
Suspended solids	51.18	73.00
Dissolved solids	1099.9	715
Total volatile solids	340.41	210
BOD	77.58	15
COD	470.42	65
Total alkalinity as $CaCO_3$	285.44	210
Total nitrogen as N	114	31
Phosphate as P	0.20	0.04
Mercury as Hg μg/l	4.42	Below detection

Source: NEERI 1990

Note: All parameters except pH, temperature, mercury, and flow are expressed as mg/l. Fecal coliform (MPN per 100 ml 10) in the influent and effluent water recorded as 46,000 and 0.91, respectively. Removal percentage of BOD was reported as 80.52% and for fecal coliform as 99.9%.

driving force is photosynthesis, which is supported by symbiotic activity between saprophytic bacteria and algae (Deb et al. 1996). The carbon dioxide is released due to bacterial decomposition of organic matter in the presence of sunlight. It is taken up and converted to algal cell material with liberation of oxygen, which is utilized by bacteria for the aerobic decomposition of organic matter (Das et al. 1990). The cultivable fish are the secondary carnivores, which thrive on the primary producers, i.e., the phytoplankton. Thus the organic nutrient content of the wastewater enters the food chain and sets up the equilibrium in this wetland system.

Water Level Management

The water level in the fishpond system is usually maintained at 1.8 m except in winter when it is reduced to about 1.5 m for augmenting water temperature to facilitate fish growth. The maintenance of water level is accomplished through a system of sluices adapting both clockwise and counterclockwise circulation of water within the system. The excess amount is released to the Hugli River through the Manikhal Canal.

Vegetation and Landscape Management

An appropriate afforestation program is accomplished by planting 30% legumi-nous plants, 30% dust and chemical absorbing plants (*Calotropis procera*,

Azadirachta indica, and *Ficus religiosa*), 30% fruit trees for attracting birds, and 10% horticultural plants to combat soil erosion and air pollution affects. Such a vegetation program also helps in fishpond fertility due to biomaturing through leaf litter decomposition of nitrogen-rich leguminous plants grown on dikes alongside the water bodies. The green patch developed in this industrial southwestern sector of congested Kolkata is a welcome relief of aesthetic importance to the urban environment. Landscape management includes the maintenance of the dikes and landscape beatification by cutting, trimming, and weeding of the uplands and wetlands after the growing season (in the post-monsoon period, October) by the MFCS. Birds attracting acidic trees such as "Triphala" are planted at a distance from the water to avoid acidification from overfertilization of the fishponds.

Aquatic Species Management

The dominant cultivable species of BBNP are the carp and *Tilapia*. The society has, however, brought 10 species of fish under culture including Catla (*Calta catla*), Rohu (*Labeo rohita*), Mirigel (*Cirrhinus mrigala*), Bata (*Labeo bata*), common carp (*Cyprinus carpio*), grass carp (*Ctenopharyngodon idella*), big head carp (*Aristichthys nobillus*), silver carp (*Hypothalamichthys molitrix*), and Tilapia (*Oreochromis mosssambica* and *Orechrmis nilotica*); Indian major carp (IMC), exotic carp and *Tilapia*. About 80% IMC is maintained for about 8 months (March–October) with a definite proportion of surface, mid-water, and bottom feeding fish while only 40% IMC is stocked for the rest of the 4 months of winter (November–February) to minimize the competitive market with sea fishes available in the local markets during winter. It has been found that the stocking density and variety of fish cultured at BBNP are also related with the quality of water. The *Tilapia*, which can endure a certain amount of toxicity, are stocked nearer to the inlet while carp as well as prawns, which are comparatively sensitive species, are grown in subsequent ponds further away from the inlet.

In general, the fingerlings are stocked for about 90–120 days before harvesting. They are stocked at a rate as high as 35,000–40,000 individuals per hectare to utilize the naturally produced plankton. The selection of species as well as their proportion depends on the plankton production of the fish ponds which includes useful varieties of both phytoplankton and zooplankton species (Deb and Santra 1995, Deb et al. 1994, Santra and Deb 1995). The regulatory stocking system, drying and weeding of fishponds, and prevention of predatory fish from the wetlands are all measures adopted for plankton production and fisheries management.

Organizational Development

Based on sewage-fed wetlands owned by the CPT, the society (MFCS) has developed an ingenious process based on eco-engineering principles to perform three important functions:

- Improving the wastewater quality using lime and macrophyte filter
- Using wastewater nutrients as inputs to grow fish food (plankton), fish
- Development of an ecologically balanced system to accommodate a number of animal and plant species for environmental and aesthetic values

Since 1985, undertaking environmental enhancement activities has become part and parcel of the overall objectives of the organization. Besides evolving a scientific systematic approach to manage the environment, the chief executive officer (CEO) initiates steps to improve upon the expertise of the MFCS members undertaking such plans by

- sending some of the members to the State Agricultural University to learn scientific pisciculture;
- deweeding and desilting of tanks or lagoons;
- intensive fish culture;
- scientific training for developing a balanced ecosystem;
- preparing the members for undertaking future development work of the society out of their own savings.

The work principles of the organization are based on absolute equality across members with respect to the workload, pay structure, and other facilities. Both casual laborers and associate members can go up the ladder and eventually become permanent members. The organizational structure rules out specialization and every member is required to do all kinds of work on rotation to avoid over specialization. Even though there is a system of functional hierarchy, the workers as well as their supervisors and commanders all enjoy the same pay scale. Maintenance of democratic principles, strict work discipline, autonomy and work ethic, openness to scientific investigation, and rapport with local residents to tackle external affairs/interference are all the major building blocks toward the success for the society (MFCS). There is also a package of welfare benefits such as medical aid, education aid, education aid, old age pension, marriage grant, housing loan, funeral expense (see Table 5.10) that provide incentives for integrity and improved performance for members of the organization.

The organization has achieved recognition from different areas at home and abroad. The comments made by Indian and foreign scientists in the register of this society about their activities and work programs relate to their remarkable success in wetlands and water quality management. A number of detailed evaluations were carried out by national agencies such as the National Environmental Engineering Research Institute (NEERI 1990), Indian Institute for Management (Dutta and Rapoor 1992), National Wasteland Development Board (Mukherjee 1991), Zoological Survey of India (Ghosh and Chattopadhyay 1990), which verifies the skills developed by this organization. The MFCS has also earned a number of awards as follows:

Table 5.10 Welfare activities for members of MFCS, Kolkata

Categories of welfare expenses	Welfare expenses	Remarks
Medical aid	Rs. 160.04–5543.82	Full reimbursement
Educational aid	Rs. 3544.00–9551.74	" " up to Class IV
Old age pension	Rs. 400/month/ member	Members or widow
Marriage grant	Rs. 2000	Members' daughter
Funeral expense	Rs. 1000	Members' death
Housing loan	Rs. 35, 000	Without interest
Consumption loan	Rs. 500	One time
Janata Insurance Policy		Coverage of accident
Drainage/sanitation/drinking water		Servicecharge paid by MFCS

Source: MFCS records.
Note: Expenses relate to the years 1980–1981 to 1989–1990.
Expenses on charity ranged from Rs. 479.00 to Rs. 53,182.75 a year.
Expenses on sports reaches as high as Rs. 61,175.00 in 1989–1990.

- National Productivity awards for fish production in 1985–1986 and 1987–1988 from the National Productivity Council, Government of India.
- Best award on productivity from Fish-Cofed and National Cooperative Union of India in 1992.
- National Film Festival award in 1992 for the documentary film "Mudialy Alternative" on the activities of this organization in the Nature Park (BBNP) being considered as the best film in the field of Environmental Conservation and Preservation in 1992.
- Best Fisherman's Cooperative Society awards in 1993 from the State Cooperative Bank and State Cooperative Union.
- Indira Priyadarshini Vrikshanitra (Friends of Trees) awards for Forestry in 1995 from the Government of India.
- Several awards from various flower and vegetable shows organized in the state of West Bengal.

Controversy Surrounding the Protection of BBNP

The urban wetlands of Kolkata have undergone critical changes amidst serious controversies. Both government and non-governmental organizations (NGO) have played their roles in arousing controversies as well as moving toward conservation concerning the (i) Brace Bridge wetlands (BBNP) and (ii) the east Kolkata wetlands (ECW). It is worth mentioning that the ECW was recently declared as national wetlands by the Department of Environment (DOE), Government of India.

The east Kolkata wetlands movement was supported by an NGO – People United for Better living in Calcutta (PUBLIC) – which provided resistance to the real estate business carried on in eastern Kolkata. Meanwhile, another movement was initiated to save the wetlands of BBNP in southwest Kolkata. With the movement surrounding the protection of BBNP, various public interest, environmental and people's science groups have taken interest. The local state government had taken note of plans of the Calcutta Trust (CPT) to reclaim the wetlands of BBNP for real estate development.

The CPT, in a very sudden move, decided to take possession of the land and water bodies on July 23, 1992. The state government intervened in favor of MFCS. Section 144 was promulgated per order of the Alipore Court, Kolkata, and stopped the CPT from developing the wetland area. The Fisheries Department, Government of West Bengal had proposed to take hold of the wetlands from CPT, but the CPT refused the request for transfer of the wetlands on August 4, 1992. They secured a stay order against Section 144 from the High Court on August 7, 1992. Three days later the MCFS also got a stay order against any further action from the same court on August 19, 1992 (Uttarpara Vigyan Sanstha et al. 1992).

Outside the court, various people's science groups, environmental groups, health movement groups like Nagrik Manch, the Scientific Workers forum, the Vigyan Vikas, the Institute of Engineers, the People's Science Coordination Center have jointly taken up the challenge to confront the CPT (Mukul 1992). So far this matter has not been settled.

Recently, the Slim Group of Indonesia has proposed a new Kolkata International Development Project which includes expressways, bridges, special economic zones, industrial hubs, plus wealth and knowledge centers. This proposed 85 km expressway starts from Baraset and will pass along the northern edge (see Fig. 5.6) of the east Kolkata wetlands, even though an exact

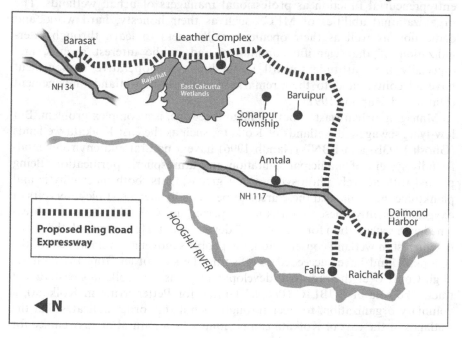

Fig. 5.6 Route of the proposed Eastern Expressway Source http://www.telegraphindia.com/1050509/images/09zzroadbig.jpg

alignment has not been chosen. But many are concerned about potential impacts to this Ramsar wetland.

Summary

The urban wetlands in Kolkata have many features for waterfront recreation. They also have lent themselves well to use by local communities. But besides these functions the Brace Bridge Nature Park has various other important roles to play in the municipal sanitation for purifying wastewater as well as in fisheries. The indigenous technology adopted to wastewater management by MFCS, a community-based fisherman organization, efficiently utilizes the algal species for performing the dual task of nutrient recovery and food chain support for pisciculture seems to be a precedent for the future of sanitation technology. In so doing, such a system could assume global significance as a "tutorial system" (Bhagat 1993). Following this example, some developing country communities could develop their apparently unproductive waterlogged city fringes into uses for environmental protection, food production, and employment generation.

In terms of success indicators such as application of modern environmental management principles, style of function, experience, socio-economic proximity, access to leadership to members, and building patronage membership, the present CBO (MFCS) has made remarkable advances in entrepreneurial function as professional managers of urban wetlands. The organizational abilities of MFCS such as their honesty, hard work, and dedication as well as their openness, willingness to learn through internalization of the scientific community and public interest groups, and especially their attitude toward leadership via cooperative management have all contributed to their remarkable success in wetland management (Dutta and Rapoor 1992).

Managing urban wastes, both solid and sewage, is a complex problem. But low-lying sewage-fed wetlands of Kolkata, such as the east Kolkata wetlands (Ghosh 1993b) and BBNP (Nandh 1996) have a natural and environmental-friendly system of municipal sanitation and atmospheric purification. Being packed with actively photosynthesizing green plants, both macrophytic and planktonic, have allowed these areas to be virtually inexhaustible reservoirs of oxygen. Presently, these two urban and periurban wetland systems are threatened with reclamation for "real estate" development. However, with the growing interest in wetland conservation, the people's movement reminiscent of the "Chipko" kind have succeeded in affecting the stay order from the Kolkata High Court over the proposed development plans for Kolkata's eastern wet tract. The role of PUBLIC (People' United for Better living in Kolkata), a voluntary organization, to steer through such a stay order indicates that the wetlands in the east of Kolkata can no longer be a natural or easy choice for urban expansion.

Aside from the traditional grassroots-inspired social engineering of the east Kolkata wetlands (ECW), there has been continual pressure on the ECW, which may affect their long-term sustainable utilization. A recent World Bank-sponsored audit and valuation of ECW was done by Dr. Chattopadhyay (2001). Key findings or discussion items include the following:

- Continual pressure for land use conversion and encroachment on wetland use. Also ownership of fisheries is highly skewed between large owners (160 ha) and small owners (0.4 ha).
- Kolkata tanneries (550 in the area) put increasing threat of toxic contamination on fisheries and vegetable production due to heavy metals and organic compounds.
- Decreasing biodiversity due to lack of reed beds for bird shelter, nesting, and roosting sites. The only birds that have adapted to this situation are the colonial water birds like herons and egrets. Studies suggest a reduction of 84% in bird species diversity (A. Ghosh 1990 and 1997).
- Change in hydrological regime affecting ecological balance and function.
- Inundation of periphery causing loss of property and life from monsoon storms.
- Loss of agriculture and fish production leading to unemployment.
- Rise in urban pollution and social unrest due to public health conditions.

El Harake (1998) also comments that there are a number of political, socio-economic, and religious factors that complicate or slow down needed changes such as watershed/waste management and land use control measures. D. Ghosh, a key proponent of this innovative project, admits that "A serious challenge is to coordinate the various activities that are now being taken up by different agencies (lists 7 agencies) ... It may be appropriate to create a separate coordinating agency to synchronize the required study" (D. Ghosh 1990).

In this case study, we have seen the roles of three types of NGOs. First there is the management-oriented local CBO, Mudialy Fisherman's Cooperative Society (MFCS) that is concerned with the day-to-day management and utilization of the wetland system of BBNP. This CBO is probably one of the most developed examples we have seen. The second NGO-People United for Better Living in Calcutta (PUBLIC) is concerned with stewardship or wetland protection of ECW against development interests. Clearly, the second NGO has been successful to date. The third type of NGO is the World Wildlife Fund or "wwfindia" which is an international NGO, which came on to the scene in late 1990s and has a web site on the ECW as a Ramsar site (http://www.wwfindia.org/calcutta_29.php?fileid = 29).

It is the dual function of these NGOs together without significant external support from international NGOs or other organizations, which is significant for conservation of wetlands in the Indian Kolkata region.

Acronyms

BBNP: Brace Bridge Nature Park
CMC: Calcutta Municipal Corporation
CMPO: Calcutta Metropolitan Planning Organization
CPT: Calcutta Port Trust
DOE: Department of Environment
ECW: East Calcutta Wetlands
FFDA: Fish Farmers Development Agency
MFCS: Mudialy Fisherman's Cooperative Society
NEERI: National Environmental Engineering Research Institute
PUBLIC: People United for Better Living in Calcutta

References

Antos, M. J., G. C. Ehmlee, C. L. Tzarios, and M. A. Weston. 2007. Unauthorized human use of an urban coastal wetland sanctuary: Current and future patterns. *Landscape and Urban Planning*, 80(1–2): 173–183.

AWB and WWF. 1993. *Directory of Indian Wetlands*. Asian Wetland Bureau and World Wildlife Fund for Nature, India, pp. 1–252.

Bhagat, R. K. 1993. Uses of Calcutta's wetlands. *Yojana*, 36(24): 17–19.

Chattopadhyay, K. 2001. *Environmental Conservation and Valuation of East Calcutta Wetlands*. Wetlands and Biodiversity EER Working Paper Series: WB-2 prepared for the World Bank. Mumbai, India: Indira Gandhi Institute of Development Research, Goregaon (East), 67pp.

Chattopadhyay, S. 1985. *Ecological Reconnaissance of the Brace Bridge Swamp – A proposed bird sanctuary of Calcutta*, Appendix II. Calcutta: Zoological Survey of India, pp. 1–8.

Costa-Pierce, B. A., A. Desbonnet, P. Edwards, and D. Baker. 2005. *Urban Aquaculture*, CABI Publishing.

Das, K. K., A. K. Biswas, A. K. Ganguly, N. Chattopadhyay, S. H. Mollah, P. B. Sanyal, and S.C. Deb. 1990. Recycle and reuse of industrial effluents for aquaculture – a case study. *Proceedings of National Seminar on Utilization of Resources*, India, pp. 73–78.

Deb, S. C., K. K. Das, and S. C. Santra. 1994. Studies on the productivity of sewage fed pond ecosystem. *Proceedings of the Academy of Environmental Biology*, 3(1): 33–42.

Deb, S. C., J. S. Pandey, and S. C. Santra. 1996. A practical approach to water pollution control through ecologically balanced wastewater management: A case study from Calcutta, India. In S. R. Mishra (ed.) *Assessment of Water Pollution*, pp. 63–79. Delhi: APH Publishing Company.

Deb, S. C. and S. C. Santra. 1995. Plankton ecology of sewage fed aquatic ecosystem in Calcutta. In R. C. Mohaznty (ed.) *Environment, Change and Management*, pp. 78–89. New Delhi: Kamal Raj Enterprise.

Dutta, S. and S. Rapoor. 1992. *Collective Action, Leadership and Success in Agricultural Cooperatives- A Study of Gujarat and West Bengal – Draft report*. Ahmedabad: Indian Institute of Management, 368pp.

El Harake, M. 1998. The future of Calcutta sewage, Waste yes, want not. At http://darwin. bio.uci.edu/~sustain/suscoasts/melharake.html

Emerton, L. (ed.). 2005. *Values and Rewards: Counting and Capturing Ecosystem Water Services for Sustainable Development*. Colombo: IUCN Water, Nature and Economics Technical Paper No. 1, IUCN Ecosystems and Livelihoods Group Asia, 93pp.

Fasela, M. 2002. Coastal and inland wetlands in China and Pakistan: Colonial water birds as bioindicators of pollutant levels and effects. Project report http://www.unipv.it/webbio/labweb/ecoeto/cina/index.htm

FAO. 2003. Urban Forestry in Asia-Pacific Region status and prospects, Annex 2 Roles and Importance of urban Trees and Forests. FAO Corporate Document Repository at http://www.fao.org/decrep/003/x1577e/x1577E12.htm

Gerrad, P. 2004. *Integrating Wetland Ecosystem Values into Urban Planning: The Case of That Luang Marsh, Vientiane, Lao PDR.* IUCN-The World Conservation Union, Asia Regional and Environmental Economics Programme and WWF Lao Country Office, Vietnam.

Ghosh, A. K. 1990. Biological resources of wetlands of East Calcutta, India. *Journal of Landscape and Ecological Studies*, 13: 10–23.

Ghosh, A. 1997. *Management of East Calcutta Wetlands and Canal System*, Dept. of Environment, Govt. of West Bengal assisted by the UK Overseas Development administration.

Ghosh, D. 1990. Wastewater fed aquaculture in the wetlands of Calcutta – an overview. In P. Edwards and R. S. V. Pullin (eds.) *Wastewater Fed Aquaculture*, pp. 50–56. Bangkok, Thailand: Environmental Sanitation Center, Asian Institute of Technology.

Ghosh, D. 1993a. Uncertainty over Mudialy Nature Park in CPT wetlands near Brace Bridge Railway Station. *Environment*, 1: 45.

Ghosh, D. 1993b. Towards sustainable development of Calcutta wetlands, India. In T. J. Davis (ed.) *Towards Wise Use of Wetlands*, pp. 107–112. Gland, Switzerland: Ramsar Convention Bureau.

Ghosh, D. 1998. The Calcutta wetlands, Turning bad into good. *Ashoka Changemakers Journal* (Oct. 1998) and http://www.changemakers.net/journal/98October/ghosh.cfm

Ghosh, S. K. 2004. Traditional commercial practices in sustainable development and conservation of man and wetlands. *Knowledgeable Marketplace Reports.* Bangkok, Thailand: The Third IUCN World Conservation Congress (17–25 Nov. 2004).

Ghosh, D. and S. Sen. 1992. Developing waterlogged areas for urban fishery and a waterfront recreation project. *Ambio* 21(2): 185–186.

Ghosh, A. K. and S. Chattopadhyay. 1990. Biological Resources of Brace Bridge wetlands. *Indian Journal of Landscape System and Biological Studies*, 13(2): 189–207.

Ghosh, D. and S. Sen. 1992. Developing waterlogged areas for urban fishery and a waterfront recreation project. *Ambio*, 21(2): 185–186.

Guntenspergen, G. R. and C. Dunn. 1998. Introduction: long term ecological sustainability of wetlands in urbanizing landscapes. *Urban Ecosystems*, 2(4): 187–188.

Hart, T. H. F. 2006. An overview of water sensitive urban design practices in Australia. *Water Practice and Technology*, 1(1): 1–8.

Irvine, K. N. 2007. The Role of Phnom Penh's Wetlands in Sustainably Treating Sewage Discharges to the Mekong/Bassac River System. Final Report. Department of Geography and Planning, SUNY Buffalo State College, 22pp.

Kada, 1991. Understanding the state of lake environment from a socio-cultural perspective-an example from Lake Biwa, Japan. In M. Hasimoto (ed.) *Guidelines of Lake Management*, pp. 7–18. Japan: ILEC/UNEP.

Kumar, D. and B. S. Keddy. 2000. *Ecology and Human Well Being.* Beverly Hills, CA: Sage Publications.

Lee, W.-I. 2006. A study of ecological spatial patterns of Tianain City, Taiwan. Paper presented at IFoU 2006 Beijing International Conference at http://dspace.lib.ksu.edu.tw:8080/dspace/handle/123456789/2700

Mukul. 1992. To save the wetlands – struggle by a fisherman's cooperative in Calcutta. *Frontline*, November 20, 1992, pp. 76–84.

Mukherjee, R. 1991. *Rapid Assessment of Flora and Fauna in Urban Fishery and Waterfront recreation Ecosystem (Near Brace Bridge Railway Station).* Calcutta: Report of Eastern Regional Centre, National Wasteland development Board, pp. 1–59.

Mukherjee, A. 1998. Fisheries in Eastern Calcutta, An Economic Study. In *West Bengal: A Social and Economic Profile*. Kolkata, India: Institute for Studies in Social and Economic Development.

Munshi, S. K. 1991. Calcutta, Land, land use and land market. In B. Dasgupta et al. (eds.) *Calcutta's Urban Future*, Calcutta: Government of West Bengal, pp. 48–59.

Nandh, N. C. 1996. *Draft Management Plan for Brace Bridge Nature Park (BBNP)*. Prepared at the International Course of Wetland Management (1996), the Netherlands: Wetland Advisory and training centre (WATC), pp. 1–52.

NEERI. 1990. *Water Quality Studies of the Jheel in Calcutta Port Area*. National Environmental Engineering Research Institute, Nagpur, India.

Patnik, A. L. 1990. An action plan for the development of the Calcutta sewage fed fishpond system. In P. Edwards and R.S.V. Pullin (eds.) *Wastewater Fed Aquaculture*, pp. 120–122. Bangkok, Thailand: Environmental Sanitation Center, Asian Institute of Technology.

Patnaik, D. C. and P. Srihari. Undated. Wetlands-A Development paradox: the dilemma of South Chennai, India, at http://ssrn.com/abstract = 591861.

Prain, G., B. Arce, and N. Karanja. 2006. Horticulture in urban ecosystems: some socio-economic and environmental lessons from studies in three developing regions. *ISHS Acta Horticulture 762*: XXVII International Horticulture Congress 1 HC 2006: International Symposium on Horticultural Plants in Peri-Urban Life.

Ramachandra, T. V. 2001. Restoration and management strategies of wetlands in developing countries. *Electric Green Journal*, Issue #15.

Ratner, B. D., D. Than Ha, M. Kosal, A. Nissapa, and S. Champhengxay. 2004. *Undervalued and Overlooked, Sustaining Rural Livelihoods Through Better Governance of Wetlands*. CABI Publishers.

Santra, S. C. and S. C. Deb. 1995. Hydro geochemistry and hydrobiology of sewage fed jheels in the tropics: A Case study. *Indian Hydrobiology*, 1: 5–17.

Smardon, R. C. 1989. Human perception of utilization of wetlands for waste assimilation, or how do you make a silk purse out of a sows ear? In D. A. Hammer (ed.) *Constructed Wetlands for Wastewater Treatment: Municipal, Industrial and Agricultural*, pp. 287–295. Chelsea, Michigan: Lewis Publishers.

Smardon, R. C. 2008. A comparison of Local Agenda 21 implementation in North American, European and Indian Cities. *Management of Environmental Quality: An International Journal*, 29(1): 118–137.

Streeter, W. J. 1998. Kooragong Wetland rehabilitation Project: opportunities and constraints in an urban wetland rehabilitation project. *Urban Ecosystems*, 2(4): 205–218.

Tapsuwan, S., G. Ingram, and D. Brennan. 2007. Valuing urban wetlands of the Gnangara Mound: A hedonic property price approach in Western Australia. Paper presented at Australian Agricultural and Resources Economics Society 2007 Conference (Feb. 13–16, 2007 Queenstown, New Zealand) at http://purl.umn.edu/10418/

Uttarpara Vigyan Sanstha et al. 1992. Now affected Mudialy: Wetland-Nature Park Cooperative – A Documentary paper (in Bengali), Sept. 1992, pp. 1–16.

Vymazel, J. 2005. *Natural and Constructed Wetlands: Nutrients, Metals and Management*. The Netherlands: Backhuys Publishing.

Wong, M. H. 2004. *Wetland Ecosystems in Asia*. The Netherlands: Elsevier, 451pp.

Wong, T. H. F. 2006. An overview of water sensitive urban design practices in Australia. *Water Practice and Technology*, 1(1): 1–8.

Zedler, J. B. and M. K. Leach. 1998. Managing urban wetlands for multiple use: Research, restoration and recreation. *Urban Ecosystems*, 2: 189–204.

Zhao, S., C. Peng, H. Jaing, D. Tain, X. Lei, and X. Zhou. 2000. Land use change in Asia and the ecological consequences. *Ecological Research*, 21(6): 890–896.

ZongMing, W., Z. ShuQing, and Z. Bai. 2004. Effects of land use change on values of ecosystem services of Sanjiang Plain, China. *China Environmental Science* 24(1): 125–128.

Chapter 6
Restoration of the Tram Chim National Wildlife Preserve, Vietnam

Introduction

Stretching about 200 km between the border of Kampuchea and the South China Sea, nine branches of the mighty Mekong River (meaning nine dragons) spread across a wide and fertile delta where three crops of rice can be harvested each year (see Figs. 6.1 and 6.2). The Mekong originates in the mountains of western China where spring melts and rains, together with summer monsoons of the tropics, combine to flood the banks from June through October. The annual cycle, interacting with the daily ebb and flow of the tides, created ideal habitats for both fresh and saltwater wildlife. But today the Mekong Delta is one of the heavily populated regions of the earth. Most of the forests and wetlands have been transferred into cities and farms. The delta is Vietnam's food basket (Torrell and Salamanca 2005).

The Mekong River basin is of truly exceptional significance to international biodiversity conservation in comparison with other parts of tropical Asia. The area supports a very large number of bird species identified as globally threatened or globally near threatened (Buckton and Safford 2004) including the famous Eastern Saurus Crane (*Grus antigone sharpii*), giant ibis (*Pseudibis gigantea*), white-shouldered ibis (*Pseudibis davison*), and the Bengal Florican (*Eupodotis bengalensis*). A recent study by IUCN lists the Mekong River as one of the nine richest watersheds for fish biodiversity globally, with 298 recorded species, including the endemic giant catfish (*Pangasianodon gigas*) and the giant Mekong bard (*Catlocarpio siamensis*) and several species of giant stingray (Hans 2000).

The Mekong wetlands also have a critical role as a staging post in flyways for a number of migratory birds (Scott 1989). The best known example is Tram Chim National Park in Vietnam, which hosts the entire population of Eastern Saurus Crane (*G. a. sharpii*) during the dry season. The freshwater wetlands are also important for migratory egrets and shorebirds.

All living resources in the Mekong River are increasingly under threat. Fishing practices in all countries include use of batteries, poison, and explosives as well as small mesh nets. Other animals that are collected in wetlands include frogs, snakes, and turtles, and their numbers are dropping. This is partially the result of trading wildlife products, particularly the manufacturing of traditional Chinese medicines. Birds have been hunted and are often victims of agro-chemical pollution.

R.C. Smardon, *Sustaining the World's Wetlands*,
DOI 10.1007/978-0-387-49429-6_6, © Springer Science+Business Media, LLC 2009

The Mekong River Basin

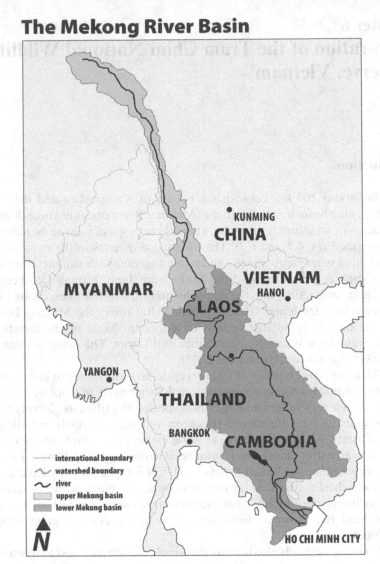

Fig. 6.1 The Mekong River, watershed, and sources: Drawn by Samuel Gordon and adapted from http://corrtho.cool.ne.jp/mekong/outline/mekong_river_c.html

Other threats to the Mekong Delta as a region are enumerated. The whole basin is under consideration for hydroelectric generation so the question is how this might affect downstream hydroelectric regime (Frappart et al. 2006, Quang 2002, Tanaka 2003, White 2002) and especially fisheries (Kite 2000, Van Zalinge et al. 2003). Much of the delta is moving toward integrated rice and fisheries production (Berg 2002, Ringler and Cai 2006, Rothius et al. 1998, Torrell and Salamanca 2003) and intensive shrimp aquaculture production at

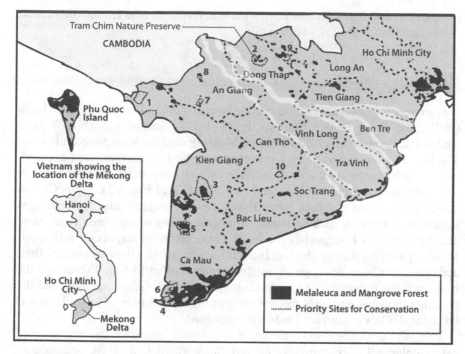

Fig. 6.2 Mekong Delta and major habitat sites within the Mekong Delta, Vietnam. Tram Chim site is indicated by number 10 within Can Tho Province. Redrawn by Samuel Gordon and adapted from Torrell and Salamanca (2005)

the southern edge of the delta. Concerns include pesticide impacts from rice culture inland (Phuong and Gopalakrishhan 2003, Torrell and Salamanca 2003) plus acid soils (Van Mensvoort 1996, Husson et al. 2000) for the inland portions and saltwater intrusion and use of aqua-cultural chemicals for the southern delta edges (Kam et al. undated). Even with these stresses on the biota within the Mekong Delta, a current ecological assessment of the Mekong basin for primary productivity and phytoplankton composition suggests the sampling sites have not suffered from ecosystem degradation.

Long-term climate change and rising sea levels may aggravate both saltwater intrusion plus chemical contamination problems (Torrell and Salamanca 2003, Wassermann et al. 2004). This may cause increasing vector-borne diseases such as malaria and West Nile. On the other hand current demographic studies of mosquito-borne malaria within rural districts of the delta have been going down (Erhart et al. 2004).

This is the story of the International Crane Foundation and Vietnamese Conservationists efforts to restore a badly damaged delta wetland habitat. Although the International Crane Foundation has had an international reputation for their efforts to restore crane populations and habitat, this effort stretched them in new directions. Also involved were a number of other private foundations and academicians from Vietnam, Germany, the United States, and Australia. That is what is so interesting.

Historical Context

Key sources for historical context are Doug et al. (2003), Duc et al. 1989, Thanh (2003), and Trong (1990). Prior to the 1961–1975 Second Indochina War, a wetland wilderness survived just east of Mekong near the Kampuchean border. The area now known as Tram Chim (meaning bird swamp) covered approximately 50,000 acres over a shallow basin, which supported wide expanses of sedge marshes and clumps of Melaleuca forests (see Fig. 6.2). Sarus Cranes nested in the open marshes, while ibises, herons, cormorants, and anhingas littered the trees with stick nests and whitewash. Deep water, dense vegetation, and proximity to Kampuchea rendered Tram Chim an important refuge for Vietcong soldiers during the last Indochina war. In an effort to control their activities, two huge drainage channels were excavated like an "X" across the basin and the inflammable Melaleuca was napalmed. Gunboats patrolled the channels. Enemies and large birds were shot from helicopters. The wetland was devastated. Only a denuded landscape remained.

Excessive drainage desiccated native vegetation, increased the frequency of catastrophic wild fires, severely altered the wetland hydrologic regime, and virtually eliminated the complex food web that the floodplain supported. As plant and animal species vanished, indigenous human populations were also replaced. Acid sulfate soils underlying the formerly waterlogged wetland substrate underwent oxidation and hydrolysis reactions, which lowered the surface water pH below 3 and released toxic concentrations of iron and aluminum ions. Thus at the onset of each rainy season floodplain waters became non-potable, non-swimmable, and biologically sterile (Pantulu 1981).

Wetland Soils

The acid sulfate soil problem either was developed or was aggravated to its present level of magnitude during the late nineteenth century. The reason is that the pyrite deposits, which react to deep drainage and consequent oxygen intrusion by acidification, underlay broad depressions in the delta; the pyrite is oxidized to iron (III) hydroxide and sulfuric acid with jarosite as the typical intermediate product (van Mensvoort 1996, Husson et al. 2000). As a direct result of anthropogenic interference, large areas (1.8×10 ha) are barren due to the acid sulfate soil problem (Van Mensvoort 1996, Husson et al. 2000). The only plant coverage on these soils is provided by weeds (e.g., *Hellicharis equisetina*) and sedges (*Carex* spp.) and acid-resistant herbaceous vegetation, mainly *Eleocharis dulis, Ischaemum aristatum, Phragmites karka*, and *Saccharum spontaneum*. They form a dense cover and reach heights of 1.5–3.0 m. The waterside forests predominately include *Melaleuca cajuputi* trees. The pH of the water in highly affected areas drops, and in some locations, to 2 during critical times of the annual cycle, namely at the outset of the rainy season. High concentrations of iron and aluminum in drainage waters in the area are toxic to crops in otherwise fertile areas and fish kills occur in canals and streams. Some scientists feel the ecological changes have reached the stage where they are irreversible (Pantulu 1981).

Wetland Flora

The original vegetation over most of Tam Nong district was the Mixed Swamp forest with Melaleuca trees as dominants intermingled with many herbaceous grasslands covered with water almost throughout the year (see Fig. 6.3a and b).

Fig. 6.3a and b Overview of Tram Chin area. Photo credit: Thanh Vo

However human disturbance (as we have reviewed) has disturbed the original forest, which is replaced at the present time by rice fields and secondary vegetation, types which still cover large areas of acid sulfate soils. There are both natural and disturbed vegetation associations present.

The following associations characterize the natural vegetation (see Pham 2003, Kiet 1991, 1994); there is Melaleuca forest, which is a relict of the degraded primitive forest including many afforested areas of new plantations.

Fig. 6.4a and b Typical current vegetation patterns in the Tram Chim Reserve. Photo credit: Thanh Vo

The remaining natural vegetation consists of grasslands differentiated into many associations dictated by ecological conditions:

- the *Panicum repens* association occurring on old alluvial soils emerging within depressions or having a relatively high elevation.
- the *Eleocharis dulcis, forma nana* association found on the edge of the alluvial terraces, in contact with the recent alluvium.
- the *Eleocharis dulcis* association confined to strongly acid sulfate soils of the deep back swamp with poor drainage.
- the *Paspalum* complex association represents a transitional type between the *Heleocharis* and the *Panicum* associations.
- the *Ischaemum indicum* association occurring in areas higher then those of *Heleocharis dulcis* and less acid.
- the *Sacciolepis–Nelumbium* association is indicative of the permanently waterlogged depression with submerged and floating plants as abundant.
- the *Eleocharis–Oryza* association is related with the improvement of dikes to retain water and sediment.

The majority of cultivated lands are rice fields reclaimed on various types of soils; therefore the weed vegetation is indicative of soil fertility as well as degree of intensive cultivation and can be classified into the following associations:

- the *Leptochloa shensis* association is dominant on recent alluvial soils with two crops per year;
- the *Pentapetes* association characterizes the same soil condition but with only one crop of rice due to prolonged flooding period;
- the *Digitaria* spp. association is confined to old alluvial soils with relatively high elevation; and
- the *Cyperus polystachyos* association is indicative of rice fields reclaimed on strongly acid sulfate soils.

Within the Tram Chim Reserve, there are about 5,000 ha, thanks to the building of levees retaining floodwater. The *Eleocharis–Oryza* association is replacing the *Eleocharis dulcis* association, indicative of the acid and toxic soils. Also large areas are being reforested with *Melaleuca leucadendron* trees.

Within the wetland association areas there is luxuriant growth of floating, submersed, immersed, and marginal aquatic macro-vegetation (see Figs. 6.5 and 6.6). Most common among such plants are the following:

Floating: Water hyacinth (*Eichomia crassipes*), water lettuce (*Pistia stratiotes*), water fern (*Salvinia cucullata*), water velvet (*Azolla pinnata*), and duckweed (*Lemna minor* and *L. polyrhiza*).

Submerged species include *Blyxa* (*Blyxa echinosperma* and *Ecclinusa lancifolia*). Florida elodia (*Hydrilla verticillata*), coontail (*Ceratophyllum demersum*), Lymnophyta (*Lymnophila heterophyla*), grass naiad (*Najas gramina*), and bladderwort (*Utricularia flexous*).

Fig. 6.5 Eastern Saurus Cranes. Photo credit: George Archibald in the ICF Bugle Vo. 3, No. 1, Feb. 1987, p. 5

Fig. 6.6 Eastern Saurus Cranes in Tram Chin Nature Preserve. Photo credit: The ICF Bugle, Vo. 13, no. 4, Nov. 1987, p. 4

Immersed and marginally wet plants include *Coix* (*Coix aquatica*), common reed (*Phragmites communis*), bulrush (*Scirpus lacustris*), bamboo (*Bambosa* spp.), water primrose (*Jussiaea repens*), water morning glory (*Ipomoea aquatica*), and water smart weed (*Polygonum tomentosa*).

Wetland Fauna

Fish fauna of the wetlands (see Kite 2000, Van Zalinge et al. 2003) are locally characterized as "Poissens blancs" or "Poissens noirs" (white or black fish) depending on their migratory patterns. The so-called white fish usually enter the wetlands during the flood season to spawn. The few that are stranded in these wetlands grow there and provide a rich fishery. Such white fish include carps and catfishes of the genus *Pongasius*, and in the Mekong plain, clupieds, threadfins, and drums. The blackfish such as the murrels (*Channa* spp.), anabantids (*Anabas testudineus*), catfish of the genera *Saccobrachus* and *Clarius*, and the spiny *Mastacemebelus* spp. are more or less permanent residents of the wetlands. Such fisheries have a very high economic value for local food subsistence and for cash (see Do and Bennett 2005, Ringler and Cai 2006).

Other vertebrates of economic importance that inhabit wetlands are frogs, snakes, crocodiles, large water lizards, and waterfowl such as grebes, pelicans, darters, herons, ducks, cranes, ibises, storks, and snipes. All these species are permanent residents with the exception of the migratory waterfowl. A number of very rare species have been observed including white-winged wood duck, Greater Adjutant Stork, black-necked stork, Bengal Florican (*E. bengalensis*) as well as the Eastern Sarus Crane (*G. a. sharpii*) (see Figs. 6.7 and 6.8) (Beilfuss 1996). The Siamese crocodile (*Crocodylus siamenis*) was formerly common in the Mekong Delta but is one of the world's most endangered crocodilians (Platt and Tri 2000).

Early Restoration Efforts

Following the war, provincial leaders in the Plain of Reeds were faced with an influx of thousands of displaced people and a collapsing economy. Their solution was to re-establish important natural resources of the wetland that could be used economically by local people. These resources include rear mangrove (*M. cajuputi*) which provided medicine, fuel, and rot-resistant wood; wild rice (*Oryza* spp.); marsh sedges which provided fodder for water buffalo; and protein-rich wetland animals such as fish, snakes, eels, turtles, shrimp, and crabs (Ngan 1989, Trong 1991). To alleviate the overpopulation and resource pressure, peasants were moved to undeveloped regions such as Tram Chim. To the chagrin of the settlers, but to the benefits of wildlife, the soil in parts of Tram Chim basin had heavy concentrations of sulfates, compounds that retarded agricultural usage.

Fig. 6.7 Typical dike and surrounding rice fields and Melaleuca forest in the Mekong Delta. Photo credit: Thanh Vo

Enter Mr. Moui Nhe, who was born in a village beside Tram Chim. Following the reunification of Vietnam in 1975, Mr. Muoi Nhe became the leader of Dong Thap Province, one of Vietnam's 38 provinces and is the one that contains what used to be the wetland area of concern. Mr. Moui did not have the luxury of a natural history or ecology studies and much of his life was spent in the armed services fighting the Japanese, the French, and the Americans.

Realizing the former importance of the Tram Chim both for aquatic wildlife and for fishing and lumbering, Mr. Moui Nhe decided to convert one-quarter of the Tram Chim basin back to its former condition. After the war 32 km of dikes were built around 14,000 acres of former wetland to prevent monsoon rains from spilling into the drainage channels during the dry season. Groves of Melaleuca were planted and the restoration began and canals were constructed around a 5800-ha portion of the former Plain of Reeds.

Eastern Sarus Cranes Rediscovered

The Sarus Crane formerly inhabited an enormous range stretching from northern India to the Philippines. There are two subspecies, the Indian Sarus (larger and with a white neck color and light tertials) and the Eastern Sarus (uniformly gray). Indian Sarus Cranes are protected by the Hindu religion in many regions, where they persist among farmlands with small natural and artificial wetlands; they perhaps number about 25,000 birds on the Indian subcontinent.

Fig. 6.8a and b Overhead photos of the water control structure and dikes engineered to recreate the hydrology. Photo credit: Thanh Vo

In contrast, the Eastern Sarus has not been protected across its Southeast Asian range and is endangered on the mainland, although since 1964 when the subspecies was first observed in northern Australia, its numbers have increased into the thousands on the island continent. Researchers in China, Thailand, and the Philippines have not recently found any Eastern Sarus Cranes and asked for the International Crane Foundation's assistance in helping to restore these enormous birds. The discovery of the Eastern Sarus at Trim Tram, in 1986, was the only confirmed record of the subspecies in its traditional range at that time and peaked the ICF's interest in Tram Chim. George Archibald, executive director of ICF, first heard about the Eastern Sarus Crane in the delta from a paper presented at a conference in China by Le Dien Duc from the Center for Natural Resources and Environmental Studies in July 1987.

At this point George Archibald connected with Charles Luthin, who formerly worked with ICF, went to graduate school at the University of Wisconsin and then served 5 years with the Brehm Fund for the International Conservation of Birds in West Germany. While working with the Brehm Fund, Mr. Luthin was able to join forces with Vietnam's leading conservationist, Professor Vo Quy at the University of Hanoi, in implementing a plan to locate and conserve a whole spectrum of endangered water birds including the Giant Ibis, Lesser Adjutant Stork, and Eastern Sarus Crane.

Under the leadership of Professor Vo Quy's deputy, Professor Le Dien Duc, a three-man Wetland and Waterbird Working Group (WWWG) started a search that resulted in the 1986 discovery of Mr. Muoi Nhe and "his" cranes. The Brehn fund has continued to provide financial support vital to development and conservation in Vietnam.

In January 1988, a three-person team from the United States – George Schaller of the New York Zoological Society, ICF Trustee Abigail Avery, and George Archibald – joined the WWWG on a weeklong exploration to Tram Chim. During this expedition, the team surveyed the vast expanses of wetlands, did rough crane counts, surveyed the status of the dikes, and did local conservation education programs in schools. During this trip, plans were laid for an education center that was constructed in Tam Nong at the edge of the Tram Chim Sarus Crane reserve, with the support of the Brehm Fund. The following program was also developed:

- Research needs were identified such as the need to determine the natural hydrology of the wetland and map the distribution of sulfate soils and then the need to identify what other areas should be restored.
- Sarus Cranes should be carefully studied throughout the year to determine habitat needs.
- Need for dike repair and sluice gates along the dikes, which could then be managed to restore the original hydrologic conditions of the wetland.

Plans were also laid for a visit of the three members of the WWWG and a local official from Tan Nong, Mr. Ngo Quoc Thang, for a one-month study visit to the United States. This was done and the WWWG with Mr. Moui Nhe in 1988 did include two field trips to Horicon Marsh in 1988 and 1989. Members of the WWWG and Ngo Quoc Thang visited ICF in 1988, and Muoi Nhe and Le Dien Duc visited ICF in 1989. Plans were also made for a co-sponsored (ICF and Vietnam) Crane Workshop near Tram Chim, with delegates invited from all Asian Nations, which had or still have cranes. By 1990, another 1200 ha of impoundment was added to the reserve. The dikes retained monsoon rainwater across the wetland and precluded forest fires while large tracts of the planted rear mangrove seedlings grew.

With the return of wetter conditions, many species of native wetland flora invaded or emerged from the dormant seed bank. Native fauna also rebounded. For example, the Eastern Sarus Crane (as mentioned, *G. a. sharpi*), black-necked stork (*Xenorhynokus asiatious*), and Bengal Florican (*E. bengalensis*) were all thought lost from southeast Asia but returned once conditions improved (Brehm Fund for International Bird Conservation 1987, Archibald 1988). With the discovery of native species, Vietnamese scientists convinced local authorities to prohibit agricultural development, outlaw wildlife hunting, and protect the area as a nature reserve called Tram Chim (Duc 1989).

Although these initial steps taken to revitalize the wetland were a tremendous success, the productivity of the wetland plateaued quickly. The diked perimeter isolated water inside Tram Chim from water outside the reserve. Stagnant standing water restricted wildlife usage, stunted vegetative growth, and prevented regeneration of flora and fauna (Kiet 1991). Finally, as the human population of the Mekong Delta swelled to more than 13.5 million people, intense pressure was placed on the wetland to either produce abundant natural resources or be converted to rice agriculture. Thus by 1990 the restoration process had reached a critical juncture.

Initiation of Joint Agreement, International Meetings Toward a Management Plan

At this point, the International Crane Foundation (ICF) initiated a long-term co-operative project with Vietnamese scientists and government officials to encourage the continuation of restoration activities of the Tram Chim. There were actually three written agreements and an initial unwritten agreement with George Archibald that the Vietnamese needed money to build dikes and restore habitat. Vietnamese scientists from the Universities of Hanoi, Can Tho, and Ho Chi Minh City teamed up with scientists from the United States to evaluate the work already completed and plan future activities. Two international conferences, the Sarus Crane and Wetland Workshop (Harris 1990) and the Tram Chim management Meeting (Barzen 1991), were convened at the reserve to generate input from delegates from 13 countries.

The International Sarus Cranes and Wetland Workshop (January 11–18, 1990) was a first in Southeast Asia. The University of Hanoi and Dong Thap Province hosted the meeting. A field trip to the preserve was conducted to see ditches being dug by three mechanical cranes (dredges), besides seeing the biological cranes. Both issues fueled discussions in the workshop. Would there be wetlands for people's or cranes' utilization? For the workshop ICF's perspective was that there was a common goal; the Eastern Sarus Crane in

Vietnam was important and needed help, but some delegates saw the reserve, its management, and their mission from a different perspective.

In 1989, there were incompatible plans and management actions displayed all over the reserve. Canals were being dug both inside and outside the preserve with no coordination and sometimes at cross-purposes. The first agreement in 1990 was a moratorium on all canal digging and other management actions so that there was time to study and understand what was going on. It was part of the funding agreement with the Brehm Fund, so there would be time to study the issues, develop a coordinated management plan, and provide funds for the water gates. There was also funding from the National Wildlife Federation.

The tensions and frictions between different parties were many and complicated. Such tensions were behind the scenes of the 1990 and 1991 meetings as well as the negotiations on how the preserve should be managed. First, there was the North Vietnam vs. South Vietnam tensions or central government vs. local autonomy. For instance, local government was upset about the costs associated with establishing the educational center with only partial funding from foundations and the central government.

There were many ideas about how the preserve should be managed. Mr. Moui Nhe wanted initially to see the area set aside as a kind of historic park. This was thwarted by the influx of people moving into the area, planting crops, and planting rear mangrove. Some of the vegetation burned and more canals were dug, and the dikes were started in 1984. Some wanted to cut the wetland into four equal parts, each with a different management goal. Some wanted to restore the entire area, and ICF was interested in wetland restoration and crane habitat. Others, such as the Mekong River Commission, were interested in water quality management and preventing the acid sulfate soils from causing problems. The University of Hanoi WWWG was interested in tying all management options to optimization of crane breeding and production.

All these different perspectives existed during the 1990 meeting and erupted full force in 1991 at the management meeting. A couple of factors were critical in preparation for the 1991 meeting for which some form of a management plan was an anticipated product (Jeb Barzen 1996). One factor was the visit of Mr. Moui Nhe to ICF back in 1989. The key experience was the visit to Horicon Marsh in Wisconsin where a drainage ditch with attendant ecological problems was shown to the Vietnamese and discussed on site.

The second factor, prior to the 1991 meeting, was intense probing discussions between ICF and the different parties as to why certain management actions (e.g., digging ditches) were done or proposed. It was often found that there was no strong rationale or basis for digging some of the ditches or other management actions. Often repeated questioning yielded different answers, no strong logic, or a history of local understandings.

The workshop program had been planned to use Tram Chim as a practical example for conservation projects and issues in southern Asia. Probably the more important goal was persistence in learning to work together and to recognize that people needed to sustain themselves as well as preserve cranes.

In the actual meeting the Vietnamese were concerned about fire and the survival of the rear mangrove trees, as well as fisheries and wildlife habitat. One of the critical issues was the proposed digging of ditches in the middle of the marsh. Mr. Moui Nhe was reminded of what he had seen and experienced at Horicon Marsh in 1989 – that the drainage ditch there did not work and caused ecological harm. This parallel logic, confirmed by the elder statesman and local leader, was also tied to the notion that once you dig the ditches some of the consequences are irreversible. But you can always try something else first and if that does not work go back to the ditches.

So the second agreement was signed in 1991, a loose management plan was in place, and detailed hydrological studies continued. Detailed hydrological studies started in 1989–1990 dry season and the first year's data were used at the 1991 management meeting. Next steps coming out of the workshop included the following:

- Continued detailed studies of the Tram Chim wetland
- Future meetings between ICF and Vietnamese officials to discuss results
- Construction of water gates to be funded by the Brehm Fund for International Bird Conservation

Hydrological Restoration

Throughout this process, hydrological restoration was deemed to be the primary mechanism by which the physical, chemical, and biological recovery of the wetland could be facilitated. Four years of hydrological restoration activity at the Tram Chim Reserve are summarized by Beilfuss and Brazen (1994). These two investigators started by trying to understand the hydrological and water quality processes of the wetland prior to degradation. Much information was obtained from interviews with long-term residents of the area, review of available literature, interpretation of archival military aerial photos, and examination of pristine remnants of the Plain of Reeds remaining in Cambodia. The latter proved difficult because of the ongoing political situation. Interviews with the local elders and reviews of published studies suggested that the dominant force for the Plain of Reeds was the same hydrological processes occurring over the entire Mekong Delta. These forces are the seasonal monsoon rainfall, combined with overbank flooding of the Mekong River plus overload sheet flow from Cambodia annually flood with 2–3 m of standing water followed by receding waters followed by the 6-month dry season. Aerial photos revealed the bisecting of numerous shallow streams and canals. Excessive drainage dropped the regional water table more than 1 m below the wetland substrate during the dry season and decreased the total period of standing water inundation from 7 to 5 months.

To supplement and verify the qualitative understanding of the pre- (natural) and post-disturbance wetland water regime, quantitative measurements of hydrology, geomorphology, and water quality were required to build a restoration model (Beilfuss and Barzen 1994). The investigators also assessed water quality parameters such as the mixing of nutrients, suspended sediment, and organic matter with Tram Chim. They examined the effect of restoration efforts on the reactivity of acid sulfate subsoils. Prior to disturbance, these water quality processes occurred as a function of the natural water-level fluctuations. Under disturbed hydrological conditions, however, severe water quality degradation has occurred in the floodplain surrounding Tram Chim.

The reserve dikes (see Fig. 6.6) altered both beneficial and detrimental mixing processes. Fine-grained sediments, dissolved nutrients, and detritus that were naturally deposited by slow-moving sheet flow across the Tram Chim area were excluded from the diked reserve (Beilfuss 1991). Export of decaying organic matter from the reserve was also impeded. Conversely, the diked perimeter excludes nutrient-poor fluvial silts that were naturally deposited on the banks of the Mekong River but are now channelized directly to the Tram Chim area with the onset of flooding (Beilfuss 1991). Acid sulfate reactions were prevented by the maintenance of anoxic conditions.

Implementation of Hydrologic and Water Quality Restoration

After modeling hydrologic and associated water quality processes of the wetland to conditions prior to the degradation of the Plain of Reeds, the scientists developed an approach to re-establish the processes at Tram Chim. Backfilling the existing canal network surrounding the reserve would be the simplest means of restoring natural overland flooding, but would prevent the extensive use of drainage canals for transportation and for irrigation of rice paddies outside the reserve. The chosen alternative was to install water gates, each consisting of four 2 × 2 m opening variable crest box culvert barrels in parallel which were designed and installed at Tram Chim (Beilfuss 1991). This design was chosen because it would

- maintain water-level equilibrium between the reserve and the surrounding floodplain during peak flooding;
- maintain water seepage loss;
- permit the natural drainage of vegetation decomposition by-products from the reserve;
- minimize the amount of physical labor and time required for water gate operation and maintenance;
- minimize operation and maintenance expenses;
- enable circumnavigation of the reserve along the perimeter dikes
- facilitate fisheries harvest.

The water gates use gravity flow to permit maximum inflow and outflow of surface water during the flooding season. Gates at Tram Chim were located at four points where large natural steam channels intersect the reserve dikes. Surface waters move to and from areas across the reserve through these stream networks.

The restoration strategy implemented at Tram Chim attempts to mimic the natural and hydrological cycle and can be described in seven steps. The timing of these steps varies among years as the timing and magnitude of flooding in the Mekong Delta vary. The following steps represent an "average" based on 30 years of region-wide data:

1. Prior to the onset of the rainy season in late April to early May the water gates are closed. No standing water and at a minimum height.
2. Water gates remain closed as the rainy season begins in May. Evapotranspiration rates decrease and water table fluctuates near the surface. Ponded waters in depressions increase and reserve water levels rise slowly to the surface, while canal waters increase rapidly as the Mekong River branches reach bankful discharge.
3. The water gates are opened when non-channelized overbank floodwaters (low in sediment and acidic material) from the Mekong River reach the reserve between June and August. The wetland water table rises sharply and most of the reserve is undated by early July.
4. From August through October precipitation increases substantially relative to evapotranspiration and the entire Plain of Reeds is inundated below 2–3 m of floodwaters. The water gates remain open throughout this period as the Mekong River flood wave is passed through the wetland.
5. By early October, floodwaters begin to recede through the stream networks and drain through the water gates to the surrounding floodplain. The rainy season diminishes through December.
6. The gates are closed when the water level above the reserve soil substrate is equal to the average depth of 20–30 cm that was ponded across the wetland prior to the channelization of the Plain of Reeds. Timing depends on the influence of tropical typhoons and annual fluctuations in the basin-wide levels.
7. From December through February, the water gates remain closed and surface water movement in the wetland ceases. Water levels in the surrounding drainage canals continue to decline, causing the degraded floodplain substrate to desiccate. Within the reserve, evapotranspiration increases steadily beneath the floating vegetation mats, while precipitation almost ceases. This completes the cycle of hydrologic restoration.

After the hydrologic study was done and the water gates installed, there was a third agreement in 1992. This agreement placed two entities in charge of the reserve, the reserve staff and a private company. The private management company had two objectives; to make money from utilization of preserve resources and habitat preservation. There was tension between these two goals and between the company and preserve staff. Things were not going well. There were massive floods in 1991 and some of the dikes were damaged. Funding was negotiated with the Mekong River Commission and the dikes were repaired.

ICF staff came back in 1992 to teach prescribed burning and vegetation management techniques. They found Vietnamese digging holes in the newly fixed dikes in order to harvest the fish sold to them by the "company". There was a substantial confrontation but the water was drained out, causing exposure to acid soils and most of the fish died before they could be harvested. The "company lost face", but the real issue was that the preserve needed national status (Barzen 1991), which in turn would provide legitimization for restoration and preservation management. The counterweight of national status was that the local provincial government did not want to have the central government tell them what to do – the north–south issue again.

In 1993 an addendum was added to the 1992 agreement, dissolving the private company, making the reserve a sole proprietorship of the reserve staff plus a major push for national status. It should be understood that the Vietnamese had been working on national recognition since 1988, and this agreement merely provided the final inputs for national recognition. The Tram Chim was declared a national preserve in 1994.

In 1993 and 1994 there were fires but the waters were high preventing much damage. In 1994, it was very dry. In 1995, the new preserve manager drew down the water too much. It should be noted that this is difficult to judge, as there were few staff gauges or accurate survey for baseline elevation conditions. There was much fire following this drawdown. The popular perception was much "loss of political face". Although no long-term ecological damage was done, there was a substantial reduction of forest cover in some areas, but the vegetation came back as did the cranes. There is a new preserve manager who now is more meticulous in terms of management but has a rougher time with interpersonal relations. So it goes.

Monitoring and Evaluation

The researchers'/investigating team's final goal was to develop an extensive monitoring and evaluation framework to test the effectiveness of implementation strategies and to be able to respond to any changes in the regional hydrological regime. This was initiated by training and collaboration with Vietnamese conservationists to refine goals and strategies of the restoration process. Scientists from Tram Chim came to the United States in 1988 and 1989 to learn wetland restoration and management techniques. International conferences held at Tram Chim in 1990 and 1991 further contributed to the input of delegates from 13 different countries to this action. Between 1992 and 1993, two Vietnamese students began full-time study in graduate programs in the

United States, with thesis research conducted at Tram Chim. Field training in Vietnam has focused upon teaching various data collection and analysis techniques to reserve staff and local villagers.

ICF realizes that in addition to working with Vietnamese conservationists and academics, the long-term hydrological restoration of Tram Chim will only be successful if it is understood and supported by the political leaders at all levels and by inhabitants living near the reserve. ICF's work with educating officials and community members about the importance of a monitoring and evaluation strategy is very important to continue restoration processes.

In the summer of 1994, the wetland hydrologist from ICF, Rich Beilfuss, traveled to Australia with four Vietnamese colleagues, Thai Van Vinh, Ngo Thinh Thang, Nguyen Huu Thien, and Phan Trong Thinh (the director, vice director, ecologist of Tram Chim National Reserve, and the ecologist for the Vietnam Wetlands Program, respectively). With financial support from the John D. and Catherine T. MacArthur Foundation and the Asian Wetlands Bureau, they spent a month learning about the Australian equivalent of the Plain of Reeds at Kadadu National Park. Training experiences were supplied by Roger Jaensch of the Asian Wetland Bureau, Dr. Max Finlayson, one of Australia's top wetland experts, Dr. Jeremy Russell-Smith, and the Australian Nature Conservation Agency Staff with three goals in mind:

- Familiarization with the ecology and hydrology of the floodplain wetlands of tropical Australia as a model for wetland restoration at Tram Chim and elsewhere in the Mekong Delta
- Firsthand experience with the practical methods of national park management
- Establishment of ongoing scientific cooperation between Vietnam and Australia

There was 1–2 weeks apiece with the three instruction groups mentioned above. During the first week AWB demonstrated how to control the invasive shrub *Mimosa*, which had destroyed thousands of acres of wetland habitat for Brolgas, geese, and countless other plant and animal species in Australia. *Mimosa* has recently infested Tram Chim as well but can be controlled if reserve managers act quickly.

There was work with Dr. Max Finlayson and his staff at the Environmental Research Institute of the Supervising Scientist (ERISS). A variety of research techniques that would aid in understanding wetland management were obtained, such as the use of fish and small invertebrates to monitor water quality as an overall measure of the health of wetlands. Other methods learned include means of studying acid wetland soils, revegetating disturbed areas, and using geographic information systems to manage tropical wetlands.

One week included work with the Australian Nature Conservation Agency for a taste of day-to-day park management as the Vietnamese participated in wildlife monitoring, problem species control, and fire management for protecting fire-sensitive communities. Throughout the training period the Vietnamese scientists/managers experienced ecotourism and educational programs appropriate to tropical wetland conservation.

One outcome of this training visit was the new relationship forged between Australian and Vietnamese colleagues. A second group of wetland managers from Laos, Cambodia, Thailand, and Vietnam traveled to northern Australia for training. This may well be the start of a permanent training program for tropical wetland management.

There has been little monitoring of vegetation except for *Mimosa pigra* control (Son et al. 2001, Thi et al. 2001), but there has been monitoring for water birds. There are now staff gauges for monitoring water levels in cooperation with the Mekong River Commission.

During the spring of 1994 it was declared that Tram Chim is a national reserve (The Central Government of Vietnam 1994, Thanh 2003) – the first wetland national reserve in the Mekong Delta and the first protected area for cranes in Vietnam. This is also an important step toward recognizing the importance of wetlands for the survival of people and wildlife. Education and training, however, are still desperately needed to help Vietnam manage the Tram Chim and its resources wisely. ICF realizes that the future of Tram Chim and other wetlands in Vietnam depends as much on wise resource management (including people) as they do on wise ecological management. Vietnam must struggle to feed and clothe its increasing population by tapping every resource available. There is till much pressure from outside the preserve for more utilization of preserve resources.

In addition, there may be new threats, which may alter the hydrology of the entire Mekong River basin. The four Mekong River Commission countries, Cambodia, Laos, Thailand, and Vietnam, are considering massive schemes for dams, barrages to harness the Mekong for flood control, and hydroelectric production for export (Lohmann 1990, Quang 2002).

The important outcome of the three agreements, and the training to date, is increasing institutional capacity to address all the management issues facing the Tram Chim National Preserve. There is still outside pressure to more intensely manage the national reserve for various products and uses. This was the subject of a fourth meeting/planning session in August 1996. This time it will be various Vietnamese factions debating issues with organizations like the ICF in the background ready to lend technical support if needed.

On December 29, 1998, the government transformed the Tram Chim Nature Reserve into Tram Chim National Park according to Decision No. 253/QD-TTg of the Prime Minister, dated December 29, 1998. A management board has been established for Tram Chim National Park.

The establishment of Tram Chim National Park has two major purposes: (1) to conserve the typical wetland ecological system of the lower Mekong Delta, the flooded zone of Dong Thap Muoi; and (2) to conserve historical, cultural, and scientific value which can sever the scientific research for wise use of wetland for national benefit and contribute to the environmental and ecological conservation of South East Asian region.

Currently the national park has 81 permanent staff and within that, 53 staff members are forest rangers, the others are scientists and administrative staff. The administration includes five functional divisions: Organizational and Administration; Planning and Accounting; the Scientific and Environmental Research; the Forest Protection Unit; and the Environmental Education and Eco-tourism Center (Tram Chim 2006).

Conservation Issues

Tram Chim now has national park status (Thanh 2003), which confers a relatively high degree of protection; however, several threats remain, such as pressure from local people's livelihood activities, pollution from surrounding rice farms, and the changes of water affecting biological dynamics of the ecosystem.

Tram Chim National Park is located in the intensive rice production areas of the Mekong Delta. It borders five communes with a total population of about 41,000 people. Most of them are rice farmers. Local livelihoods are based on rice production in the dry season, and fishing and collecting natural products in the flood season.

The establishment of Tram Chim National Park and the construction of a long enclosure dike system lead to the reduction of livelihood opportunities for the local population. The frequent encroachment of local people into the national park to hunt, collect firewood and other wild products could be considered a conservation conflict issue. The park is surrounded by intensive rice cultivation, which heavily uses pesticides and fertilizers, which have a substantial impact on the integrity of the wetland ecosystem of the national park. Examples of such impacts are pollutant discharge and alteration of natural water levels (Buckton et al. 1999).

In 2000, the national park management board began constructing six canals inside the national park, the construction of which could have fragmented the natural habitat and altered the water regime, leading to changes in habitat. However, construction of the canals was halted after only two were completed. The construction of the enclosure dyke system, about 71 km long, has isolated the national park from the whole ecological system, which directly affects the hydrological dynamic and biological links between the park and the external system of the Mekong Delta.

The construction of canals is not, perhaps, the major threat to the Sarus Crane population at Tram Chim. The most important factor in maintaining suitable habitat, for this species, is appropriate management of the water level at the site. In 2000, a partial drawdown was carried out, and, in 2001, a full drawdown took place, which facilitated natural vegetation recovery. It is hoped that appropriate water-level management will result in an increase in the crane population at Tram Chim. The most recent management effort in Tram Chim has been a focused effort to control *M. pigra*, a very aggressive exotic vegetative species (Thi et al. 2001, Son et al. 2001, Walden et al. 2002).

Tram Chim meets the criteria for designation as a site of international importance for wetland conservation under the Ramsar Convention. Currently, Tram Chim National Park is one of the most important wetland sites in Mekong Delta and the lower Mekong region.

The Role of the International Crane Foundation and Other NGOs

Initially the ICF, with the Vietnamese conservationists, focused on preservation and restoration of the Eastern Sarus Crane habitat, but this focus broadened out to encompass wetland restoration (vegetation and hydrology) as well as consideration of multiple use of the Tram Chim Reserve. Clearly, there was a foundation of local respect of wise use principles practiced by the Vietnamese, which allowed collaborative efforts to find the resources necessary to move forward with vegetation restoration, hydrologic restoration, and training and wetland management activities. In the process of doing all of the above, ICF acted as a facilitator to bring together, with the Vietnamese, foundation financial support, as well as other NGO and government technical support from the United States, Australia, and the Asian Wetlands Bureau (now part of Wetlands International).

Probably the critical event was the management workshop in 1991, where some tough negotiations and discussions had to bridge language and cultural gaps to address different perspectives on management of the Tram Chim Preserve. Whereas ICF was initially focused on crane habitat protection and restoration – the end-negotiated management plan may have reflected more of a locally defined "wise" multiple use of the wetland which is reflected in current management strategies. It is the process by which this local definition, subject to some ecological lobbying, takes place which is important to wetland management throughout the region as well as follow-up activities which are the implementation of the management plan.

Other Key Actors

Other key actors include the University of Hanoi WWWG, especially for getting support for national status for the preserve. Initially the European and US Foundations such as the Brehm Foundation, National Wildlife Federation, and MacArthur Foundation provided funding at critical points in the process. The Mekong River Commission later augmented this support, which was especially important for providing support for the management plan and national preserve status.

The University of Cantho played a significant role in the 1991 meeting by way of assisting to achieve management consensus. They have disengaged at various points in the process. Recently however Cantho University has gathered together researchers to establish long-term research agenda for the region (Claire-Ashton et al. 2005). There is some very promising student work going on. One student is working on social surveys of local perceptions of management in the Mekong Delta region and another is working on multiple use and growing rear mangrove in the buffer areas surrounding the Tram Chim Preserve.

Summary

Currently there is ongoing ecosystem assessment of the Downstream Mekong River Wetlands (Thong 2005). Stage 1 was from May 2003 to February 2004, and stage two from February to the end of 2004. The purpose of this study is to look at the state of biodiversity for the region and pressures on this biodiversity. A summary paper by Torrell and Salamanca (2005) points out that the most prominent pressures on Mekong Delta wetlands are rice production and associated large-scale water control structures, shrimp aquaculture, and the inadequacy of current institutional arrangements. The latter being that wetlands are neither land nor water and thus fall through the Vietnamese land tenure and regulatory jurisdictional system – they belong to no one (see Cai et al. 2005). It should be noted that Tram Chim National Park is second of all prioritized wetland conservation sites in the Mekong Delta.

This an extremely interesting story of many actors as we can see from the stakeholders listed above. Perhaps the main message is that different actors, ICF, WWG, and local residents, can develop meaningful wetland management strategies if all parties are actively engaged in management negotiations, learn from each other, and learn of other wetland management possibilities by seeing them work. This still a very fragile situation that can be aggravated by both natural conditions and human resource needs. The management scheme can be upset both by extreme monsoon-induced flooding and by extreme dry conditions and fires. Local people need the rice and fish production to survive; thus more pressure will be exerted in this direction. One of the keys, from this author's perspective, is to develop local management strategies that incorporate adaptive management and learning progressive management techniques from Australia or other hydrologically and ecologically relevant situations. This was done initially, thanks to the local Vietnamese community leadership and ICF. The question is, can this continue and support sustainable wetland management for Tram Chim Nature Preserve?

Acronyms

ERISS: Environmental Research Institute Supervising Scientist
ICF: International Crane Foundation
WWWG: Wetland and Waterbird Working Group

References

Alger, W. N., P. M. Kelly, and N. H. Ninh (eds.). 2001. *Living with Environmental Change, Social Vulnerability, Adaptation, and Resilience in Vietnam*. London: Rutledge, 314pp.

Archibald, G. W. 1988. Vietnam cranes thrive at Tram Chim. *ICF Bugle*, 14: 4–5.

Barzen, J. A. 1991. Restoration mixes science, people and luck in Vietnam. *The ICF Bugle*, 17: 2–3.

Barzen, J. 1996. Phone interview (April 1996).

Beilfuss, R. D. 1991. Hydrological restoration and management of tram Chim Wetland Reserve, Mekong Delta, Vietnam. Unpublished Thesis, University of Wisconsin, Madison, 259pp.

Beilfuss, R. D. 1994. ICF, Vietnam, and Australia Build Wetland Network. *The ICF Bugle*, 20: 1, 4–6.

Beilfuss, R.D. 1996. Written comments on earlier chapter manuscript.

Beilfuss, R. D. and J. A. Brazen. 1994. Hydrological wetland restoration in the Mekong Delta, Vietnam. In W. J. Mitsch (ed.) *Global Wetlands: Old and New*, pp. 453–468. The Netherlands: Elsevier Science B.V.

Berg, H. 2002. Rice monoculture and integrated rice-fish farming in the Mekong Delta, Vietnam-economic and ecological considerations. *Ecological Economics*, 41: 95–107.

Buckton, S. T., N. Cu, H. Q. Quynh, and N. D. Tu. 1999. *The Conservation of Key Wetland Sites in the Mekong Delta*. Hanoi: Bird Life International Vietnam Program.

Buckton, S. T. and R. J. Safford. 2004. The avifauna of the Vietnamese Mekong Delta. *Bird Conservation International*, 14: 279–322.

Cai, H. H., D.T. Ngo, N. An, and T. T. Giang, 2005. The legal and institutional framework and the economic values of wetlands in the Mekong River Delta. In *Millennium Ecosystem Assessment: Downstream Mekong River Wetlands Ecosystem Assessment, Vietnam*. Hanoi City, Vietnam: Institute of Geography, National Center for Natural Science and Technology, and at http://www.millenniumassessment.org/en/subglobal.mekong.aspx

Claire-Ashton, E., N. T. B. Nhi, and T. Nielsen. 2005. *Research Priorities for the Mekong Delta – Environmental Status and Future Requirements*. Workshop at Cantho University, College of Agriculture, Cantho City, Vietnam, 72pp.

Do, T. N. and J. Bennett. 2005. *An Economic Valuation of Wetlands in Vietnam's Mekong Delta: A Case Study of Direct Use Values on Camau Province*. Environmental Management and Development, APSEQ Occassional Paper no. 8, The Australian National University at http://hdl.handle.net/1885/43111

Doug, V. N., E. Maltby, R. Tafford, T.-P. Tuong, and V.-T. Xuan. 2003. Status of the Mekong Delta; agricultural development, environmental pollution and farmer differentiation, In M. Torrell, A. M. Salamanca, and B. D. Ratner (eds.) *Wetlands Management in Vietnam; Issues and Perspectives*, pp. 25–29. Penang, Malaysia: World Fish Center.

Duc, L. D. 1989. Eastern Sarus Cranes in Indochina. *Proceedings 1987 International Crane Workshop*, pp. 317–318. Qiqihar, China, International Crane Foundation, Wisconsin.

Duc, L. D., H. V. Thang, and G. W. Archibald. 1989. *Biology and Conservation of the Eastern Sarus Crane in Vietnam*. Madison, Wisconsin: Rome Publications, 9pp.

Erhart, A., N. D. Thang, T. H. Blen, N. M. Tung, N. Q. Hung, L. X. Hung, T. Q. Tuy, N. Speybroeck, L. D. Cong, M. Coosemas, and U. D'Alessandro. 2004. Malaria epidemiology in a rural area of the Mekong delta; a prospective community based study. *Tropical Medicine and International Health*, 9(10): 1081–1090.

Frappart, F., K. Do Minh, J. L'Hermitte, A. Cazenave, G. Ramillien, T. Le Toan, and N. Mognard-Campbell. 2006. Water volume change in the lower Mekong from satellite altimetry and imagery data. *Geophysical Union International*, 167(2): 570–584.

Hans, F. 2000. The biodiversity of the wetlands in the Lower Mekong Basin, The World Commission on Dams at http://www.dams.org/kbase/submissions/showsub.php?rec+ENV148

Harris, J. T. 1990. Asians meet on Behalf of cranes. *The ICF Bugle*, 16: 1–6.

Husson, O., M. T. Phung, and M.E. F. van Mesvoort, 2000. Soil and Water indicators of optimal practices when reclaiming acid sulfate soils in the Plain of Reeds, Vietnam. *Agricultural Water Management*, 45(2): 127–143.

Kam, S.P., C. T. Hoanh, T. P. Tuong, N. T. Khiem, L.C. Dung, N. D. Phong, J. Barr, and D. C. Ben. Undated. *Managing Water and Land Resources Under Conflicting Demands of Shrimp and Rice Production for Sustainable Livelihoods in the Mekong River Delta, Vietnam.* UK: Department for International Development, 11pp.

Kiet, L. C. 1991. *The Interaction Between Water Depth and Species Occurrence for Plants at Tram Chim.* Paper presented at the Tram Chim Management Meeting, Tamnong, Vietnam, 3pp.

Kiet, L. C. 1994. Native freshwater vegetative communities in the Mekong Delta. *International J. of Ecology and Environmental Sciences*, 20: 55–71.

Kite, G. 2000. *Developing a Hydrological Model for the Mekong Basin; Impacts of Basin development on Fisheries Productivity.* Colombo, Sri Lanka: International Water management Institute (IWMI), IWMI Working Paper 2, 14pp.

Lohmann, L. 1990. Remaking the Mekong. *The Ecologist*, 20: 61–66.

Ngan, P. T. 1989. Vegetation of the Tam Nong District with special reference to the Tram Chim area. *Garrulux*, 6: 3–5.

Pantulu, V. R. 1981. *Effects of Water Resources Development on Wetlands of the Mekong Basin.* Bangkok: Interim Committee for Coordination of Investigations on the Lower Mekong Basin, Bangkok, 13pp.

Pham, T. T. 2003. Managing and classifying wetlands in the Mekong River Delta. In M. Torrell, A. M. Salamanca, and B. D. Ratner (eds.) *Wetlands Management in Vietnam; Issues and Perspectives*, Penang, Malaysia: World Fish Center, pp. 31–43.

Phuong, D. M. and C. Gopalakrishhan. 2003. An application of the contingent valuation method to estimate the loss of value of water resources due to pesticide contamination in the case of the Mekong delta, Vietnam. *International Journal of Water Resource Development*, 19(4): 617–633.

Platt, S. G. and N. V. Tri. 2000. Status of the Siamese crocodile in Vietnam. *Oryx*, 334(3): 217–221.

Quang, N. N. 2002. Vietnam and the sustainable development of the Mekong River Basin. *Water Science and Technology*, 45(11): 261–266.

Ringler, C. and X. Cai. 2006. Valuing fisheries and wetlands using integrated economic-hydraulic modeling – Mekong River Basin. *Journal of Water Resource Planning and Management*, ASCE 132(6): 480-487.

Rothius, A. J., D. K. Nhon, C. J. J. Richter, and F. Ollevier. 1998. Rice with fish culture in the semi-deep waters of the Mekong Delta, Vietnam; Interaction of rice culture and fish husbandry management on fish production. *Aquaculture Research*, 29(1): 59–66.

Scott. D. A. 1989. *A Directory of Asian Wetlands.* Gland, Switzerland and Cambridge, UK: IUCN, pp. 774–780.

Son, N. H., P. V. Lam, N. V. Cam, D. V. T, Thanh, N. V. Dung, L. D. Khanh, and I. W. Forno. 2001. Preliminary studies on control of Mimosa pigra in Vietnam. *Strategic Weed*

Management in Vietnamese Wetlands; Weed control and Occupational Health and Safety Issues, pp. 110-116. Tran Chim National Park, Vietnam.

Tanaka, M., T. Sugimura, S. Tanaka, and N. Tamai. 2003. Flood-drought cycle of Toule Sap and Mekong Delta area observed by DMSP-SSM/1. *International Journal of Remote Sensing* 24(7): 1487–1504.

Thanh, N. C. 2003. Socio-economic situation, management, rational utilization and development potentials of Tram Chim, a Wetlands Ecosystem Conservation National Park. In Torrell et al. (eds.) *Wetlands Management in Vietnam, Issues and Perspectives*, pp. 75–80. Penang, Malaysia: World Fish Center.

The Central Government of Vietnam. 1994. *Decision of the Prime Minister on the Recognition for Tram Chim National Wetland Reserve of Tam Nong District, Dong Thap Province*. The Central Government, Hanoi, 2pp.

Thi, N.T. L., T. Triet, M. Storrs, and M. Ashley. 2001. Determining suitable methods of control of Mimosa pigra in Tram Chin National Park Vietnam. *Strategic Weed Management in Vietnamese Wetlands; Weed Control and Occupational Health and Safety Issues*, pp. 91–95. Tran Chim National Park, Vietnam.

Thong, M. T. 2005. *Millennium Ecosystem Assessment: Downstream Mekong River Wetlands Ecosystem Assessment, Vietnam*. Hanoi City, Vietnam: Institute of Geography, National Center for Natural Science and Technology, and at http://www.millenniumassessment.org/en/subglobal.mekong.aspx

Torrell, M. and A. M. Salamanca. 2005. Wetlands management in Vietnam's Mekong delta: An Overview of the pressures and responses. In *Wetlands Management in Vietnam, Issues and Perspectives*, Penang, Malaysia: World Fish Center, pp. 1–16 and at http://www.millenniumassessment.org/en/subglobal.mekong.aspx

Tram Chim National Park. 2006. Basic facts about Tram Chim National Park. Internal introduction document.

Trong, N. X. 1990. *History of the Dong Thap Moui*. Paper presented at the Sarus Crane and wetlands workshop, Tamnong, Vietnam, 3pp.

Trong, N. X. 1991. *The History of Creating the Tram Chim Reserve*. Paper presented at the Sarus Crane and Wetlands workshop. Tamnong, Vietnam, 3pp.

Van Mensvoort, M. E. F. 1996. Soil knowledge for farmers, farmer knowledge for soil scientists: The case of acid sulfate soils in the Mekong Delta, Vietnam. Wageningen University Dissertation abstract no. 2112, 5pp.

Van Zalinge, N., P. Degen, C. Pongsri, S. Nuov, J. G. Jensen, H, V. Hao, and X. Choulamary. 2003. The Mekong River System. *Second International Symposium on the Management of Large Rivers for Fisheries* (Phnom Phen, 11–14 Feb. 2003), 17pp.

Walden, D., C. M. Finlayson, R. van Dam, and M. Storrs. 2002. Information for risk assessment and management of *Mimosa Pigra* in Tram Chim National Park, Vietnam. In J. Rovis-Herman, K. G. Evans, A. L. Webb, and R.W.J. Pigen (eds.) *Environmental Research Institute of the Supervising Scientist Research Summary 1995-2000, Supervising Scientist Report 166*. Darwin NT: Department of Environment and Heritage, Environment Australia.

Wassermann, R., N. X. Hien, C. T. Hoanh, and T. P. Tuag. 2004. Sea level rise affecting the Vietnamese Mekong Delta; Water elevation in the flood season and implications for rice production. *Climatic Change* 66(1–2): 89–107.

White, I. 2002. *Water Management in the Mekong Delta; Changes, Conflicts and Opportunities*. International Hydrological Program, Technical Documents in Hydrology no. 61, Paris: UNESCO, 66pp.

Chapter 7
The Great Lakes Wetlands Policy Consortium – Bilateral NGO Action Aimed at the Great Lakes

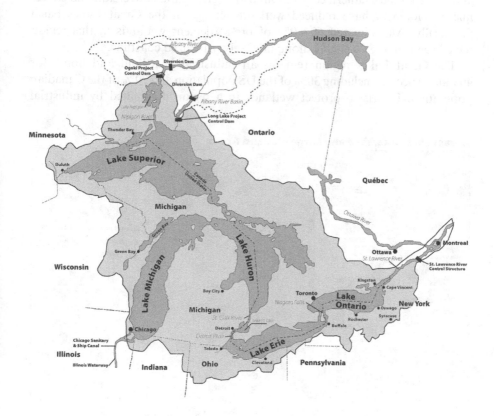

Background

This case study concerns the formation of the Great Lakes Wetlands Policy Consortium which was a bilateral action group formed in 1989–1990 to negotiate and coordinate environmental NGO policy positions on protection management and restoration of wetlands in the Great Lakes drainage basin. This includes the states of Wisconsin, Illinois, Michigan, Indiana, Ohio, and New York plus the province of

R.C. Smardon, *Sustaining the World's Wetlands*,
DOI 10.1007/978-0-387-49429-6_7, © Springer Science+Business Media, LLC 2009

Ontario in Canada. The major player in this case study is the Tip of the Mitt
Watershed Council located in Michigan. They organized the policy consortium and
obtained the grant for to do the activity and are most active with follow-up. Other
actors are individual environmental advocate groups in the United States and
Canada as well as regulatory agencies in both countries.

The Great Lakes drainage basin contains 95% of the surface water of North
America (see Figs. 7.1 and 7.2). It covers nearly 300,000 miles^2 in eight states
and two Canadian provinces and boasts a rich variety of freshwater wetland
communities. Prior to European settlement, however, there were wetlands
stretching from the western edge of Lake Erie clear across Ohio, into Indiana,
and covering the southern edge of Ontario. Agricultural conversion and shoreline development have reduced wetland acreage in the Great Lakes basin
drastically. An estimated 60–80% of pre-settlement wetlands in this region
have been lost and 80–100% along intensely urbanized coastline.

The Great Lakes region remains an industrial heartland and home for
40 million people, including 30% of the US population and 70% of the Canadian
population. Efforts to protect wetlands in a region dominated by industrial

Great Lakes Watershed and Large Wetland Areas

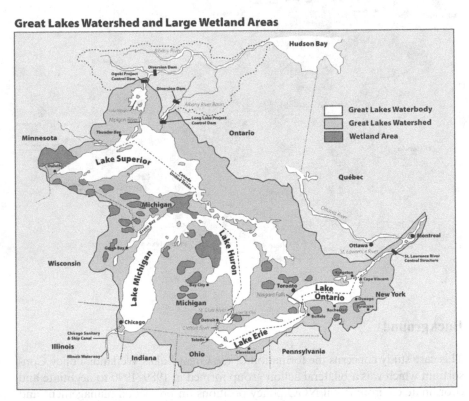

Fig. 7.1 Great Lakes watershed and large wetland areas. Drawn by Samuel Gordon and
adapted from Environment Canada and USEPA, 2000. Great Lakes Atlas

Fig. 7.2 Open lake edge marsh. Photo credit: Canadian Wildlife Service

activity and expanding development inevitably encounters regulatory, economic, cultural, and political resistance.

The Resource

Marshes, shrub/scrub, and forested wetlands occur along the margins of the Great Lakes and throughout the watersheds. Wetlands along the margins of the Great Lakes are located primarily in areas protected by wave action including bays, the entrances to rivers, and behind spits and barriers. These marshes are globally unique in that their plant communities and species composition have adapted to high water level fluctuations, wind and ice action (Bedford 1992, Burton et al. 2002, Geis 1985, Herdendorf 2004a, Keddy and Fraser 2000, Tilton and Schwegler 1978).

Some wetlands are located along the open coast in shallow water areas in semi-protected zones (see Figs. 7.2, 7.3, 7.4, and 7.5). Upper watershed wetlands are located along rivers, streams, and smaller freshwater lakes, and in isolated pockets. These wetlands tend to be coniferous or hardwood swamps, shrub carr-heaths, and bogs (see Fig. 7.6).

Wetlands in the Great Lakes drainage basin are valued (see Tilton and Schwegler 1978) for contributions to water quality (especially riverine and lacustrine marshes), hydrology and flood dissipation (especially upper watershed swamps), shoreline protection (limited), human use (especially marshes), primary production and diversity (especially marshes), rare and endangered species (especially swamps, bogs, and fens), and peat (carbon) accumulation (especially inland swamps, fens, and bogs).

There have been a number of studies assessing biophysical attributes and classifying Great Lakes wetlands and their connecting water bodies

Fig. 7.3 St. Lawrence River marsh. Photo credit: New York Sea Grant

Fig. 7.4 Emergent marsh vegetation. Photo credit: Canadian Wildlife Service

(Herdendorf 2004a, Herdendorf and Hartley 1980, Environment Canada Wildlife Service 2002, National Wetlands Working Group 1981). There have also been Great Lakes wetlands focused conferences (Champagne 1981, Kusler and Smardon 1990) as well as special journal issues (Munawar 2004, Kreiger et al. 1992) and books (Prince and D'Itri 1985). One of the best summaries of the values of wetland habitat in the Great Lakes basin is by Tilton and Schwegler (1978) and Whillans (1987) for fish, and Hecnar (2004) for amphibians. There have also been a number of resource value assessment studies for Great Lakes wetlands as well (Amacher et al. 1989,

Fig. 7.5 Typical embayment wetland along Lake Ontario shore. Photo credit: New York Sea Grant

Fig. 7.6 Northern coniferous swamp. Photo credit: Canadian Wildlife Service

Herdendorf 2004, Smith et al. 1991) and the State of the Great Lakes Ecosystem (SOLRC) reports were done by Dodge and Kavetsky (1995) and Maynard and Wilcox (1997).

In general, many coastal wetlands are subject to change over periods of time, but many emergent wetland vegetative communities are dominated by a few species. This is even truer for embayment wetlands and those along connecting waterways where water levels are controlled or very less than the past. Dominant submergent plant species include *Chara* spp., *Myriophyllum spicatum*, and *Potamogetum pectinatus*. Dominant emergent species for these wetland communities include cattail (*Typha glauca*), burreed (*Sparganium eurycarpum*), reedgrass (*Calamagrostis canadensis*), wild rice (*Zizania aquatica*), and sedge (*Carex* spp.). Where there are shrubs and flooded trees herbaceous species include common horsetail (*Equisetum arvense*), sensitive fern (*Onoclea sensiblis*), fringed loosestrife (*Lysimachia ciliata*), and Canada goldenrod (*Solidago canadensis*). Specific studies on coastal wetland vegetation ecology include work by Keough et al. (1996), Klarer and Millie (1992), Jean and Bouchard (1993), Sager et al. (1985), and Whyte et al. (1997). Studies of phytoplankton and zooplankton nearshore communities include work by Booth (2001), Cardinale et al. (2004), Hwang and Heath (1999), and Klarer and Mille (1994). Studies of invertebrate habitat include work of Burton et al. (2004), and there is insect emergence work for coastal wetlands by Mackenzie and Kasler (2004) and McLaughlin and Harris (1990). Price et al. (2005) and Environment Canada (1995) have studied coastal marsh habitat relationships to amphibians.

Herdendorf and Hartley (1980) provided a comprehensive literature review of fish and wildlife resources of the US side of Great Lakes wetlands. Many fish species are dependent on wetland habitat for part of their life cycle such as spawning, nursing, and resting. Herdendorf and Hartley list at least 24 species of fish that spawn in Great Lakes coastal wetlands. Holmes and Whillans (1984) reported 77 fish species recorded in Hamilton Bay, Lake Ontario. There are historic and recent documentation of fish inhabiting some 70 locations along the north shore of Lake Ontario, about half of which are wetlands (Whillans 1987). Thirty-seven of those species are wetland dependent (breeding, food, and cover). Stephenson (1990) found that 32 species of fish, representing 89% of all species present, use coastal Great Lakes marshes that she studied. Additional Great Lakes fisheries studies have been done by Botts (1997), Brazner et al. (2004), Chubb and Liston (1986), Hook et al. (2001), Leach (1995), and Leslie and Timmins (1991).

Although much attention has focused on exotic introduced salmonids not found in wetlands the forage base for these fish include species which spawn and/or are reared in wetlands. In shallower waters in the United States and Canada foci of fishing is on centrarchids and preclids that inhabit wetlands in all or part of their life cycles. Canadian commercial fish harvests have also been heavily composed of centrarchids and perclids (Ridgley 1985). Walleye or northern pike (*Esox lucius*) and carp (*Cyprinus carpio*) are heavily fished in Great Lakes coastal waters in both United States and Canada and the muskellunge (*Esox masquinongy*) is one of the largest and most prized game fish in Lake Erie, Eastern Lake Ontario, and the St. Lawrence River (Farrell 2001).

Approximately 3 million waterfowl migrate into or out of the Great Lakes region (Crowder and Bristow 1988, Prince et al. 1992, Tilton and Schwegler

1978). Great Lakes wetlands also provide habitat for waterfowl, especially ducks and geese, other birds, and a number of animals (see Figs. 7.7, 7.8, and 7.9). These areas are important regionally during migration, especially between the Atlantic Coast and inland locations in northern Canada. Waterfowl rest

Fig. 7.7 Ducks in flight. Photo credit: Canadian Wildlife Service

Fig. 7.8 Wood thrush.
Photo credit: Canadian
Wildlife Service

Fig. 7.9 Common tern. Photo credit: Canadian Wildlife Service

and feed in these areas, especially lake edge wetlands. According to Hummel (1981) 42 bird species are totally dependent on southern Ontario wetlands, 26 bird species are partially dependent, 16 mammal species are dependent, and 20 reptile species are heavily dependent. Glooschenko et al. (1988) found high occurrences of five out of six bird guilds in Great Lakes wetlands including divers, dabblers, waders, gulls and terns, and passerines. The largest numbers of dabbling and diving ducks use a corridor passing over south-eastern Michigan and northern Ohio. Similarly, numbers of Canada, snow and blue geese are highest in fall migration corridors, which pass through Saginaw Bay of Lake Huron, Green Bay of Lake Michigan, and Grand Traverse Bay of Lake Michigan (Crowder and Bristow 1988, Prince et al. 1992,Tilton and Schwegler 1978).

Wetland ecosystems of the Great Lakes region are also valuable as waterfowl production areas. The following birds are examples of the variety of species nesting in wetland habitats: black-crowned night herons (*Nycticorax nycticorax*), marsh hawk (*Circus cyaneous*), great blue heron (*Ardwa herodias*), short-billed marsh wren (*Cistothorus platensis*), red-winged blackbird (*Agelaius phoniceus*), as well as numerous species of ducks, and to a lesser extent, geese.

Coastal Great Lakes wetlands are important breeding grounds for muskrat (*Ondatra zibethica*), beaver, otter, and other mammals and habitat for fish eating mammals such as raccoon, otter, and mink (see Fig. 7.10). Information on mammals exists for some locations but not consistently throughout the Great Lakes wetlands (Crowder et al. 1986, Herdendorf and Hartley 1980).

Fig. 7.10 Muskrat. Photo credit: Canadian Wildlife Service

Threats

Historical threats: Historically wetlands have been destroyed or degraded by the following:

- **Fills**: Filling of wetlands has been particularly serious at the mouths of rivers and in urban areas (e.g., Chicago, Milwaukee, Toronto, Buffalo). Large areas of waterfront are typically filled wetlands. However, fills have also taken place in some other areas to facilitate development or provide protection from coastal erosion (see Fig. 7.11).

Fig. 7.11 Wetland fill. Photo credit: Canadian Wildlife Service

- **Drainage**: Drainage, primarily for agricultural purposes, has apparently been the major cause of loss of watershed wetlands in both the United States and Canada. Drainage has also taken place to facilitate development, particularly in urban areas (see Fig. 7.12).
- **Dredging**: Dredging to facilitate commercial and recreational water traffic has widely taken place in some rivers and at the mouths of rivers. It also has been widely undertaken along urban waterfronts (see Fig. 7.13).

Fig. 7.12 Wetland drainage for agriculture. Photo credit: New York Sea Grant

Fig. 7.13 Dredging for marina use. Photo credit: New York Sea Grant

- **Dikes**: Dikes to provide flood protection and to create waterfowl impoundments have been constructed in some areas along the Great Lakes. The dikes
are a mixed blessing as they affect wetlands hydroperiods or cut off wetlands
from adjacent waters but also provide them protection from storm damage.
- **Water pollution**: Water pollution has taken the form of both direct point
sources of pollution from industrial and commercial operations and non-
point sources of pollution from agricultural runoff, stormwater, and other
sources. Pollutants include sediment, excess nutrients, and toxic trace metals,
and organic pollutants (Crowe et al. 2004) (see Fig. 7.14).

Fig. 7.14 Water pollution from dumping in wetlands. Photo credit: New York Sea Grant

- **Stabilization of water levels**: Water levels have been partially stabilized in
all of the Great Lakes. This has resulted in decreased productivity and
modification of wetland aquatic plant community's structure and diversity
(Patterson and Whillans 1985, Wilcox 1993, 2004) (see Fig. 7.15).

Present Threats: Some of the old threats have been reduced and others
continue. Major present threats include the following:

- **Further stabilization of water levels**: The International Joint Commission
continues to consider a variety of proposals for stabilizing water levels,
particularly for Lake Ontario. Concerns include impact on fisheries (Liston
and Chubb 1985, Manny 1984), birds (McNicholl 1985), and plant communities (Keddy and Reznicek 1986, Lyon et al. 1986, Jaworski et al. 1999,
Wilcox 1993, Wilcox et al. 2002).
- **Drainage**: Although reduced, drainage for agriculture and other purposes
continues in both the United States and Canada.
- **Fills**: Fills have been reduced, but continue to "nibble away" at small residual
wetlands in both the United States and Canada.

Fig. 7.15 Flooding and stabilization of water levels. Photo credit: New York Sea Grant

- **Non-point**, and to a lesser extent, point sources of pollution have been reduced, but some new pollution continues to occur. In addition, non-point sources of pollution continue, essentially unabated, resulting in contaminated wetland sediment and vegetation. Work is continuing on phosphorus budgets (Mitsch and Reeder 1997, Reeder 1994). Mercury, especially, remains a concern for its availability for fish and fish eating birds (Pijanowski et al. 2002). More recent concerns arise from persistent organic chemicals and endocrine disruptor affects on wildlife (Fox 2001).
- **Maintenance dredging**: Navigational dredging continues, particularly in urban areas. A number of proposals have been made for new marinas or expanded marinas.
- **Invasion of exotic species**: The Great Lakes continue to be susceptible to unwanted invasion of exotic species from zebra mussels (Bowers and de Szalay 2004, Hudson and Bowen 2002, Leach 1995, Mills et al. 1993, Schloesser et al. 2006) to unwanted plant species such as purple loose strife (Zedler and Kercher 2004) and common reed (Wilcox et al. 2003).
- **Global warming**: Includes aggravation of water level and exotic species affects above plus vegetative community shifts and long-term carbon storage issues (Armentano and Menges 1986, Crowley 1990, Hartmannn 1990, Magnuson et al. 1998, Mortsch and Quinn 1996, Smith 1991).

Protection Policies

Until two decades ago, no or little protection was provided to wetlands of the Great Lakes. Since then a variety of measures have been adopted at national, state and provincial, and local levels, but the effectiveness of these measures vary. A detailed accounting of wetland protection policies both in Canada and the United States can be found in Loftus, Smardon and Potter (2004). The following is an abbreviated overview of these protection programs.

Regulations

In the United States, the US Army Corps of Engineers provides some control of fills and other structures in wetlands along the margins of the Great Lakes and in wetlands along the major rivers and streams and in watersheds through the Section 404 program. However, smaller fills and many types of drainage have not been regulated. This program, subject to various types of conditions, has typically issued permits. There is no comparable federal permitting program in Canada, but there is a Federal Policy on Wetlands Conservation and the Canada Fisheries Act does intend to protect fish and wildlife habitat in both inland and marine waters.

In the United States, most (Wisconsin, Minnesota, Michigan, New York, Pennsylvania) but not all the states have adopted wetland regulatory programs, which require permits for specified activities. However, drainage is not extensively regulated. In addition, many states limited the types or sizes of wetlands subject to regulation (e.g., the New York program applies only to wetlands 12.4 acres in size and larger). In Canada, a general wetland policy has been adopted at the provincial level by Ontario, which has jurisdiction over all Great Lakes wetlands except for the St. Lawrence. However, implementation is up to local governments.

In the United States many local governments have also adopted wetland protection regulations, particularly in Wisconsin, Michigan, Pennsylvania, and Minnesota; as noted above, regulation of wetlands is solely local government responsibility in Ontario (Smith et al. 1991).

Despite the rather broad scope of regulations already in place (particularly in the United States), the effectiveness of implementation is questionable. There are a fair number of exemptions to the existing regulations; regulations are in some instances, not enforced; the effectiveness of compensation (restoration/creation) measures to compensate for losses is highly questionable, the typical balancing approaches utilized for permits result in gradual, cumulative losses. Lack of manpower, staffing, and budgets are all problems for implementing agencies.

Tax Incentives

There are both provincial and federal tax incentive programs in Canada (see Loftus et al. 2004) and some tax incentive programs in the US states. There is also a tax incentive for US landowners if they donate land to a land trust or not-for-profit under the US Federal Income tax code. The increase in land trust activity in the last 20 years, partially because of this measure, in the United States has been substantial.

Acquisition/Securement

The national, state, and local governments in some instances have acquired selected wetland areas, although there seems to be more activity in non-government organizations moving toward acquisition of wetlands for habitat management, heritage values, or interpretation potential. There are many programs (see Loftus et al. 2004) in Ontario that include acquisition, dedication, agreements, co-management, land use allocation, and extension outreach. Much of the wetland acquisition in the United States is connected to major NGOs such as Ducks Unlimited, Nature Conservancy, Audubon, and local land trusts.

Restoration

Some wetland restoration efforts have occurred in both regulatory and non-regulatory contexts mainly in urban settings such as Toronto Harbor and Chicago and others are proposed as part of long-term pollution abatement such as Hamilton Harbor. In the United States, both the US Fish and Wildlife Service and Natural Resource Conservation Service have wetland creation, restoration, and enhancement programs that have been very effective through-out the Great Lakes drainage basin. There has been more ongoing research in general on wetland restoration (Barry et al. 2003, Kreiger 2003, Lundholm and Simser 1999, Wang and Mitsch 1998, Wilcox 1999) and a very concentrated research on Cootes Paradise March restoration in Hamilton Harbor (Chow-Fraser 1998, 2005, Chow-Fraser et al. 2004, Wei and Chow-Frazer 2005).

Thus we have the wetland resource, past threats, current threats, and reg-ulatory programs. Now the question is what was the role of non-government organizations in the midst of the Great Lakes wetland management context.

Significant Tri-national Wetlands Management Efforts

The other major effort by an NGO affecting Great Lakes wetlands is the role of Ducks Unlimited in their suggestion and support of the *North American Waterfowl Management Plan*. This ambitious plan was initiated in 1979 by the United States, Canada, and Mexico and was 7 years in gestation. It set out principles for the cooperative conservation and use of North American water-fowl. It established population objectives (to restore 1970s levels by the year 2000) and harvest strategies for ducks, geese, and swans. It identified and priced the habitat, research, and management initiatives required to attain the population objectives. It identified new administrative arrangements to oversee the plan's implementation. The signing of the plan by the two federal governments in 1986 did not constitute a financial commitment to its implementation. Nevertheless funding was later available for bi-national habitat restoration and improvement projects. Mexico signed the agreement in 1989. A series of subsequent implementation measures in both the United States and Canada have helped to make this a very successful effort (Loftus et al. 2004).

Tip of the Mitt Watershed Council's History

The watershed council was formed in 1979 by a group of lake associations from Cheboygan, Charlevoix, and Emmet Counties with the assistance of the University of Michigan Biological Station. The lake associations wanted to coordinate efforts in order to keep their lakes clean and protect the water resources of the region.

The series of events, which led to the formation of the watershed council, began in the early 1970s. At this time, the University of Michigan Biological Station (UMBS) provided substantial assistance to lake associations. These programs were funded through federal grants. The station did a complete water quality evaluation for 40 lakes from 1972 to 1974 and also produced publications and conferences. Additional assistance came from NEMCOG (Northeast Michigan Council of Governments) and the Northwest Michigan Regional Planning and Development Commission. These planning commissions, in turn, also depended largely on federal programs to fund their programs.

At the end of the 1970s, federal grant funding for assistance to lake commissions became unavailable. This is when the associations decided it was time to raise funds to support their own program and formed the Tip of the Mitt Watershed Council to oversee it. In 1979, the watershed council was a volunteer organization with a budget of about $1500. Support and staff have grown steadily to the point today where the council has a staff of six and a substantial operating budget. The watershed council currently has over 43 member organizations, including 27 major lake associations, Little Traverse Conservancy, and 15 other local groups who want to see water resources protected. The board of directors consists of eight representatives from member organizations and 10 members at large.

History of the Great Lakes Wetlands Policy Consortium

Two years before the wetlands policy consortium became a reality – the council was concerned with Michigan state rules for wetland permit processing. There was much controversy as an unusual joint legislative committee on administrative rule making was going to make these wetland regulations and this same joint legislative committee had been previously challenged on constitutional grounds, e.g., mixing legislative and executive functions and powers. There was a need to balance environmental and industrial interests as part of the rule-making process. The council was called upon to provide two of the four environmental representatives as part of this process. There were four environmental and four industry representatives.

One of the council's environmental representatives, Stephen Brown, stated that at that time he felt that the environmental representatives were "outgunned" and at an extreme resource disadvantage. The industry representatives had lawyers and specialists that were constantly being "cycled in" as the issues came up – the environmentalists had few resources. Stephen Brown and Gail Gruenwald (past executive director of the council) prepared papers on such issues as the "feasible and prudent alternatives test" which is a complex issue. Definitions were problematic and difficult to discuss and negotiate. Non-consensus plus the ad hoc process convinced some that the environmental community was not prepared. Environmentalists at the state level were not ready to negotiate policy. They needed a think tank to solicit policy. Most environmental advocate groups were focused on micro issues and needed resources to allow people to come together.

Stephen Brown stated that it was this feeling of being unprepared and at a resource disadvantage, which motivated him to write the grant with Gail Gruenwald to the Mott Foundation for the Great Lakes Wetlands Policy Consortium. The Mott Foundation was a little doubtful that the council was ready to take on this level of international multigroup discussions on wetland policy. But, the council's experience with wetland issues with the earlier Michigan rule-making negotiations served them well. They knew the issues, but they took a position outside the existing regulatory arenas and focused on a vision of how ecosystems management should happen. Stephen Brown and Gail Gruenwald also did lots of recruiting while developing the proposal to the Mott Foundation to get letters of support from many of the participating organizations.

Development of the Great Lakes Wetlands Policy Consortium

At the first meeting, all the issues were put on the table. For the organizers, it was a real learning experience. It was expected that everyone would have similar perspectives, as opposed to the many and varied differences of views and opinions as was actually the case. An example was the representative from the Association of Conservation Districts who said they were using the wrong approach, e.g., not having industry at the table. As a result, this group ended up in an advisory role and did not sign the agreement.

There were three major meetings plus one meeting with the regulatory people. At the first meeting there was a lot of "flexing of muscles". Some groups were new at policy making and some were not. Some of the seasoned groups were more skeptical. Smaller organizations were very pleased to sit down with the other groups and discuss wetland policy issues. Stephen Brown initially laid out what he thought were the major issues. The group as a whole decided to reorient, but ended up with some of the same issues.

The group after reviewing the major issues broke into subgroups to work on four areas:

- Regulatory issues
- Incentives
- Public outreach
- International policy linkages

Regulatory issues were not popular, yet were some of the major issues. The organizers tried to maintain working groups and to promote discussion of issues. Funds were provided for meetings, but did not provide for release time between meetings. Finding time for individuals to work on project tasks was a real problem with a resultant high degree of variation in input. Teams would develop their own style and personalities which made it difficult to coalesce material together.

Many participants had strong ideas about what should or should not be done. An example was the vision statement. Its purpose was to make a public splash and influence policy, whereas others felt that it was better to work behind the scenes and this public imaging was counterproductive.

Stephen Brown pulled much of the material for the drafts together himself. They also got a small group together to work on the regulatory issues in Chicago. They mostly worked in small groups to finalize text and was sent out to larger working groups for review. They asked regulators to respond but did not get much response.

The big issue in the end was how to release the recommendations and which groups would sign. As was stated before the National Association of Conservation Districts did not sign and the Freshwater Foundation was also concerned about their role and the implications of signing off on the recommendations. Much time has spent stating the role of the organizations in the development of guidelines and recommendations.

The Final Report of the Great Lakes Wetlands policy Consortium ended up with a *Vision Statement, Wetland policy Issues and Recommendations for Change* and an *Action Agenda* (Gruenwald 1990). The recommendations had the following structure and are included in summary form as an appendix to this chapter:

A. Improving regional coordination and planning

    1. International recommendations
    2. Recommendations for provinces, states, and local government
    3. Recommendations for non-governmental organizations

B. Increasing incentives for preservation

    1. Direct and indirect payments
    2. Income and property tax incentives
    3. Other incentives

C. Strengthening wetland regulatory programs
   1. Guidelines for all regulatory programs
   2. Recommendations for local regulatory programs in the United States
   3. Recommendations for state regulatory programs
   4. Recommendations for the US federal program
   5. Recommendations for wetland regulatory programs in Canada
D. Ending government funding of wetland destruction
E. Expanding restoration efforts
   1. Management needs
   2. Research priorities
G. Supporting acquisition programs
H. Extending education outreach

All three parts were produced in a report as well as the Great Lakes wetlands newsletter. Also a number of videos were produced and supported by separate grants from the Mott Foundation. The Mott Foundation wanted evidence that someone would use the results of the consortium work. As part of the third year of the initial grant, it was proposed that there be developed a Michigan Action Coalition to assist in implementation for one state. Also there was grant funding for the Great Lakes wetlands newsletter for the first 2 years. This newsletter became almost self-sufficient according to Gail Gruenwald, executive director of Tip of the Mitt Watershed Council. The consortium recommendations also had an impact on Wisconsin's wetland programs, the National Wildlife Federation Great Lakes Program and in Ontario from a Canadian perspective.

The author interviewed two individuals connected with NGOs at the time the policy consortium recommendations were developed and continued on with key implementation roles. These individuals were Cam Davis who was with the Lake Michigan Federation and is now with the National Wildlife Federation, and Nancy Patterson who was with the Ontario Federation of Naturalists and is now with the Canadian Wildlife Service. Both individuals were asked about implementation progress.

Cam Davis, indicated on behalf of the Lake Michigan Federation (LMF) that LMF did everything they were supposed to do and that the GL wetlands policy consortium was a great effort and very "forward thinking" at the time. He felt personally that there was an overemphasis on tax credits and financial incentives, but otherwise the balance of effort was good. The biggest drawback, he felt, was that many of the participants did not meet for one and one-half year's time after the recommendations were released and currently on the US side, no one knows what other groups did or did not do. To this end he has four recommendations:

• Implement monitoring to find out which action items were done by which groups
• Invite regional commentary on regulatory actions and issues. There was a governor's conference in Indiana, which did such

- A lot of the recommendations and action agenda are policy oriented – where is the enforcement? or could some of the agenda be more enforcement oriented
- Do pilot projects with individual landowners with hands-on application to understand some of the implications of the recommendations

So in essence, a typical US side response is we did what we were supposed to, but aspects could be taken further with monitoring and enforcement. Let us see what the Canadian side did with the recommendations and action agenda.

After the policy consortium recommendations and action agenda were released Nancy Patterson left the Federation of Ontario Naturalists Association, a Canadian NGO, and went to work for the Canadian Wildlife Service, a Federal Canadian agency. Her job was to implement the recommendations and action plan from a Canadian perspective (Patterson 1992). Step 1 according to Nancy was to take the policy consortium's results as a framework and find out what "real" actions and the major actors could agree upon attendant resource costs. In this case NGOs and government had to achieve consensus on priorities and proposed legislation. Government had to take a hard look at what they were prepared to do, especially at the provincial and local levels regarding compliance. To do this Nancy Patterson organized a network of committees (she used the term "nightmare of committees") to negotiate these issues, which took approximately 3 years. The result was the *Great Lakes Wetlands Conservation Action Plan* (GLWCAP). The plan specified actions, resource costs, and milestones to measure accomplishments for the period of 1994–2000. Parties represented in the committees and negotiations included farmers groups, sportsman's groups, naturalists, NGOs, and government. Aspects of the plan include

- regulation and compliance;
- secure and protection (easements and management);
- priorities for wetland restoration;
- education and outreach.

After the committees were dissolved step 2 was implementation, which contains some interesting strategies. One such strategy is the creation of an implementation committee made up of two government agencies, the Canadian Wildlife Service and the Ontario Ministry of Natural Resources, and two NGOs, the Federal of Ontario Naturalists and the Nature Conservancy. The committee is chaired be someone from an NGO and its job is to track the process on deliverables utilizing the milestones and actions in the plan.

The other strategy for implementation was to formalize progress on the action plan within legislation, the *Canada-Ontario Agreement*, which in turn formalizes progress on the International Joint Commission's Water Quality Agreement between the United States and Canada. A third strategy was to market Great Lakes Wetlands Conservation Plan as not a new program, but an ongoing program. As part of this strategy, use of existing resources and partnerships between NGOs and government was used to underwrite the plan. During the critical phase of implementation, Canada, like the United States was undergoing a phase of deregulation – in Canada called land use planning reform and "harmonization". So implementation of any program, which affects use of wetlands, would be under a good deal of scrutiny. It is a testament to the leadership ability of those who carried the spirit of the GL Wetlands Policy Consortium through a tough negotiation process to the present program implementation in Canada. It is also important, as pointed out by Nancy Patterson herself, that someone from an NGO orientation carry through this same perspective and negotiation style, even though the leadership role is with a Canadian federal agency.

Stephen Brown, who wrote the Mott Foundation Grant and the Consortium Report, feels that whether all the recommendations were implemented is not the issue. He suggests that the results in this case is getting the environmental advocate groups to talk to each other – to broaden their horizons whereas most groups tend to function in isolation. The networks that were set up to do consortium work are still functional. Other results still evolving are that (1) groups involved are better prepared to address their own wetland policy strategic needs and (2) are not so much at a resource disadvantage as they were previously. This also needs to be put into a regulatory context that there has been backlash and regression of some federal, state, and provincial wetland programs in both Canada and the United States during the last 2 years. The latter point means that anyone serious about implementation needs to be innovative in developing action and funding strategies combined with endurance needed for long protracted negotiation on tough resource and local government decision-making issues.

But, the "proof is in the pudding". To date in 1994, the Great Lakes wetlands Conservation Action Plan (GLWCAP) brought together both governmental and non-governmental partners in an effort to conserve and rehabilitate remaining wetlands. The Action Plan complements the goals and objectives of Canada's Federal Wetlands Policy of 1991 and the Ontario Wetlands Policy Statement of 1992.

The first Plan of Action (1994–2001) was produced under the auspices of the 25-year Strategic Plan for wetlands of the Great Lakes basin. Launched in 1993, the strategic plan involves several public and private agencies working together individual citizens and landowners. The long-term goal of the original plan is to protect the area and function of 30,000 ha of existing wetlands in the Great Lakes basin by the year 2020.

In July 1994, the Canadian federal and provincial environment ministers signed the Canada-Ontario Agreement Respecting the Great Lakes Ecosystem (COA), a 6-year agreement that set specific targets and time frames for restoring, protecting, and sustaining the basin's ecosystems. GLWCAP was to be the delivery system for COA's goal of rehabilitating and protecting 6,000 ha of wetland habitat by the year 2001. This target was met and surpassed with over 5,000 ha of wetland receiving protection and over 12,000 ha being rehabilitated.

GLWAP's strategies and associated milestones are implemented by representatives from Environment Canada, the Ontario Ministry of Natural Resources, Ducks Unlimited Canada, The Federal of Ontario Naturalists, and the Nature

Conservancy of Canada. Other major partners include the Eastern Habitat Joint Venture off the North American Waterfowl Management Plan and the Great Lakes Stewardship Fund.

In addition to protection of several thousand hectares of wetlands in the Great Lakes basin, accomplishments of the first Action Plan include the production of wetlands publications, displays and facilitation of workshops, and the development of Temperate Wetlands Restoration Workshop and Training course. All of this activity is detailed in four progress reports for 1994–2001, 1997–2001, 2001–2003, and 2003–2005 which are available at the Great Lakes Wetlands Conservation Action Plan web site at http://www.on.ec.gc./wildlife/wetlands/glwcap-e.cfm. The latest progress report from 2003 to 2005 is summarized in Table 7.1. As one can see there has been substantial progress made under each of the eight strategies within the GLWCAP. Also notice the similarities of these strategies to the initial recommendations of the Great Lakes Wetlands Policy Consortium (Brown 1990) and pp.199–200 of this chapter.

Table 7.1 Great Lakes wetlands conservation action plan (GLWCAP) progress as of 2006[1]

Strategy 1: Publicize information concerning wetland protection, rehabilitation, policies, and regulations and encourage involvement by individuals, groups, corporations, and industries in all aspects of Great Lakes wetlands protection and rehabilitation

Milestones	Progress
1.1 Publicize wetland values to society, to water, and to wildlife in order to encourage wetlands conservation. This may involve developing, publishing and distributing brochures, educational packages, and status reports	75%
1.2 Produce and distribute communication packages targeted to corporations, agriculture (landowners), industry and development interests, school curriculum, and municipal and regional governments	75%
1.3 Expand distribution network through web-based information and like links	75%
1.4 Provide a publicly accessible, web-based basic wetland attribute and mapping resource (e.g., provide Ontario Coastal Wetlands Atlas online)	100%

Strategy 2: Conduct and facilitate study of wetland functions, status, and trends to improve understanding, communicate values, and set priorities for protection and rehabilitation. Develop an accessible, computerized database for coastal Great Lakes wetlands

Milestones	Progress
2.1 Establish an ad hoc interagency data management group or technical coordination team	50%
2.2 Create/maintain an integrated computer database for coastal wetlands of the lower Great Lakes and expand to include the remainder of the Great Lakes basin (e.g., Ontario Great Lakes Coastal Wetland Atlas, plans for interior Ontario wetlands, bi-national coastal outcomes from Great Lakes Coastal Wetlands Consortium)	75%
2.3 Continue wetland health monitoring at a variety of spatial and temporal scales including maintenance and enhancement of a bi-national Great Lakes wetland monitoring program (e.g., community-based Marsh Monitoring Program and Great Lakes Coastal Wetlands Consortium indicators work)	50%

Table 7.1 (continued)

2.4 Investigate and report on targets, status, and trends in wetland area and other attributes	50%
2.5 Investigate and report on loss of wetlands (area and function) due to agricultural drainage and other causes in selected watersheds	50%
2.6 Investigate the science of wetlands, including the relationship between wetland hydrology and groundwater discharge/recharge; features that define faunal wetland preferences; wetland function within a landscape mosaic-hydrology, connections to uplands, buffers: exotics; species at risk; species toxicology, sensitivity to climate change; relationship between wetlands and water quality; and economic values.	25%
2.7 Use up-to-date science to develop a more cost-effective method-ology for evaluating wetland functions and values, while maintaining the scientific rigor of the provincial wetland evaluation system	25%

Strategy 3: Determine priority securement sites and the most effective techniques to secure those sites. Undertake wetlands securement at priority sites involving publicly owned lands to demonstrate innovative securement strategies. Undertake extension and stewardship activities with private landowners to protect the area and function of existing Great Lakes basin wetlands and achieve the "no loss" long-term goals

Milestones	Progress
3.1 Secure 6,000 ha of wetland (8890 ha pervious plus 3993 ha in 2003–2005 = 12,883 ha overall)	100%
3.2 Promote and facilitate improved responsible wetland protection and management (strategy 4) on Crown lands by all provincial and federal government agencies/owners. Identify opportunities by documenting location and ownership of all provincially owned lands with wetlands to complement existing federal report	25%
3.3 Convene an expert's workshop to identify, map, and describe biodiversity investment areas and to develop basin-wide conservation blueprint for priority securement	75%
3.4 Identify, promote, and assist activities of conservation authorities and municipalities to maintain and improve, where necessary, the security and management of other publicly owned lands	25%
3.5 Promote and facilitate responsible wetland protection and management (Strategy 4) on private lands by landowners through extension and stewardship programs such as organizing workshops to promote local initiatives	50%

Strategy 4: Undertake rehabilitation projects and priority sites. Pursue opportunities for wetland rehabilitation/creation through existing programs, including Remedial Action Plans and the Eastern Habitat Joint Venture. In the long term consider ecological and watershed-based goals to achieve an overall increase in the area and function of wetlands in the Great Lakes basin

Milestone	Progress
4.1 Rehabilitate/create 6000 ha of wetland	75%
4.2 Strengthen and enhance wetland rehabilitation and management expertise through training and technology transfer to rehabilitation practitioners	50%
4.3 Establish management plans on 6000 ha of secured or rehabilitated wetland, based on federal, provincial or non-government guidelines as appropriate. Develop and refine guidelines as needed	75%

Table 7.1 (continued)

Strategy 5: Define and improve compliance with existing regulatory programs, strengthen wetland conservation and protection through ongoing regulatory/agreement/policy review opportunities

Milestone	Progress
5.1 Influence official plans through stewardship and efforts to promote wetlands being designated and zoned for conservation in local planning documents	50%
5.2 Periodically review the effectiveness of the provincial wetlands policy as part of the provinces (Ontario) 5-year process and recommend any changes and resources required to improve effectiveness of the policy	100%
5.3 Evaluate and implement Parks and Forest Management Guidelines where appropriate for wetland management on provincially owned lands	25%
5.4 With appropriate agencies, review the application and effectiveness of the Federal Wetlands Policy, Fisheries Act, Canada Environmental Assessment Act, Migratory Birds Convention Act, Agriculture Act, Species at Risk Act, Drainage Act, Lakes and Rivers Improvement Act, Conservation Authorities Act, and Ontario Farm Practices, Protection and Promotion Act with regard to wetlands protection and rehabilitation (see Loftus et al. 2004 for act specifics)	50%
5.5 Conduct workshops involving conservation authorities, the MNR municipalities and other government and non-government stakeholders to review the effectiveness of current wetland conservation practices such as impact assessment and mitigation and provide necessary follow-up and information exchange	25%
5.6 Review and evaluate grants, loans, and other financial incentives / disincentives to determine their impact on wetlands resources (e.g., Conservation Land Tax Incentive Program)	50%
5.7 Optimize implementation of GLWCAP through the Canada- Ontario Agreement Responding to the Great Lakes basin ecosystem	75%

Strategy 6: Ensure that all new plans such as resource management plans, watershed management plans, local land use plans, official plans, and habitat management plans incorporate wetland protection and rehabilitation strategies. Also encourage recognition and designation of appropriate adjacent and upstream land uses

Milestone	Progress
6.1 Update the MNR's natural heritage strategies and guidelines for coa stal areas (Crown lands) as required	50%
6.2 Identify, promote, and assist activities of conservation authorities and municipalities to maintain current watershed plans/strategies, integrated resource management plans, zoning, and other activities for wetland protection	50%

Strategy 7: Coordinate and integrate all action plan protection, rehabilitation, and other creation initiatives with other ongoing programs that affect Great Lakes wetlands, in particular activities associated with relevant international conventions and agreements

Milestones	Progress
7.1 Through linkages with strategy 1, maintain a current GLWCAP web site with regular updates to share progress with wetlands stake-holders	75%
7.2 Build alliances with new and existing wetlands and other wildlife hab itat conservation initiatives to ensure coordination and efficiency as well as facilitate reporting on the full range of wetland activities in the Great Lakes basin	75%

Table 7.1 (continued)

7.3 Coordinate bi-national Great Lakes wetland activities (including Lakewide Management Plans, International Joint Commission Lake Ontario – St. Lawrence River Study, etc.)	75%
7.4 Coordinate bi-national Great Lakes wetlands meetings to complement initiatives such as the North American Bird Conservation Initiative, Great Lakes Conservation Blueprint, and SOLEC	50%

Strategy 8: Evaluate the action plan components, including a careful assessment of individual techniques and their application

Milestone	Progress
8.1 Share partners (e.g., Nature Conservancy, Conservation Ontario, Ducks Unlimited, Ontario Nature, Ontario Ministry of Natural Resources, Environment Canada) annual work plans within the implementation team	100% +
8.2 Report on program progress at least twice during the lifespan of the action plan	100%
8.3 Regular review of the program by all implementation team partners	100% +

[1]Source: Environment Canada 2006.

On the US side the Great Lakes Legacy Act was passed in 2005 which provides funding for cleanup of contaminated sediment site "hit spots" as part of remedial action planning process (see Hartig and Thomas 1988) for the Great Lakes. There is legislation pending in the US Congress for restoration of the Great Lakes ecosystem, which is similar to those efforts proposed for the Chesapeake and San Francisco Bay or Columbia or Kissimmee Rivers. Such legislation would be strengthening of existing programs plus some new programs. Some of these programs would address the wetlands stresses addressed in the beginning of this chapter.

Summary

Great Lakes wetlands protection activities have sometimes taken the same paths – the North American Waterfowl Treaty – and different paths – the Great Lakes Wetlands Policy Consortium and outcomes. Both examples illustrate different strategies and tools used by NGOs – international, national, and state/local. For the North American Waterfowl Treaty, both in the United States and in Canada, Ducks Unlimited played a key role in developing and getting the treaty adopted. Once adopted in the United States. Canada and Mexico, federal, provincial, and state agencies played key roles in implementation such as the USFWS's "Partners in Wildlife" Program or NRCS's Wetland Reserve Program. It also helped that specific waterfowl populations stopped declining and started increasing, essentially validating the treaty and attached support programs.

The contrast of outcome results from the Great Lakes Wetlands Policy Consortium is interesting. In Canada, we have a steady and focused implementation of the Great Lakes Wetlands Conservation Action Plan with great partnering between government agencies and national and regional NGOs. In the United States, on the surface, we have lots of wetland outreach activity from both NGOs and federal/state agencies. We also have fragmentation and diversion of federal/state regulatory activity due to adverse court decisions and state wetland regulatory funding and personnel problems. So although some of the reforms pushed by the Great Lakes Wetlands Policy Consortium have been implemented, some of the wetland regulatory programs have actually gone backward. The prime example being the SWANC Supreme Court case which has directed the US Corps of Engineers that do not have jurisdiction over hydrologically "isolated wetlands" under Section 404 of the US Clean Water Act.

On the other hand Canada has never had strong wetland regulatory programs at the federal or provincial level, so much of the wetland protection has been through incentives and partnerships (see Loftus et al. 2004) as opposed to a strong regulatory focus in the United States. On the horizon there is current action in the US Congress to pass a Great Lakes Restoration Act that would focus regulation, incentive, and research programs for the Great Lakes much as Great Lakes Wetlands Conservation Action Plan in Canada. There is also progress in adopting and implementing Lakewide Management Plans or LaMPs for the Great Lakes. A good example is the Lake Ontario Management Plan, which included four agencies: Environment Canada, Ontario Ministry of Natural Resources, US Environmental Protection Agency, and New York State Department of Environmental Conservation. With such agreements in place there is more imputes for wetland protection as a key piece of ecosystem maintenance.

The other bit of good news is that there is lots of connected activity through out the Great Lakes basin, especially by NGOs and researchers. Groups like the Alliance for the Great Lakes and the Great Lakes Aquatic Network and Fund have extensive web pages on Great Lakes wetlands. The USEPA-sponsored GLIN web page has an extensive listing of NGO and state agencies with wetland programs at http://www.great-lakes.net/envt/air-land/wetlands.html

A number of wetland researchers are active with the Great Lakes Coastal Wetlands Consortium. The consortium consists of scientific and policy experts drawn from key US and Canadian federal agencies, state and provincial agencies, non-governmental organizations, and other interest groups with responsibility for Great Lakes coastal wetlands monitoring. In doing this work they are attempting to develop common methods/protocols for assessing coastal wetlands health, e.g., vegetation, fish, amphibian, and benthic sampling techniques. There are also researchers who are very concerned with wetland health monitoring and the effects of global warming effects on Great Lakes wetlands as well (see Bourdaghs et al. 2006, Danz et al, 2005, Finklestein et al. 2005, Lougheed and Chow-Fraser 2001, Mortsch et al. 2006, Niemi et al. 2004, Shear et al. 2003, Uzarski et al. 2004). There are other scientists using geographic information systems to study wetland change over time (Gottgens et al. 1998, Host et al. 2005, Jean and Bouchard 1991, Williams and Lyon 1991).

Acronyms

COA: Canada-Ontario Agreement
GLWCAP: Great Lakes Wetlands Conservation Action Plan
LaMP: lakewide management plan
NEMCOG: Northeast Michigan Council of Governments
NRCS: Natural Resources Conservation Service
SWANCC: Solid Waste Agency of Northern Cook County vs. US Army
    Corps of Engineers
UMBS: University of Michigan Biological Station
US F&WS: US Fish and Wildlife Service

References

Interview with Cam Davis March 11, 1996.
Interview with Nancy Patterson March 15, 1996.
Interview with Gail Gruenwald February 1996.
Interview with Stephen Brown, February 14, 1996.
Amacher, G. S., R. J. Brazee, J. W. Bulkey, and R. A. Moll. 1989. *Application of Wetland Valuation Techniques: Examples from Great Lakes Coastal Wetlands*. National technical Information Service, Springfield, VA as PB90-112319/AS, E. Lansing: Michigan Institute of Water Research.
Armentano, T. V. and E. S. Menges. 1986. Patterns of change in the carbon balance of organic soil wetlands of the temperate zone. *The Journal of Ecology*, 74(3): 755–774.
Barry, M. J., R. Bowers, and F. A. de Szalay. 2003. Effects of hydrology, herbivory and sediment disturbance on plant recruitment in a Lake Erie coastal wetland. *The American Midland Naturalist*, 151(2): 217–232.
Bedford, K. W. 1992. The physical effects of the Great Lakes on tributaries and wetlands, a summary. *Journal of Great Lakes Research*, 18: 571–589.
Booth, R. K. 2001. Ecology of testate amoebae (protozoa) in two Lake Superior coastal wetlands: Implications for paleoecology and environmental monitoring. *Wetlands*, 21(4): 564–576.
Botts, P. 1997. Spatial pattern, patch dynamics and successional change: Chironomid assemblages in a lake Erie coastal wetland. *Freshwater Biology*, 37(2): 277–286.
Bourdaghs, M., C. A. Johnston, and R. R. Regal. 2006. Properties and performance of the floristic quality index in Great Lakes coastal wetlands. *Wetlands*, 26(3): 718–735.
Bowers, R. and F. A. de Szalay. 2004. Effects of hydrology on Unionids (Unionidae) and Zebra mussels (Dreissenidae) in a Lake Erie coastal wetland. *The American Midland Naturalist*, 151(2): 286–300.
Brazner, J. C., S. E. Campana, and D. K. Tanner. 2004. Habitat fingerprints for Lake Superior coastal wetlands derived from elemental analysis of yellow perch otoliths. *Transactions of the American Fisheries Society*, 133(3): 692–704.
Brown, S. 1990. *Preserving Great Lakes Wetlands: An Environmental Agenda: The Final Report of the Great Lakes Wetlands Policy Consortium*. Conway, MI: Tipp of the Mitt Watershed Council, 78 pp.
Burton, T. M., C. A. Stickler, and D. G. Uzarski. 2002. Effects of plant community composition and exposure to wave action on invertebrate habitat use of lake Huron coastal wetlands. *Lakes and Reservoirs: Research and Management*, 7(3): 255–269.

Burton, T. M., D. Uzarski, and J. Genet. 2004. Invertebrate habitat use in relation to fetch and plant zonation in northern Lake Huron coastal wetlands. *Aquatic Ecosystem Health and Management*, 7(2): 249–267.

Canada Wildlife Service. 2002. *Where Land Meets Water: Understanding Wetlands of the Great lakes*. Downsview, ON: Environment Canada, Canadian Wildlife Service.

Cardinale, B. J., V. J. Brady, and T. M. Burton. 2004. Changes in the abundance and diversity of coastal wetland fauna from the open water/macrophyte edge toward shore. *Wetlands Ecology and Management*, 6(1): 59–68.

Champagne, A. (ed.). 1981. *Proceedings of the Ontario Wetlands Conference*. Toronto: Federation of Ontario Naturalists.

Chow-Fraser, P. 1998. A conceptual ecological model to aid restoration of Cootes Paradise marsh, a degraded coastal wetland of Lake Ontario, Canada. *Wetlands Ecology & Management*, 6(1): 43–57.

Chow-Fraser, P. 2005. Ecosystem response to changes in water level in lake Ontario marshes: Lessons from the restoration of Cootes Paradise Marsh. *Hydrobiologia*, 539(1): 189–204.

Chow-Fraser, P., V. Lougheed, V. le Thiec, B. Crosbie, L. Simser, and J. Lord. 2004. Long-term response of the biotic community to fluctuating water levels and changes in water quality in Cootes Bay Marsh, a degraded coastal marsh of Lake Ontario. *Wetlands Ecology and Management*, 6(1): 19–42.

Chubb, S. L. and C. R. Liston. 1986. Density and distribution of larval fishes in Pentwater marsh, a coastal wetland on Lake Michigan. *Journal of Great lakes Research*, 12(4): 332–343.

Crowley, T. E. II. 1990. Laurentian Great Lakes double – CO_2 climate change hydrological impacts. *Climate Change*, 17: 27–47.

Crowder, A. A. and J. M. Bristow. 1988. The future of waterfowl habitats in the Canadian lower Great Lakes wetlands. *Journal of Great Lakes Research*, 14: 115–127.

Crowder, A. A., B. McLaughlin, R. D. Weir, and W. J. Christie. 1986. Shoreline fauna of the Bay of Quince. *Canadian Special Publication of Fisheries and Aquatic Sciences*, 86: 190–200.

Crowe, A. S., S. G. Shikaze, and C. J. Ptacek. 2004. Numerical modeling of groundwater flow and contaminant transport to Point Pelee marsh, Ontario, Canada. *Hydrologic Processes*, 18(2): 293–343.

Danz, N. P., R. R. Regal, G. J. Niemi, V. J. Brady, T. Hollenhorst, L. B. Johnson, G. E. Host, J. M. Hanowski, C. A. Johnston, T, Brown, J. Kingston, and J. R. Kelly. 2005. Environmentally stratified sampling design for the development of Great Lakes environmental indicators. *Environmental Monitoring and Assessment*, 102(1–3): 41–65.

Dodge, D. and R. Kavetsky. 1995. *Aquatic Habitat and Wetlands of the Great Lakes*. 1994 State of the Lakes Ecosystem Conference (SOLEC) Background paper. Environment Canada and US Environmental Protection Agency EPA 905-R-95-014.

Environment Canada. 1995. *Amphibians and Reptiles of the Great Lakes Wetlands; Threats and Conservation*. Environment Canada, 12 pp.

Environment Canada 2006. *Great Lakes Wetlands Conservation Action Plan Highlights Report 2003-2005*. Toronto, Ontario: Environment Canada, 24 pp. and at http://www/on.ec.gc.ca/wildlife/publications-e.html

Farrell, J. M. 2001. Reproductive success of sympatric Northern Pike and Muskellunge in an Upper St. Lawrence River Bay. *Transactions of American Fisheries Society*, 130(5): 796–808.

Finklestein, S. A., M. C. Peros, and A. M. Davis 2005. Lake Holocene paleoenvironmental change in a Great Lakes coastal wetland: integrating pollen and diatom data sets. *Journal of Paleolimnology*, 33(1): 1–12.

Fox, G. A. 2001. Wildlife as sentinels of human health effects in the Great Lakes/St. Lawrence Basin. *Environmental Health Perspectives*, 109(Supp. 6): 853–861.

Glooschenko, V., J. H. Archibold, and D. Herman. 1988. The Ontario wetland evaluation System: replicability and bird habitat selection. In Hook et al. (eds.) *The Ecology and*

Management of Wetlands, Vol. 2, pp. 115–127. Portland, OR: Timber Press and Croom Helm Ltd., Publishers.

Geis, J. W. 1985. Environmental influence on the distribution and composition of wetlands in the Great Lakes. In H. H. Prince and F. M. D'Itri (eds.) *Coastal Wetlands*, pp. 15–31. Chelsea, MI: Lewis Publishers.

Gottgens, J. F., B. F. Swartz, R. W. Kroll, and M. Eboch. 1998. Long-term GIS-based records of habitat changes in a Lake Erie coastal marsh. *Wetlands Ecology and Management*, 6(1): 5–17.

Gruenwald, G. 1990. Recommendations of the Great Lakes Wetlands policy Consortium. In J. Kusler and R. Smardon (eds.) *Wetlands of the Great Lakes; Protection and Restoration Policies: Status of the Science*, pp. 17–18. Berne, NY: Association of Wetland Managers.

Hartig, J. H. and R. L. Thomas. 1988. Development of plans to restore degraded areas in the Great Lakes. *Environmental Management*, 12: 327–347.

Hartmannn, H. C. 1990. Climate change impacts on Laurentian Great Lakes levels. *Climatic Change*, 17: 49–68.

Hecnar, S. J. 2004. Great Lakes wetlands as amphibian habitats. *Aquatic Ecosystem Health and Management*, 7(2): 289–304.

Herdendorf, C. E. 2004a. Morphometric factors in the formation of Great Lakes Coastal wetlands. *Aquatic Ecosystem Health & Management*, 7(2): 179–198.

Herdendorf, C. E. 2004b. Great Lakes estuaries. *Estuaries*, 13(4): 493–503.

Herdendorf, C. E. and S. M. Hartley (eds.). 1980. *A Summary of the Knowledge of Fish and Wildlife Resources of Coastal Wetlands of the Great Lakes of the United States. Vol. 1: Overview.* U.S. Fish and Wildlife Service, Twin Cities, MN, 468 pp.

Holmes, J. A. and T. H. Whillans. 1984. Historical review of Hamilton Harbour Fisheries. *Canadian Technical Report Fish and Aquatic Science*, No. 1257, 117 pp.

Hook, T. O., N. M. Eagan, and P. W. Webb. 2001. Habitat and human influences on larval fish assemblages in northern Lake Huron coastal marsh bays. *Wetlands*, 21(2): 281–291.

Host, G. E., J. Schuldt, J. H. Ciborowski, L. B. Johnson, T. Hollenhorst, and C. Richards. 2005. Use of GIS and remotely sensed data for a priori identification of reference areas for Great Lakes coastal ecosystems. *Journal of Remote Sensing* 26(23/10): 5325–5342.

Hudson, P. L. and C. A. Bowen. 2002. First record of *Neoergasilus Japonicus* (*Poecilostomatoida: Ergasilidae*), a parasitic copepod new to the Laurentian Great lakes. *Journal of Parasitology* 88(4): 657–663.

Hummel, M. 1981. Wetland wildlife values. In A. Champagne (ed.) *Proceedings of the Ontario Wetlands Conference*, pp. 27–32. Toronto: Federation of Ontario Naturalists.

Hwang, S. and R. Health. 1999. Zooplankton bacterivory at coastal and offshore sites of Lake Erie. *Journal of Plankton Research*, 21(4): 699–719.

Jaworski, E: C. N. Raphael, P. J. Mansfield, and B. B. Williamson. 1999. *Impact of Great Lakes Water Level Fluctuations on Coastal Wetlands*. National Technical Information Service, Springfield VA as PB-296403, Institute of Water Research, Michigan State University.

Jean, M. and A. Bouchard. 1993. Riverine wetland vegetation: Importance of small-scale environmental variation. *Journal of Vegetation Science*, 4(5): 609–620.

Jean, M. and A. Bouchard. 1991. Temporal changes in wetland landscapes of a section of the St. Lawrence River, Canada. *Environment Management*, 15(2): 241–256.

Keddy, P. and L. H. Fraser. 2000. Four general principals for the management and conservation of wetlands in large lakes: The role of water levels, nutrients, competitive hierarchies and centrifugal organization. *Lakes and Reservoirs: Research and Management*, 5(3): 177–185.

Keddy, P. and A. A. Reznicek. 1986. Great Lakes vegetation dynamics; the role of fluctuating water levels and buried seed. *Journal of Great Lakes Research*, 12(1): 25–36.

Keough, J. R., M. E. Sierszen, and C. A. Hagley. 1996. Analysis of a Lake Superior coastal food web with stable isotope techniques. *Limnology and Oceanography*, 41(1): 136–146.

Klarer, D. M. and D. F. Millie 1992. Aquatic macrophytes and algae at Old Woman Creek estuary and other Great Lake coastal wetlands. *Journal of Great Lakes Research*, 18(4): 622–633.

Klarer, D. M. and D. F. Millie. 1994. Regulation of phytoplankton dynamics in a Laurentian Great lakes estuary. *Hydrobiologia*, 286(2): 97–108.

Kreiger, K. A. 2003. Effectiveness of a coastal wetland in reducing pollution in a Laurentian Great Lake: hydrology, sediment, and nutrients. *Wetlands*, 23(4): 778–791.

Kreiger, K. A., D. A. Klarer, R. T. Heath and C. A. Herdendorf (eds.). 1992. Special Issue on Coastal wetlands of the Laurentian Great lakes. *Journal of Great Lakes Research*, 18(4): 521–768.

Kusler, J. and R. C. Smardon. 1990. Introduction and Key Recommendations. In J. Kusler and R. Smardon, (eds.) *Wetlands of the Great Lakes: Protection and Restoration Policies; Status of the Science*, pp. 2–5. Berne, NY: Association of Wetland Managers.

Leach, J. H. 1995. Non-indigenous species in the Great Lakes: were colonization and damage to ecosystem health predictable? *Journal of Aquatic Ecosystem Stress and Recovery*, 4(2): 117–128.

Leslie, J. K. and C. A. Timmins. 1991. Distribution and abundance of young fish in Chenal Ecarte and Chematogen Channel in the St. Clair River delta, Ontario. *Hydrobiologia*, 219(1): 135–142.

Liston, C. R. and S. Chubb. 1985. Relationships of water level fluctuations and Fish. In H. H. Prince and F. M. D'Itri (eds.) *Coastal Wetlands*, pp. 121–140. Chelsea, MI: Lewis Publishers.

Loftus, K. K., R. C. Smardon, and B.A. Potter. 2004. Strategies for the Stewardship and conservation of Great Lakes Coastal wetlands. *Aquatic Ecosystem Health and Management*, 7(2): 305–330.

Lougheed, V. L. and P. Chow-Fraser. 2001. Development and use of a zooplankton index for wetland quality in the Laurentian Great Lakes Basin. *Ecological Applications*, 12(2): 474–486.

Lundholm, J. T. and W. L. Simser. 1999. Regeneration of submerged macrophyte populations in a disturbed Lake Ontario coastal marsh. *Journal of Great Lakes Research*, 25(2): 395–400.

Lyon, J. G., R. D. Drobney, and C. E. Olson. 1986. Effects of Lake Michigan water levels on wetland soil chemistry and distribution of plants in the straits of Mackinac. *Journal of Great Lakes Research*, 12(3): 688–700.

Mackenzie, R. A. and J. L. Kaster. 2004. Temporal and spatial patterns of insect emergence from a lake Michigan coastal wetland. *Wetlands*, 24(3): 688–700.

Mills, E. L., J. H. Leach, J. T. Carlton, and C. L. Secor. 1993. Exotic species in the Great Lakes: a history of biotic crises and anthropogenic introductions. *Journal of Great Lakes Research*, 19: 1–54.

Munawar, M. (ed.). 2004. Special Issue: Coastal Wetlands of the Laurentian Great Lakes; Health, Integrity and Management. *Aquatic Ecosystem Health and Management*, 7(2): 169–333.

Magnuson, J. J., K. E. Webster, R. A. Assel, C. J. Browser, P. J. Dillon, J. G. Eaton, H. E. Evans, E. J. Fee, R. I. Hall, L. R. Mortsch, D. W. Schindler and F. H. Quinn. 1998. Potential effects of climate changes on aquatic systems: Laurentian Great lakes and Precambrian shield region. *Hydrological Sciences*, 11(8): 825–871.

Manny, B. A. 1984. Potential impacts of water diversions on fishery resources in the Great lakes. *Fisheries*, 9(5): 19–23.

Maynard, L. and D. Wilcox. 1997. *Coastal Wetlands*, Background paper for the State of the Lakes Ecosystem Conference 1996. Environment Canada and US Environmental Protection Agency EPA 905-R-95-015b

McLaughlin, D. B. and H. J. Harris. 1990. Aquatic insect emergence in the Great Lakes marshes. *Wetlands Ecology and Management*, 1(2): 111–121.

McNicholl, M. K. 1985. Avian wetland habitat functions affected by water level fluctuations. In H. H. Prince and F. M. D'Itri (eds.) *Coastal Wetlands*, pp. 87–98. Chelsea, MI: Lewis Publishers.

Mitsch, W. J. and B. C. Reeder. 1997. Nutrients and hydrologic budgets of a Great Lakes coastal freshwater wetland during a drought year. *Wetlands Ecology and Management*, 1(4): 211–222.

Mortsch, L., J. Ingram, A. Hebb, and S, Doka (eds.). 2006. Great lakes Coastal Wetland Communities: Vulnerability to Climate Change and Response to Adaptation Strategies, Final Report. Coastal Zone Project A592-A. Faculty of Environmental Science. University of Waterloo, Canada.

Mortsch, L. D. and H. F. Quinn 1996. Climate scenarios for Great Lakes ecosystem studies. *Limnology and Oceanography*, 41(5): 903–911.

National Wetlands Working Group. 1981. *Wetlands of Canada*. Map, Ecological Land Classification Series No. 14, 1:7,500,000. Ottawa: Environment Canada, Lands Directorate.

Niemi, G., D. Wardrop, R. Brooks, S. Anderson, V. Brady, H. Paerl, C. Rakocinski, M. Brouwer, B. Levinson, and M. McDonald. 2004. Rationale for a new generation of indicators for coastal waters. *Environmental Health Perspectives*, 11(9): 979–986.

Patterson, N. 1992. Great Lakes Protection and Rehabilitation: Rising to the Challenge. *Great Lakes Wetlands*, 3(4): 1–3, 11.

Patterson, N. J. and T. H. Whillans. 1985. Human interference with natural water level regimes in the context of other cultural stresses on Great lakes wetlands. In H. H. Prince and F. M. D'Itri (eds.) *Coastal Wetlands*, pp. 209–251. Chelsea, MI: Lewis Publishers.

Pijanowski, B. C., B. Shellito, S. Pithadia, and K. Alexandridis. 2002. Forecasting and assessing the impact of urban sprawl in coastal watersheds during eastern Lake Michigan. *Lakes and Reservoirs: Research and Management*, 7(3): 271–285.

Price, S. J., D. R. Marks, R.W. Howe, J. M. Hanowski, and G. J. Niemi. 2005. The importance of spatial scale of conservation of anuran populations in coastal wetlands of Western Great Lakes, USA. *Landscape Ecology*, 20(4): 441–454.

Prince, H. H. 1985. Avian communities in controlled and uncontrolled Great lakes wetlands. In H. H. Prince and F. M. D'Itri (eds.) 1985. *Coastal Wetlands*, pp. 99–119. Chelsea, MI: Lewis Publishers.

Prince, H. H., P. I. Padding, and R. W. Knapton 1992. Waterfowl use of the Laurentian Great Lakes. *Journal of Great Lakes Research*, 18: 673–699.

Prince, H. H. and F. M. D'Itri (eds.). 1985. *Coastal Wetlands*. Chelsea, MI: Lewis Publishers.

Reeder, B. C. 1994. Estimating the role of autotrophs in nonpoint source phosphorous retention in a Laurentian Great Lakes coastal wetland. *Ecological Engineering*, 3(2): 161–169.

Ridgley, R. 1985. *Ontario Commercial Fishing Industry, Statistics on Landings*. Toronto, Canada: Ontario Ministry of Natural Resources.

Sager, P. E., S. Richman, H. J. Harris, and G. Fewless. 1985. Preliminary observations on the seiche-induced flux of carbon, nitrogen and phosphorus in a Great Lakes coastal marsh. In H. H. Prince and F. M. D'Itri (eds.) *Coastal Wetlands*, pp. 59–68. Chelsea, MI: Lewis Publishers.

Schloesser, D. W., J. L. Metcalfe-Smith, W. P. Kovalak, G. D. Longton, and R. D. Smithee. 2006. Extirpation of freshwater mussels (*Bivalia: Unionidae*) following the invasion of Dreissenid mussels in an interconnecting river of the Laurentian Great Lakes. *The American Midland Naturalist*, 155(2): 307–320.

Shear, H., N. Stadler-Salt, P. Bertram, and P. Horvatin. 2003. The development and implementation of indicators of ecosystem health in the Great Lakes Basin. *Environmental Monitoring and Assessment*, 88(1–3): 119–151.

Smith, J. B. 1991. The potential impacts of climate change on the Great Lakes. *Bulletin of the American Meteorological Society*, 72(1): 21–28.

Smith, P. R. G., V. Glooschenko, and D. A. Hagen. 1991. Coastal wetlands of the Canadian Great lakes; inventory, current conservation initiatives and patterns of variations. *Canadian Journal of Fisheries and Aquatic Sciences*, 48(8): 1581–1594.

Stephenson, T.D. 1990. Fish reproductive utilization of coastal marshes of Lake Ontario near Toronto. *Journal of Great Lakes Research* 16(1): 71–81.

Tilton, D. L. and B. R. Schwegler. 1978. The values of wetland habitat in the Great Lakes basin. In P. E. Greeson, J. R. Clark, and J. E. Clark (eds.) *Wetland Functions and Values: The State of Our Understanding*, pp. 267–277. Minneapolis, MN: American Water Resources Association.

Tip of the Mitt Watershed Council. Undated (a). *Michigan Wetlands: Yours to Protect.* Conway Michigan: Tip of the Mitt Watershed Council, 16pp. appendices.

Tip of the Mitt Watershed Council. Undated (b). *Our Valuable Wetland Resource.* Conway Michigan: Tip of the Mitt Watershed Council, VHS color video 27 minutes

Tip of the Mitt Watershed Council. Undated (c). *Citizens the Essential Link in Wetland Protection.* Conway Michigan: Tip of the Mitt Watershed Council, VHS color video 28 minutes.

Tip of the Mitt Watershed Council. Undated (d). *Wetlands of the Great Lakes.* Conway Michigan: Tip of the Mitt Watershed Council, VHS color video 13 minutes.

Tip of the Mitt Watershed Council. Undated (e). *Wetland Regulation in Michigan: The Citizens Role.* Conway Michigan: Tip of the Mitt Watershed Council, VHS color video 27 minutes.

Uzarski, D., T. Burton, and J. Genet. 2004. Validation and performance of an invertebrate index of biotic integrity for Lakes Huron and Michigan fringing wetlands during a period of lake level decline. *Aquatic Ecosystems Health and Management*, 7(2): 269–288.

Wang, N. and W. J. Mitsch. 1998. Estimating phosphorus retention of existing and restored coastal wetlands in a tributary watershed of the Laurentian Great Lakes in Michigan, USA. *Wetlands Ecology and Management*, 6(1): 69–82.

Wei, A. and P. Chow-Fraser. 2005. Untangling the confounding effects of urbanization and high water level on the cover of emergent vegetation in Cootes Paradise Marsh, a degraded coastal wetland of lake Ontario. *Hydrobiologia*, 54(1): 1–9.

Whillans, T. H. 1987. Wetlands and aquatic resources. In W. C. Healy and R. R. Wallace (eds.) Canadian Aquatic Resources, *Canadian Bulletin of Fisheries and Aquatic Sciences*, 215: 321–256.

Wilcox, D. 1993. Effects of water level regulation on wetlands of the Great Lakes. *Great Lakes Wetlands* 4: 1–2, 11.

Wilcox, D. 2004. Implications of hydrologic variability on the succession of plants in Great lakes wetlands. *Aquatic Ecosystem Health And Management* 7(2): 223–231.

Wilcox, D., J. E. Meeker, P. L. Hudson, B. J. Armitage, M. G. Black, and D. G. Uzarski. 2002. Hydrologic variability and application of index of biotic integrity metrics to wetlands: A Great Lakes evaluation. *Wetlands*, 22(3): 588–615.

Wilcox, D. and T. H. Whillans, 1999. Techniques for restoration of disturbed coastal wetlands of the Great lakes. *Wetlands*, 19(4): 835–857.

Wilcox, K. L., S. A. Petrie, L. A. Maynard, and S. W. Meyer. 2003. Historical distribution and abundance of *Phragmites australis* at Long Point, Ontario. *Journal of Great lakes Research*, 29(4): 664–680.

Williams, D. C. and J. G. Lyon. 1991. Use of geographic information system database to measure and evaluate wetland changes in the St. Marys River, Michigan. *Hydrobiologia*, 219(1): 22–26.

Whyte, R. S., D. A. Francko, and D. M. Klarer. 1997. Distribution of floating-leaf macrophyte *Nelumbo lutea* (American water lotus) in a coastal wetland in lake Erie. *Wetlands*, 17(4): 567–573.

Zedler, J. and S. Kercher. 2004. Causes and consequences of invasive plants in wetlands: Opportunists, and outcomes. *Critical Reviews in Plant Sciences* 23(5): 431–457.

Web Sites Used

Alliance for the Great Lakes. 2004. An advocate Field Guide to Protecting Lake Michigan.
    http://www.lakemichigan.org/field_guide/habitat_wetlands.asp
Environmental Canada. 2005. Great Lakes Wetlands Conservation Action Plan. http://www.
    on.ec.gc./wildlife/wetlands/glwap-e.cfm
Great Lakes Commission. 2004. Great Lakes Coastal Wetlands Consortium. http://www.glc.
    org/wetlands/
Great Lakes Aquatic Habitat Network & Fund. 2005. Great Lakes Wetlands/Great Lakes
    Directory. http://www.greatlakesdirectory.org/wetlands/wetlands.htm
GLIN. 2006. Wetlands in the Great lakes Region. http://www.great-lakes.net/envt/air-land/
    wetlands.html
Bird Studies Canada. Undated. The Great lakes Monitoring Program. http://www.bsc-eoc.
    org/mmpmain.html
US Geological Survey. 2004. Effects of Global Climate Change on Great Lakes wetlands
    http://www.nrel.colostate.edu/projects/brd_global_change/proj_31_great_lakes.html

Chapter 8
Estuaries on the Edge, Yucatan Peninsula, Mexico

Introduction

This case study addresses the different roles of NGOs in management to two of the most important coastal estuaries in the Yucatan Peninsula, Mexico. These two biosphere reserves sustain an estimated 24,000 Caribbean flamingos that migrate between Ría Celestún (wintering site) and Ría Lagartos (breeding site). Both sites are also wintering sites for thousands of migratory waterfowl, which explains much of the international interest. Both sites are at the Neotropical edge and illustrate management issues typical of subtropical North and Central America. Although the Mexican government agencies play the dominant role in day-to-day management of these two biosphere reserves; international, national, and regional NGOs play major roles in research management decisions as well as management support. It should be noted that Amigos de Sian Ka'an, a regional Mexican NGO, collaborates in the management of the Sian Ka'an Biosphere Reserve on the Caribbean side of the Yucatan Peninsula, but the author is much more familiar with the other two estuaries, Ría Celestún and Ría Lagartos.

Regional Context

The Yucatan Peninsula branches off the eastern coast of Mexico to form the southern coast of the Gulf of Mexico (Fig. 8.1). Mexico, Belize, and Guatemala all possess territory on the peninsula. The peninsula's location and a historic lack of transportation and communication links kept the Yucatan well isolated from mainland Mexico until relatively recently. The isolation accounts for many of the peninsula's natural and cultural differences from mainland Mexico. Transportation facilities on the northern coast of the Yucatan Peninsula have modernized rapidly with the growth of several economic activities. In the past, henequin or sisal was the most profitable product. More recently, however, tourism has overtaken all

R.C. Smardon, *Sustaining the World's Wetlands*,
DOI 10.1007/978-0-387-49429-6_8, © Springer Science+Business Media, LLC 2009

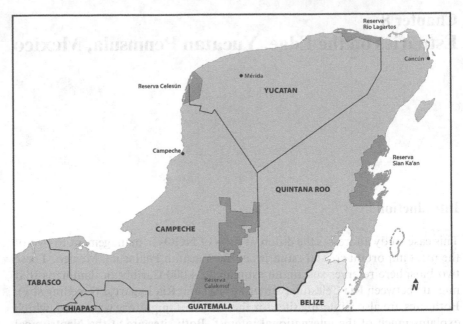

Fig. 8.1 Map of the Yucatan Peninsula and major biosphere reserves. Drawn by Samuel Gordon and adapted from Cartographic Laboratory, Department of Human Ecology, CINVESTAV, Merida, Yucatan, Mexico

other activities, with the emergence of Cancun as a world-class beach resort.

Most of the residential population on the Yucatan Peninsula concentrates in several main cities. In the northern Yucatan these cities are Mérida, Cancun, Progresso, Tizimin, and Valladolid. Mérida is the largest city on the peninsula and has a population of 600,000. Merida's economy is primarily based on agriculture, commerce, and more recently tourism. Mérida has all of the typical amenities of a city of its size. Cancun is the tourist mecca of Mexico's Caribbean coast. Cancun also acts as the gateway for many tourists who visit sights farther inland on tours from the larger hotels, or on their own.

Progresso is a port city to the north of Mérida. It has a small population (30,000) and depends on cargo transshipment as its economic mainstay. There is a dock facility that can accommodate ocean cruise ships. This dock has been extended so that it can accommodate larger cargo ships. Tizimin is a smaller city located between Merida and Cancun. The inhabitants of Tizimin rely on farming and ranching as their primary economic support. This city sees some tourist activity as tourists pass on their way to Cancun, Mérida, or Ría Lagartos. Valladolid is another small agricultural-based city located between Mérida and Cancun that is beginning to see an increase in tourism due to its proximity to Chechen Itza.

Mexican Coastal Zone Management History

Although Mexico has a huge coastline, much of it remains underpopulated. Harsh coastal climates, disease, lack of protection from coastal storms, and lack of freshwater historically limited Mexican coastal communities (Chavarria 1988). Of all the 17 native populations of Mexico, only the Maya used the coastline to any degree. Today, only 12.7% of the Mexican population lives on the coast, and Mexico's three most important economic centers, Mexico City, Guadalajara, and Monterey, are all located far inland.

A reason for Mexico's past lack of interest in ecological management of the coastal zone is the fact that it does not want to decrease any opportunities for its four main coastal activities; oil and gas extraction, fisheries, tourism, and marine transportation (Valdes 1988). Recently there have been new efforts to acquire key databases for coastal zone planning at Centro de Investigacions Y Estudios Avandos CINVESTAV (Euan-Avila and Witter 2002, Clark 1991, Rivera-Monroy et al. 2004, Yanez-Arancibia et al. 2004). Mexico has historically let these industries manage themselves, and such a policy has resulted in use conflicts and lack of ecological consideration in coastal development decisions.

Legal constructs that would be useful for coastal zone management (CZM) began in Mexico via an early Spanish influence that coastal areas are lands of public trust (Chavarria 1988). This concept was further referenced in the Independence Act of 1821 and further articulated in the National Constitution of 1917 in Article 27, The General Law on National Welfare and Public Trust (Chavarria 1988). Foreign natural area programs began to exert influence on Mexican policy at this time and many national parks and forestry reserves were designated under the administration of Lazaro Cardenas. These areas were given protective legislation in 1934 under article 5 of the Second Law of Forestry in the Mexican Constitution. In 1958 the International Union for the Conservation of Nature and Natural Resources (IUCN) prompted Mexico to begin a new approach to natural area conservation and protection by instituting the International Commission on National Parks.

In 1961 the Mexican Institute for Renewable Natural Resources studied the status of Mexican protected areas and recommended redefining them based on their goals, recreation facilities, flora and fauna, outstanding characteristics, and technical requirements. Natural areas were then assigned designations as national parks, natural reserves, natural monuments, and pristine region reserves (Bourdelle 1956). In 1971 two laws aimed at coastal protection were passed and addressed "public health and pollution prevention and control matters" (Chavarria 1988). Unfortunately, these laws did not have the power necessary for strict enforcement policies (Chavarria 1988).

In 1982 the Environmental Protection Law addressed "problems of marine ecosystem protection and recognition of more restrictive use within some of the existing protected coastal areas" (Chavarria 1988). Mexican policy on environmental protection though was still low on the political agenda. When the de la Madrid administration took over in late 1982 it expressed an interest in environmental matters and in 1988 introduced a revised version of environmental quality regulations which strengthened Mexico's environmental policy.

While Mexico has reached a stage where it is beginning to consider ecological concerns in coastal management, it has only recently addressed the need for cooperation between its federal agencies and the need for a comprehensive plan for its coastal resources (Silva and Delvestre 1986). This is due in part to Mexico's recent adoption of sectoral planning (Chavarria 1988).

The Ría Lagartos and Ría Celestún Preserves were designated Protected Wildlife Refuges in 1979 and then upgraded to special biosphere reserves in 1988 under Mexico's Environmental Protection Law of 1982 (Chavarria 1988).

Current Coastal Zone Planning in Mexico

Coastal zone management in Mexico is primarily accomplished by means of "national sectoral planning, a concept in which each (economic) activity is considered a separate category deserving specific and separate development planning" (Valdes 1988:3). Valdes goes on to assert that this type of management is too single activity minded and "fails to make necessary next step of integrating plans of these sectors into a coherent coastal development plan" (Valdes 1988:3).

Valdes (1988) also notes that the numerous specialized agencies, which deal with managing activities in the coastal zone often, lack coordination and sometimes their goals are contradictory. This all leads to a rather ineffective coastal zone management program with the inception of the Ministry of Urban Development and Ecology (SEDUE but now called SEMARNAP) in 1982. However, coastal zone planning in Mexico began to take a more systems-oriented approach (Chavarria 1988). Methods involving "sustained use", ecosystem management, and integration of regional priorities are slowly beginning to replace sectoral management policies.

Power for most of the activity in Mexico's coastal zone is wielded by the federal government through agencies such as Secretary of Environmental, Natural Resources and Fisheries (SEMARNAP), Ministry of Fisheries (SEPES), and Ministry of Tourism (SECTUR) (Valdes 1988). State and local governments often do not have the power or funds to enact their own programs. Government at the state level though is beginning to ask for, and receive; more power and local government considerations are beginning to be heard (Valdes 1988).

One of the primary reasons for Mexico to begin managing its coastal zone has been the rapid economic activity in this zone. Most of these activities are resource consumptive and have been exploiting the coastal resources for decades. Recent worry over the decline in the ocean's ecological and economic importance, however, has made the management of these activities a primary concern in many countries. Mexico, in order to protect both its economic and its ecological resources, has become quite active in coastal management (Euan-Avila 2002, Clark 1991, Rivera-Monroy et al. 2004, Yanez-Arancibia et al. 2004).

Conflicting resource uses and their impact of coastal estuarine wetland complexes are some of the reasons why we have seen increasing activity of international, national, and regional NGOs in Mexico as well as academic institutions. Many of the NGO actors can be characterized by their roles. For instance CINVESTAV (Centro de Investigacions y Estudios Avanzados) is the research university that ends up doing many of the field studies on wetland systems as well as impacts studies within the coastal zone. Program for Nature (PRONATURA Peninsula de Yucatan AC) has been involved with assisting with both ecotourism development and reserve management support both at Ría Celestún and Ría Lagartos as well as other reserves like Calakmul in the Campeche. Ducks Unlimited (DUMAC) has been involved in on-site research on migratory waterfowl both in Ría Celestún, where they have a research station, and in other coastal sites largely with funding from the North American Waterfowl Treaty (NAWT). The NAWT also supports research activities of CINVESTAV and PRONATURA. Also involved in technical assistance is the USDI, Fish and Wildlife Service, and the National Park Service's Office of International Affairs. Other International NGOs like Wetlands International, IUCN, Audubon Society, and the Nature Conservancy support PRONA-TURA's activities for three parks with the "Parks In Peril" Program (Andrews et al. 1998). Biosphere Management Plans have been formulated for many of these two biosphere reserves with funds provided by the World Bank. So there are many actors – international, national, and regional – but little coordination or collaboration at times.

The Case Study Areas Ría Celestún and Ría Lagartos

The two biosphere reserve wetland complexes at the edge of the Yucatan Peninsula are treated within this case study because they are linked in function. They are both used by the Caribbean flamingo, which migrates back and forth between these two areas. Both areas are stopovers or destinations for migratory waterfowl and support significant water birds. They also share similar climate, geomorphology, geology, soils, hydrology, and to some degree flora and fauna. Key background documents for the case studies include Andrews et al. (1998), Fraga et al. (2006), Moan (1992, ParksWatch (2002), and Conroy (1998).

Climate

The Koppen climate classification system lists the northern part of the Yucatan as tropical wet/dry, AW (Wilson 1980). This area receives approximately

100–150 cm of rain annually, with most of the rain falling during the wet season between June and November (Murgia 1989b). Rainfall on the peninsula is heaviest in the south. Temperatures vary little, remain between 23 and 28°C (Wilson 1980). The warm climate and the timing of the dry season with the peak tourist season in the temperate northern countries help make the coasts of the Caribbean Yucatan popular tourism destinations. North winds (nortes) affect the Rio Lagartos area from November to March (Anon 1989a). Hurricane season lasts from June to November (Murgia 1989b, Anon 1989a) and develop at 13° north latitude. When warming has started in the insular region pf the Antilles, hurricanes are formed, some of long duration and extraordinary power, if formed in the months of August, September, and October. Some cross the Yucatan Peninsula through Cozumel or Cancun, or Chetamal or through the north coast from where they reach the states of Tamaulipas, Veracruz, and the southwest coast of the United States.

Geomorphology

From the Maya mountains in Belize, surface elevation decreases as the Yucatan Peninsula stretches northeast. The landform is relatively flat and the northern tip is only slightly above sea level (Wilson 1980, Valdes 1988). Due to the peninsula's predominately limestone composition there is little surface water (Wilson 1980, Wilson and Williams 1987). What little there is exists in sinkholes or coastal lagoons (Wilson 1980, Wilson and Williams 1987, Perry 1991). There are eight large lagoons on the coast of the northern Yucatan Peninsula, of which Ría Celestún and Ría Lagartos are the largest two (Wilson and Williams 1987, Correa et al. 1989).

A coastal barrier island called the Ría Lagartos Peninsula borders the northern coast protecting inland areas from the physical forces of the gulf and storms. The same type of formation occurs on the gulf side of Ría Celestún. This land is composed of white unconsolidated calcareous beach sand, which is deposited by ocean and gulf currents along the coast (Sauer 1967, Wilson 1980, Wilson and Williams 1987, Perry 1991). This sand is derived from the erosion of coral reefs parallel to the eastern coast of the Yucatan Peninsula (Valdes 1989). The sand is deposited by a westward long shore current (Perry 1991). This movement of sand seems to have reached an equilibrium, the result of which is that the Ría Lagartos Peninsula has not changed much in centuries (Ibid.). The same general phenomena can be said for the Ría Celestún.

Northern Yucatan Hydrology

Seasonal rains deposit approximately 1000–1500 mm annually on the northern Yucatan coastline (Murgia 1989b, Correa et al. 1989, Wilson and William 1987). Although this is considered a moderate amount of rain, the seasonality of rain episodes combined with high porosity of the area's limestone geology explains why there is little surface water on the northern portion of the peninsula. What little that exists is found in sinkholes or coastal lagoons.

L. J. Cole (1910) suggests, and it has been proven, that because the higher interior zones in the southern portion of the peninsula receive more rain, this produces a hydrostatic pressure, which causes subsurface water to flow northward toward the lower coasts. This freshwater aquifer flows through a system of rock fractures and is filtered through the calcareous ground occurring at a depth no greater than 12.5 m in the northern Yucatan (Cole 1910). Closer to the coast, fresh groundwater can be found in sinkholes at depths of 1 m or less below ground level (Ibid.) This water flows out of springs at or below sea level (Wilson 1980). It is believed that up to half this groundwater aquifer is confined near the northern coast and is protected from saltwater intrusion by a thin nearly impermeable calcareous layer (Perry 1991). This confining layer (called a coastal aquitard) is believed to form the landward edge of the barrier beach lagoon system that protects the northern Yucatan coast. It is also believed that the edges of this layer move with fluctuations in mean sea level (Lee 1995, Sklar and Browder 1998).

The sources of freshwater found in the northern Yucatan which are readily available to humans and wildlife are cenotes, aguadas, and petenes (Wilson 1980, Wilson and Williams 1987). Cenotes are sinkholes caused by the action of subterranean water collapsing the weak limestone surface (see Fig. 8.2a). This leaves natural steep-walled open wells containing freshwater. The size and depth of these wells varies from a few meters to over 60 m (Wilson 1980). Aguadas are shallow pools of water formed by the action on surface limestone. The sides are usually more gentle and sloped than a cenote (Fig. 8.2b).

Petenes are pools of freshwater forced from the limestone by the subsurface hydraulic gradient (Fig. 8.2c). Petenes are distinct more for their vegetation than for their hydro-geologic structure, however, that affects the vegetation (Wilson 1980, Wilson and Williams 1987). Upwelling of freshwater forms Petenes, which are surrounded by dry land, or saline water. A petene is usually flooded during the rainy summer months. This flooding is not due to the amount of rain in that area, however, but due to the water pressure caused by heavier rainfall in higher inland areas to the south (Cole 1910, Wilson 1980, Wilson and Williams 1987). This pressure forces water from lower coastal sinkholes causing an island of freshwater in an area that is dry or inundated with saline water (Wilson 1980, Correa et al. 1989). This change in hydrology causes a shift in the area's vegetation. Such vegetation makes the petene area appear as mounds in the landscape because the petene's vegetation grows taller than surrounding saline-influenced vegetation (Lara-Dominguez et al. 2005, Rejmankava et al. 1995, Wilson 1980, Wilson and Williams 1987).

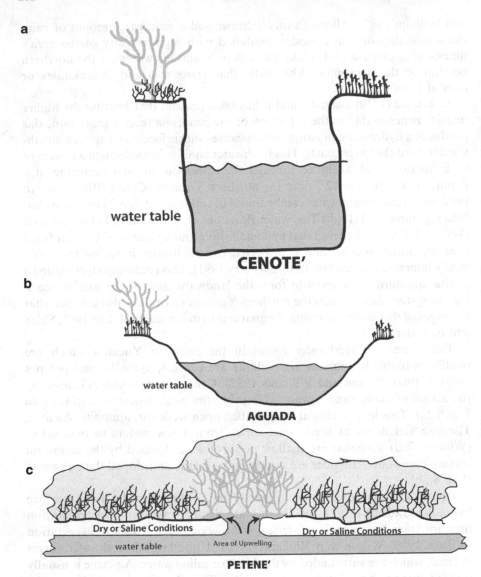

Fig. 8.2 Cross sections of cenote (a), aguada (b), and pentene (c). Drawn by Samuel Gordon and adapted from Scott Moan, 1992. Ecotourism in the Yucatan Peninsula; Ecotourism potentials for the Ria Lagartos Wildlife Reserve. Unpublished master's project, SUNY College of Environmental Science and Forestry, Syracuse, NY, p. 106

Northern Yucatan Vegetation

Wilson (1980) states that due to a declining range of rainfall from east to west on the peninsula, there is a gradation of forest types. It should be noted that the east side of the north coast is more humid and the west side is drier. Using J. S.

Beard's forest classification system (1944), Wilson classifies the northern Yucatan as primarily deciduous seasonal forest with some scrub forest areas present. The deciduous seasonal forest is typically composed of two low tree stories, one reaching 20 m, the second 3–10 m. Epiphytes are scarce, probably due to low rainfall (Wilson 1980). Scrub forest is primarily found closer to the coast where there is less rain. Trees reach approximately 7 or 8 m, with a dense understory of evergreen and deciduous shrubs (Ibid.). Much of the coastal vegetation is unique in its composition because of the environmental stresses along the Yucatan Peninsula (Rejmankova et al. 2007).

Ría Lagartos Preserve

The Ría Lagartos Wildlife Preserve was named for the many crocodiles that once were present in the lagoon's water. The preserve is located on the northern tip of the Yucatan Peninsula in the state of Yucatan, Mexico (see Fig. 8.3). The site is 210 km from Mérida. It lies between 21 26′ and 21 38′ northern latitudes and 87 30′ and 88 15′ eastern longitudes (Murgia 1989b).

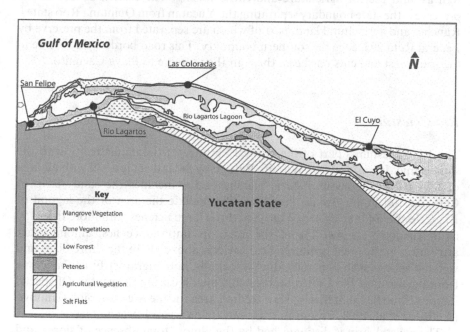

Fig. 8.3 Vegetative communities and towns within Ria Lagartos Wildlife Preserve. Drawn by Samuel Gordon and adapted from Scott Moan, 1992. Ecotourism in the Yucatan Peninsula; Ecotourism potentials for the Ria Lagartos Wildlife Reserve. Unpublished master's project, SUNY College of Environmental Science and Forestry, Syracuse, NY, p. 111

The preserve encompasses approximately 47,820 ha, with a coastal lagoon extending over 80 km in length (west–east) and varying between 0.02 and 3.5 km in width (north–south) (Murgia 1989b, Correa et al. 1989). The true boundaries encompass an area of 55,350 ha (Andrews et al. 1998). The Ría Lagartos lagoon is the second largest of the northern Yucatan's eight coastal lagoons and is approximately 285 km from the largest lagoon, Ría Celestún. The Ria Lagartos lagoon is approximately 3 m in depth at its deepest point (Correa and Boege 1989, Correa et al. 1989). The lagoon is separated from the Gulf of Mexico by a vegetated barrier island (Murgia 1989b, Correa et al. 1989). This barrier has two major inlets, the Boca San Felipe and the Ría Lagartos inlet. The Boca San Felipe is the natural mouth of the lagoon, while the Ría Lagartos inlet was constructed in order to allow the fishermen of Ría Lagartos town easier access to the gulf. These inlets are major points of water exchange between the lagoon and the Gulf of Mexico.

The preserve is bounded on the northern side by the Gulf of Mexico "Yucatan Channel". The preserve's western limits begin at San Felipe, a small fishing village that exists within the reserve. There are other villages in the preserve, Rio Lagartos, Las Colorados, and El Cuyo (see Fig. 8.3). There is also a salt extraction plant and a gravel mining facility within the preserve's boundaries as well as land use for agriculture and cattle raising. The eastern border of the preserve is the state boundary separating the Yucatan from Quintana Roo states. Ranches and agricultural land, part of which are separated from the preserve by Federal Ruta 295, edge the southern boundary. This road borders the reserve in the southwest and cuts northeast through the preserve to Playa Cacunito.

Ría Celestún

The Ría Celestún is part of the natural heritage of the municipality with the same name. It was declared Wildlife Refuge by federal act in 1979 with an area of 59,130 ha and about 79% is marsh, 11.5% includes the sandy strip of Celestún and Punta Arenas, less than 8% include the area of the lagoon of Ría Celestún and the estuary Yalton and the large petenes with more than 150 m in diameter are just 2% of the area. Its importance lies, among other attributes, in the great ornithological variety, above all, in the shore birds and wetland species such as herons, ducks, seagulls, and migratory birds that come from the north of the United States and Canada during the winter. Of special interest is the fact that this is also a feeding area for the Mexican pink flamingo (Andrews et al. 1998, Espino-Barrios and Baldassarre 1989a and b).

The coastal strip is distinguished by the almost total absence of slopes and topographic contrasts, besides the minimal undulations of small coastal dunes in the sandy strip. They are flat and low terrain, which allow infiltration of the saline mantles from the sea, the freshwater springs, and the rainfall to accumulate. The slope of the land within an average distance of 6 km inland is 0.013% in Ría Celestún; to the south of this strip the landforms become slightly undulated.

The length of the lagoon is approximately 22.5 km and its average width is approximately 1.25 km with a maximum of 2.24 km and a minimum of 0.48 km. Its area is 28.24 km². Its rectangular shape is long with a northeast–southwest orientation. The linkage with the Gulf of Mexico is through an opening (mouth), 0.46 km wide, located in the southernmost part of the lagoon. It consists of a tidal channel that goes through all its length. Its depth varies from 3.5 m near the mouth to 0.5 m in the inner area, with an average of 1.5 m, which is the navigable portion of the lagoon.

In the middle of its length, there is a bridge that connects the village of Celestún with Merida. Out on the tidal channel, there are very shallow areas that become exposed when the tide is low, showing the existence of microalga and seaweeds.

Ría Lagartos and Ría Celestún Preserves Hydrology

Lagoons are often grouped with estuarine systems in the study of coastal and wetland dynamics (Clark 1974). However, lagoon systems are often much more closed than conventional estuarine systems (Bianchi et al. 1999). Lagoons often receive more input from both terrestrial and ocean systems than they contribute back to these systems, and therefore act as sinks for organic material (Bianchi et al. 1999, Chmura et al. 2003, Lee 1995, Twilley et al. 1992). In this case, inputs from the Ria Lagartos lagoon come from upland systems in the lagoon's watershed and from the gulf through the dune system. These effects are enhanced by the substrata's porous composition. This closed system aspect of the lagoon makes it susceptible to human-induced impacts in the watershed of the lagoon (Alonzo 2007, Cable et al. 2002, Day et al. 1995, Mulholland et al. 1998, Sklar and Browder 1998, Young et al. 2005). Activities, which increase erosion and pollution or affect water inputs and circulation, will affect the preserve's biota and human activities.

While lagoon systems tend to be more closed than other systems, they are not self-contained (Clark 1974, Sprunt et al. 1988). Productivity in these systems is high and contributes to neighboring systems via seasonal or episodic communication and through the nutrient cycle and food web. This high productivity is important to aquatic species in the Gulf of Mexico as well as to terrestrial and avian species (Lee 1995, Sklar and Browder 1998, Twilley et al. 1992, Vega-Cendejas and Arregun-Sanchez 2001, Young et al. 2005).

The Rías Lagartos and Celestún lagoons exhibit varying degrees of salinity ranging from ocean levels (35 ppt) near the lagoon's mouth and largest inlet (Fig. 8.10) to hypersaline (100 ppt) in the easternmost section of the lagoon (Sprunt et al. 1989, Anon 1989a). High salinity is due to seasonal rainfall patterns and high temperatures which cause rapid evapotranspiration, few surface or subsurface inputs of freshwater, limited surface water exchange with the Gulf of Mexico (Correa et al. 1989, Sprunt et al. 1989), and the presence of a "confining layer" in the lagoon's water column which separates the saline water from the fresher water beneath (Perry 1991). Although there is some tidal flushing at the western end of the lagoon near the Boca San Felipe, water circulation in the lagoon is sluggish and occurs primarily via wind action (Murgia 1989b).

Water input into the lagoons is low but enters in several ways; water from the gulf enters as ground water through the barrier beach (Perry 1991): freshwater enters from the coast through fractures in the limestone subsurface pushed by a hydraulic gradient (Cole 1910, Wilson 1980, Perry 1991), and strong trade winds associated with the "nortes" forces water from the gulf into the lagoon through the inlets (Murgia 1989b). Surface runoff as a source of freshwater is probably insignificant and is limited to periods following high rainfall events (Sprunt et al. 1988).

These inputs of water act to lower water temperature and salinity, raise the lagoon's oxygen content, and help in cycling of organic material. It is also surmised that the freshwater flows help to keep the Boca in San Felipe from being closed by sand carried by the coastal current (Perry 1991). While these inputs help to recycle water in the lagoon, it continues to exhibit hypersaline conditions all year long (Sprunt et al. 1989). There are several petene areas in the reserve and one cenote near the town of Ría Lagartos and in Ría Celestún also. These areas provide freshwater for the four towns and the salt works within the preserve.

Ría Lagartos Vegetation

The biota of the Ría Lagartos Preserve is very diverse due to the Yucatan Peninsula's location at the overlap of the Neartic and Neotropical zones (Correa et al. 1989, Boo 1990). There are at least nine vegetative community types (mangrove, low thorny forest, coastal dune scrub, petenes, savanna, hammocks, cattails, sawgrass, and mudflats). For a typical cross section of vegetation units within the preserve see Fig. 8.4. Approximately 280 plant species have been identified in the Ría Lagartos Preserve (Correa et al. 1989; Murgia 1989a, Sprunt et al. 1989). Due to the preserve's unique environment

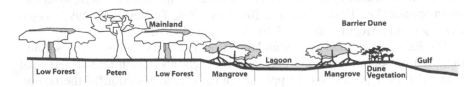

Fig. 8.4 Cross section of vegetation within Ria Lagartos Wildlife Preserve.
Source: Scott Moan, 1992. Ecotourism in the Yucatan Peninsula; Ecotourism potentials for the Ria Lagartos Wildlife Reserve. Unpublished master's project, SUNY College of Environmental Science and Forestry, Syracuse, NY, p. 113

some of these species are endemic to the area and have highly specialized adaptations (Anon 1989a, Comtreras-Espimosa and Warner 2004, Correa et al. 1989, Lara-Dominguez et al. 2005, Lugo et al. 1988, Murgia 1989a, Rejmankova et al. 1995).

The protection from the gulf's waves and tidal action offered by the Ría Lagartos Peninsula has allowed a low halophytic mangrove vegetation to establish itself on the shores of the Ría Lagartos lagoon (see Fig. 8.4). This vegetation unit fringes much of the lagoon and is composed predominately of red mangrove (*Rhizophora mangle*), black mangrove (*Avicennia germinans*), white mangrove (*Laguncularia racemosa*), and buttonwood (*Conocarpus erectus*). The mangrove is one of the principle sources of organic production in the lagoon (Bianchi et al. 1999, Navarrete and Olivia-Rivera 2002, Reyes and Merino 1991, Rivera-Monroy et al. 1998, Rivera-Monroy and Twilley 1996, Sprunt et al. 1989, Young et al. 2005). The particulate and dissolved organic matter from the mangrove is utilized by the second principal source of organic production in the preserve, microorganisms suspended in the water column. Both of these sources provide organic matter for the source of production in the lagoon, the benthic algal mat. This mat acts to store reduced organic matter and provides mineral nutrients for primary production and it is surmised that this is the primary recycling mechanism in the lagoon (see Fig. 8.5).

Fig. 8.5 Close up of mangrove root systems. Photo credit: Rick Newton

In places the mangrove vegetation blends into haloxerophytic dune vegetation composed of sea grape (*Coccoloba uvifera*), prickly pear cactus (*Opuntia dillenii*), sea rocket (*Cakile lanceolata*), and the threatened Chit (*Thrinax radiata*) and Kukaa palm (*Pseudophoenix sargentii*) (Sauer 1967). This coastal dune vegetative component of the region has been called the most floristically complex section of vegetation on the Gulf coast of Mexico (Sauer 1967). This dune vegetation is important because it stabilizes coastal landforms and establishes the conditions for a successional progression of vegetation types on the Ría Lagartos Peninsula. Dune vegetation stabilizes the substrate base, collects organic material, adds organic material to the developing soil layer, and provides protection from wind and salt spray.

Moving inland from the shore of the lagoon, glycophytic forest species displace mangrove vegetation along the decreasing salinity gradient (Sprunt et al. 1989). Mounds of taller vegetation mark the presence of freshwater petenes among the mangroves or low thorny forest typical to the preserve (Murgia 1989b, Wilson and Williams 1987). Species, which are important food sources for many waterfowl in the Yucatan's coastal lagoons, are muskgrass (*Chara* spp.) and widgeongrass (*Ruppia maritima*). These plants, however, are found only in limited quantities in the Ría Lagartos Preserve due to high salinity and wind scouring.

Wetland Vegetation Ría Celestún

The vegetative composition of this region, and most of Yucatan, is complex and different from the rest of the Gulf of Mexico due, in great extent, to its semi-arid climate. There is a mixture of coastal dune vegetation, petenes, savanna, marshes, reedbeds, flooded scrub forest, and deciduous scrub forest with cactus. The combination of barrier islands with coastal lagoons produces a mixture of dune vegetation and mangrove toward the mainland. Inland from the mangrove area is old growth forest and its resplendent variety. Its species are mixed with the mangrove in combination with dune scrub.

1. Coastal dune vegetation. Ría Celestún has some species that are endemic to the Yucatan Peninsula such as *Echites yucatanensis*and *Coccothrinax readii*. The distribution of the species follows the tolerance to some factors, mostly *edaficos*that follow a gradient from the beach to the mangrove (Espejel 1984). There are two large vegetation communities in this area: the pioneers and the scrub. The first community occurs in the beach areas and mobile dunes and is characterized by herbaceous plants and short shrubs tolerant to extreme environmental conditions such as high salinity, strong winds, sand movements, and high tides. Some of them are *Sesuvium portulacastrum*, sea rocket (*C. lanceolata*), *Suaeda linearis*, and Poinsettia (*Euphorbia buxifolia*) (Espejel 1984).

Within the scrub area species grow that are less tolerant. There are primarily shrubs or very ramified trees surrounded by clear areas with forbs and grasses. In Ría Celestún there is an interesting alternation of dune vegetation and mangrove. Within the profile of dune vegetation, the first 30 m of non-flooding soil has species such as *Bravaisia tubiflora*, *S. linearis*, *Cyperus artitatus*,logwood (*Haematoxylon campechianum*), and some *Schomburgkia tibicinis*, while the next 22 m of flooding soils has mangrove species (Espejel 1984). This happens successively along the barrier beach strip toward the estuary where red mangrove (*R. mangle*) is established.

In the disturbed areas such as the edges of the roads and popular beaches there are species such as balsam (*Croton punctatus*), *Flaveria linearis*, and *Ambrosia hispida*.

2. Mangrove vegetation. Around the lagoon, Trejo (1986) says that there exist two structures: the first represents the solid shell sandy soils of the strip that protects the lagoon and the second includes the inland forests. The mangrove in the inland edge has severe spatial limitations and it is also affected by competitive dune vegetation that has established all along the littoral even in the ledges of the lagoon. This association of black mangrove (*A. germinans*), *Batis maritima*, white mangrove (*L. racemosa*), buttonwood (*C. erectus*), sea purslane (*S. portulacastrum*), red mangrove (*R. mangle*), dwarf glasswort (*Salicornia bigelovii*), *Nopalea gaumeri*, and dropseed (*Sporolobus virginicus*) decreases toward the "mouth" of the lagoon. The size and foliage density of the trees decrease toward the point of the strip at which vegetation is scarcer, trees are shorter, and interspersed with the dune vegetation.

On the other side of the lagoon, toward the mainland, there are extensive mangrove forests. Black mangrove (*A. germinans*) exists within 7 km of the lagoon. These mangroves develop along protected sides and they are affected by the tide fluctuation that floods the soils. They are dominated, in the external areas, by red mangrove (*R. mangle*) with its supportive roots that allow it to survive on slightly consolidated and unstable sediments. Black mangrove (*A. germinans*) is established in the internal areas of low relief where the removal of the waters occurs much more slowly.

3. The vegetation of higher less flooded lands has been less studied. The flooded scrub forest that spreads in the western side of the lagoon develops in a narrow strip of recent sediments to the north and a belt of swamps of permanent flooding surrounds it on the west side. The plant community's characteristic of these areas is the *marshes*, natural pastures, *tintales*, and petenes, with species such as southern cattail (*Typha domingensis*), spikerush (*Eleocharis cellulosa*), saw grass (*Cladium jamaicense*), and common reed (*Phragmites communis*) (Espejel 1984).

In the transition area between limestone and the flood plains there are small-size trees such as *Byrsonima crassifolia*, *Crescentia cujete*, *Metopium brownei*, *Achras zapota*, and logwood (*H. campechianum*). The dominant species are buttonwood (*C. erectus*) and *B. tubiflora*. There are other species intermixed where ponds are formed with rainfall water: white water lily (*Nymphaea ampla*), water lettuce (*Pistia stratiotes*), water hyacinth (*Eichhornia*sp.), pennywort (*Hydrocotyle ranunculoides*), and water clover (*Marsilea mexicana*) among others (Espejel 1984).

4. The petene (see Fig. 8.2c) is a type of vegetation of medium forest that exists like an island in the mangrove. This association is located in the southeast of

Ría Celestún in the area known as Los Petenes in Campeche. Petene vegetation is usually related to a cenote or freshwater spring. The ecosystem balance is maintained between the contribution of freshwater and the intrusions of saline water from the bottom. Its main characteristic is the presence of species that are less common in areas with saline water such as *Manilkara achras*, gumbo limbo (*Bursera simaruba*), *Malvaviscus arboreus*, fig (*Ficus tecolutensis*), pond apple (*Annona glabra*), and *Sabal yapa*among others, and they can reach up to 25 m or more in height.

Inland from the lagoon, on limestone, there are extensive communities of old scrub forest with *cactaceascandelariformes* (Miranda, 1958) with species such as *Beaucarnea pliabilis*, *Thevetia ovata*, Frangipani (*Plumeria obtusa*), gumbo limbo (*B. simaruba*), beard grass (*Gymnopodium ovalifolium*), *Ceiba aesculifolia*, *Cordia dodecandra*, *Guaiacum sanctum*, *Enterolobium cyclocarpum*, *Lemaireocereus griseus*, and *N. gaumeri* among others.

Inside the lagoon, the underwater vegetation is very interesting, in that the macrobenthos cover approximately 80% of the lagoon. Unlike any other water bodies of the state, the microalga dominates (more than 70% of the biomass), whereas seaweeds are about 10% of the biomass. The most prominent species are muskgrass (*Chara fibrosa*) and *Batophora oerstedii* up estuary. In the middle and down estuary there are *Diplantera wrightii*, Manatee grass (*Syringodium filiforme*), and *Chaetomorpha linum*on the lagoon edges form very dense "carpets" (Herrera-Silveira 1987). Near the mouth, the dominant species is turtle grass (*Thalassia testudinum*).

Ría Lagartos Fauna

The Ría Lagartos Preserve has many diverse migratory and some endemic animal species (Murgia 1989b, Sprunt et al. 1989). This diversity is due to geographic location, diversity of habitats, and interspersion of vegetative communities. It is also a coastal area where saline and freshwater systems meet terrestrial systems. Adding to the abundance of wildlife in the preserve is the fact that the preserve acts as an island of refuge and a corridor for wildlife movement along the Yucatan's developing northern coast. Many species have been driven from their natural ranges by logging, agriculture, and residential development concentrated in the Ria Lagartos Preserve.

Mammals: There are over 20 species of mammals occurring in the preserve, of which the jaguar (*Felis onca*) is hunted for its pelt and alleged attacks on livestock and people. Also the once plentiful white-tailed deer (*Odocoileus osceola*), which has been extensively hunted in the region, is of special concern (Correa et al. 1989). Other species include the raccoon (*Procyon lotor*), gray fox (*Urocyon cinereoargenteus*), armadillo (*Dasypus novemcinctus*), and peccary (*Tayassu* spp.). None of these species are wholly wetland dependent. Also of concern is the Tapir population levels which have been heavily hunted in the past (Brooks et al. 1997). Bat populations are found throughout the peninsula in forest islands (Montel et al. 2006).

Reptiles: The lagoons and beaches of the Ría Lagartos Preserve offer habitat to over 20 species of reptiles (Barron and Correa 1989). Of special note, the hawksbill (*Eretmochelys imbricata*) and green sea turtles (*Chelonia mydas*) are endangered species, which use the coastal waters to feed and the preserve's beaches to lay their eggs. The Morelett's crocodile (*Crocodylus moreletti*) and the American crocodile (*Crocodylus acutus*), for which the preserve was named, are also present although their numbers are greatly reduced due to hunting and habitat loss (Hurley 2005, Platt and Thorbjarnason 2000). The American crocodile is extremely rare. Some amphibians such as the Rio Grande Leopard frog are of concern (IUCN 2000).

Birds: The most abundant and spectacular group of wildlife in the preserve is the avifauna. Over 260 species of birds occur in the region and some of these species – migratory and resident – use the lagoon (Andrews et al. 1998, Murgia 1989b, Anon 1989b, Barron and Correa 1990, Correa and Garica 1991, Correa et al. 1989, Withers 2002, Woodin 2004). The quantity and diversity of birds using the preserve is the primary reason for the Ría Lagartos lagoon system being named a wildlife preserve by the Mexican government in 1979 and a wetland of international significance under the Ramsar Treaty in 1986.

It is believed that the high primary productivity of the lagoon system, the close proximity of so many varied vegetative communities, and its strategic location on the eastern flyway account for the large number of birds using the preserve. Indeed, the entire coastal lagoon region of the Yucatan's north coast has been called one of the most strategically located waterfowl habitats in all of Mexico (Andrews et al. 1998, Batilori 1990). Sprunt et al. (1988) have stated, "Without access during the spring migrations to sites such as Ria Lagartos, whole species may be lost" (1989:3). However, it has been noted that Ría Lagartos harbors relatively small percentage of certain types of waterfowl (most notably ducks and geese) as the scouring action of the wind and the lagoon's high salinity inhibits the growth of favorable plant foods (Batilori 1990).

The various vegetative communities of the Ría Lagartos Preserve provide food and habitat for many species of migratory birds. The Jaribu stork (*Jaribu mycteria*), the turquoise-browed mot mot (*Eumomota superciliosa*), and the occasional Peregrine falcon (*Falco peregrinus*) are three of the more rare species, which use the preserve seasonally (Murgia 1989a, Anon 1989a, Correa et al., 1989). The preserve also is home to many species of water birds (Ramo and Busto 1993, Thompson and Baldassarre 1990, Withers 2002, Woodin 1994) including the brown pelican (*Pelecanus occidentalis*) (see Fig. 8.6), white pelican (*Pelecanus erythrorhynchos*), roseate spoonbill (*Ajaia ajaja*), and two species of cormorants (*Phalacrocorax* spp.), to name just a few. The preserve is also the only place in the continental Americas where the Greater Flamingo (*Phoenicopterus ruber*) nests (Espino-Barrios and Baldassarre 1989a,b, Murgia 1989b, Sprunt et al. 1989, Correa et al. 1989, Hernandez and Barron 1989, Schmitz and

Fig. 8.6 Brown pelicans at Celestún outlet. Photo credit: Richard Smardon

Baldassarre 1992a,b). The lagoon provides the physical conditions necessary for the flamingos to feed and nest. The flamingos are also economically important to the area because they draw most of the tourists who visit the preserve (Andrews et al. 1998, Batilori 1990, Murgia 1989b, Correa et al. 1989).

Concentrations of flamingos can be found from Campeche to Quintana Roo (Hernandez and Barron 1989). Most of the population, however, winters in Ría Celestún and flies 285 km to the Ría Lagartos lagoon every year in April and stays until late August in order to nest and fledge their young (Hernandez and Barron 1989). The number of flamingos in Ría Lagartos during breeding season has recently been surveyed at about 24,000 birds (peak numbers occurring during June) (Espino-Barrios and Baldassarre 1989a,b, Schmitz and Baldassarre 1992a,b) (see Figs. 8.7a, b, and c).

Primary feeding areas on the peninsula are Ría Celestún, Dzilam de Bravo, San Felipe, Rio Lagartos, Los Colorados, El Cuyo, and Isla Holbox (Hernandez and Barron 1989). Old breeding grounds included Punta Marco, Vidal, and Sac-Boc; these sites however were destroyed by a hurricane in 1951 (Anon 1989a). Most nesting is now done in Mulsunic and Los Colorados within Ría Lagartos (Hernandez and Barron 1989, Schmitz and Baldassarre 1992a,b). Peak courtship activity occurs during April. May and June is the peak-nesting season. The flamingos choose an exposed sand bar or small island and build their nests out of sand, shells, feathers, etc. The flamingo is a social bird and there may be many nests within a small area (Schmitz and Baldassarre 1992a,b). The top of the cone-shaped nest is built high enough so that, given normal water

Fig. 8.7a Flamingos at Celestún Wildlife Preserve. Photo credit: Richard Smardon

Fig. 8.7b Flamingos plus gulls taking off. Photo credit: Richard Smardon

Fig. 8.7c Flamingos in flight. Photo credit: Rick Newton

conditions, the egg will not be submerged. Each flamingo then lays one egg on top of its nest (Barron and Correa 1989, Hernandez and Barron 1989, Schmitz and Baldassarre 1992a,b). After the fledglings mature they leave with adult flamingos for Rio Celestún in August.

In addition to the aforementioned species, many species of fish, lobster, crab, shrimp, and shellfish use the preserve's lagoon and coastal waters and mangrove as habitat during some point of their cycle.

Fish of Ría Celestún and Ría Lagartos

The benthic macro fauna is mostly mollusks and also fish, crustacean, and *anelidos*. The most common families of bivalves are Verenidae and Mesodesmatidae and the most rare are Diplodontidae and Arcidae among others. In the gastropod family the most common are Margdae, Columbellidae, Calyptrocidae and the most rare are Retrucidae, Burcidae, and Tricotrophidae. The most common species of fish are *Acanthostraciom quadricornis*, *Orthopristis chrysoptera*,*Haemulon aurolineatum*, and *Pomadacidae*. The most common crustacean is *Penaeus aztecus*, *Emerita*sp., *Callinectes sapidus*and *Hammarus*sp. and is found mostly near the mouth.

The fish include such species as *Archosargus rhomboidalis*, *Lagodon rhomboides*, *Serranus atrobronchus*, *Sparisoma radians*, *Spheroides testudineus*, *Lutjanus griseus*, *Monacanthus hispidus*, *Chloroscombrus chrysurus*, *Caranx hippos*, *Chilomycterus schaeffi*, *Syngnathus lousianae*, *O. chrysoptera*, *Arius melanopus*, *Eucinostomus gula*, and *E. argentus*. There have been identified 53 species inside the lagoon and most of them are considered resident species (Arellano-Torres et al. 2006, Flores-Verdugo et al. 1988, Ramos-Miranda et al. 2005, Vega-Cendejas et al. 1994, Vega-Cendejas and DeSantilliana 2004, Vazquez et al. 2005, Yanez-Arancibia et al. 1988, 1993).

Fauna of Ría Celestún

The area is very important in terms of the variety of shorebirds, resident and migratory (see Fig. 8.8). Among the nesting birds are olivaceous cormorants (*Phalacrocorax olivaceus*) and black-billed whistling duck (*Dendrocygna autumnalis*). All year-round species are brown pelican (*P. occidentalis*), white pelican (*P. eythrorhynchos*), darters (*Anhinga*), least bittern (*Ixobrychus exilis*), wood stork (*Mycteria americana*), roseate spoonbill (*A. ajaja*), and lttle blue heron (*Egretta caerulea*) (Thompson and Baldassarre 1990, Withers 2002). The lagoon is an important feeding area for the American flamingo (*Phoenicopterus ruber ruber*) with a population of 15,000–20,000 individuals from the nesting colony of Ría Lagartos (Espino-Barrios and Baldassarre 1989a,b).

There are more than 13 species of migratory ducks and two local species: blue-winged teal (*Anas discors*), lesser scaup (*Aythya affinis*), *Nareca americana*,

Fig. 8.8 Egret in Celestún mangrove. Photo credit: Richard Smardon

northern pintail (*Anas acuta*), shoveler (*Spatula clypeata*), ring-necked duck (*Aythya collaris*), American coot (*Fulica americana*), bufflehead (*Bucephala albeola*), North American ruddy duck (*Oxyura jamaicensis*), merganser (*Mergus serrator*), *Cairina moschata*, green-winged teal *Anas carolinensis*, cinnamon teal (*Anas cyanoptera*), and wood duck (*Aix sponsa*) (Withers 2002, Woodin 2004).

The most important reptiles are Morelett's crocodile (*Crocodylus moreletii*), loggerhead turtle (*Caretta caretta*), hawksbill turtle (*E. imbricata*), slider turtle (*Trachemys scripta*), painted turtle (*Chrysemys picta belli*), and mud turtle (*Kinosternon subrubrum*).

Mammals such as *Felis wiedii*, jaguar (*F. onca*), ocelot (*F. pardalis*), white-tailed deer (*Odocoileus virginianus*), *Dicotiles tajacu*, and *Ateles goeffroyi* are not common. Only the white-tailed deer can be found, which is not extensively hunted.

Human Use of the Preserves – Historical

Unlike the rest of Mexico, early indigenous populations on the Yucatan Peninsula settled in coastal areas as well as in inland sites. The lack of freshwater and good land, and the high percentage of wetlands, however, historically limited population growth in the northern coastal settlements (Chavarria 1988). Most of these settlements centered around cenotes or petenes in order to obtain freshwater (Wilson and Williams 1987, Chavarria 1988). Also wetlands were critical to early Mayan agricultural systems (Heimo et al. 2004, Rejmankova et al. 1995, Smardon 2006).

Petenes were also important because they provided suitable conditions for desirable plant species (Wilson and Williams 1987). Canals were often built to reach petene areas. Remnants of canal systems and the existence of certain non-native plant species such as banana (*Musa* spp.), sapodilla (*Manilkara zapota*), avocado (*Persea americana*), tropical red cedar (*Cedrela mexicana*), and mahogany (*Swietenia macrophylla*) are good indicators of early settlements (Wilson and Williams 1987).

Archaeological evidence suggests that the Ría Lagartos area has been inhabited for more than 1,500 years (Anon 1989a). The porous limestone subsurface and low amount of rainfall however made it difficult for early settlers to produce anything but salt and honey (Wilson 1980). Salt production was an important activity and during the pre-Columbian period, the area became an important salt supplier and successful trading center (Correa et al, 1989). Isla Cerritos, an island off the tip of what is now the Ría Lagartos Preserve, became an important seaport for the empire of Chechen Itza because it was a defensible place to store large quantities of salt (Garret 1989and personal conversation with Raul Murgia 1989).

With the arrival of the Spanish, the exploitation of Mexican resources for the old world began. Logging and some mining were the most economically important activities carried out under Spanish influence. Exploitation of natural resources continued as Mexico gained independence. Henequin production and other agricultural crops were important (Wilson 1980). Campeche wood (*H. campechianum*) was logged for the chemical substance hemantin, used in the dying and the chicle tree (*A. zapota*), was important for supplying chicle for chewing gum (Cesar-Dachary and Arnaiz 1984, Correa et al. 1989, Anon 1989a). Mining continued and grew in importance with addition of a newfound resource, oil that is being developed just north of the Yucatan Peninsula in Campeche and Tabasco states in wetland areas.

Land Use/Current Economic Activity at Ría Lagartos

There are several prominent economic activities occurring in the Ría Lagartos Preserve. During the 1950s there was a large growth in the fishing, salt, agriculture, and cattle ranching industries (Murgia 1989b). This growth is attributed to an increase in the demand for the area's products (shrimp, lobster, and salt) and an economic diversification of the Yucatan State (Murgia 1989b). Other factors, which contributed, were better communication and transportation facilities and an increase in foreign markets (Valdes 1988, Murgia 1989b).

Fishing: What began as small Mayan fishing villages have evolved into town-run cooperatives with sizable fishing fleets. This process began in the early 1950s as lobster harvesting became profitable (Murgia 1989b). Before this, lobster was primarily considered a nuisance species. In the early 1980s, the fishermen in the preserves formed town cooperatives in order to obtain newly required permits to fish for lobster and to receive low-interest government loans to purchase fishing boats and freezer facilities (Murgia 1989b). This greatly increased fishing activity within the preserve for octopus, lobster, shrimp, sharks, mojaria, mullet, and drum (Andrews et al. 1998). Most fishing is conducted in the open sea.

Today, Rio Lagartos and San Felipe are the two most productive fishing towns in the preserve (see Figs. 8.9a and b). In Rio Lagartos 84% of the economically active population earns its living from fishing (SPP 1989). In San Felipe 64% of the population fishes (SPP 1989), El Cuyo is third in fishing activity, and Las Colorados a distant fourth (SPP 1989). Presently, fishing in the preserve supports about 1,400 families (Andrews et al. 1998, Correa et al. 1989, Sprunt et al. 1988).

Fig. 8.9a Fishing fleet in Celestún. Photo credit: Richard Smardon

Most of the fishing is done from 16 to 24 foot skiffs with outboard motor (Arellano-Torres et al. 2006, Faust and Sinton 1991). There are usually no more than two men to a boat, and they fish the gulf close to shore. During the nortes (strong northerly winds) fishing must be limited to the lagoon for safety (Andrews et al. 1998, Anon 1989a). Approximately 60 tons of shrimp is caught a year (Correa et al. 1989, Sprunt et al. 1989). This activity is done exclusively on the lagoon and may be done with smaller boats (Murgia 1989b). Lobster fishing occurs farther offshore in the Gulf of Mexico and the fisherman use larger boats.

The usual catch is pollack, red grouper, octopus, lobster, mojaria, mullet, drum, and shark (Andrews et al. 1998, Murgia 1989b). In 1984, 3,231 tons of

Fig. 8.9b Fishing boats in San Felipe on Ria Largartos. Photo credit: Richard Smardon

fish was caught by the preserve's fishermen, generating an income of $678,861,564 dollars (Anon 1989a). This amount does not include local consumption, which is estimated at 35% of the overall catch (SPP 1989). It does not include the profit from lobster and octopus fishing. Lobster is the most lucrative catch and was the impetus for the forming of early cooperatives in order to obtain loans to buy larger boats (Anon 1989a).

Salt processing: It should be noted that artesanal saltpans historically covered significant areas. After the collapse of the henequin plantations in the 1940s, the Roche family bought much of the area around the Ria Lagartos lagoon (Andrews et al. 1998, Murgia 1989b). The Roches then moved Mayan campesinos that had been working on the henequin plantations into the preserve and established the town of Las Colorados, which was built to house the salt workers. Las Colorados is a company town; most of the residents work the salt operations. The salt company, Industria Salinera de Yucatan, SA (ISYSA), owns workers' houses and the local stores are company owned. Workers can buy goods in these stores and have money taken out in their weekly pay. ISYSA is Mexico's second largest salt producer.

In the late 1970s SEMIP (Ministry of Industry and Mining) gave permission for the salt industry in the lagoon to form ISYSA, the Yucatan's Industrial Salt Society (Correa and Boege 1989). The Roches, using modern technology and campesinos for labor, then produced salt to sell to US chemical firms. By 1987, 50,000 tons of salt a year was being produced (Murgia 1989b, Correa et al. 1989, Sprunt et al. 1988). As the facility was further modernized, fewer workers were needed. The work force dropped from 600 employees in the 1950s to approximately 400 in the 1970s and has continued to drop to just over 200 employees (Faust 1991). In 1988, Hurricane Gilbert destroyed most of the salt processing plant and the town of Las Colorados. Presently, the Roches have re-established the salt factory (Anon 1989a, Murgia 1989b, Faust 1991) and have constructed a pier to load cargo vessels for shipment to Japan.

Modern methods for the solar extraction of salt entail closing off large sections of the lagoon with walls made of sand and wood creating pools of standing water called salt "charcas" (Andrews et al. 1998, Perry 1991). These pans allow evaporation to create a salt gradient, which increases toward the bottom of the pool. The less saline water is then pumped off the top layers and the salt is collected and put into piles for the final drying stages (personal conversation with Raul Murgia 1989). Besides the industrial salt operation there is also a traditional co-op style salt extraction enterprise in the Ría Lagartos Preserve, which supports approximately 40 families (Murgia 1989b, Sprunt et al. 1988).

Tourism: Early tourism in the Yucatan Peninsula focused on the beaches of the Caribbean coast, primarily Cancun and Cozumel. Any visits farther inland were usually short side trips from resort areas. However, as transportation becomes more convenient and as interest in the environment grows, increasing numbers of tourists are traveling beyond the beaches, seeking out the peninsula's natural and cultural attractions.

The Ría Lagartos Wildlife Preserve, with its endemic flora and fauna, has begun to attract people interested in nature experiences (see Fig. 8.10). Birders

Fig. 8.10 View of Ria Lagartos outlet from Hotel San Felipe. Photo credit: Richard Smardon

often pass through the area looking to fill their bird lists with some rare species, which inhabit the preserve. In fact fisherman often earn extra money by bringing tourists by boat to see the flamingos and other bird species which inhabit the mangroves. In an attempt to keep tourists in the area longer, a 20-room hotel, the Maria Nefertiti (now closed), was built in the town of Rio Lagartos and there is now a hotel in San Felipe. Along with the foreign tourists there are many Mexican tourists who visit from the interior regions of the peninsula. Many of these tourists have summer homes in the Rio Lagartos Preserve. Most of these homes are along the lagoon's shore and are only used during the hot summer months when school is out. The primary attractions for these tourists are the beaches and the rural atmosphere (Meyer-Arendt 1991, Moan 1992, Murgia 1989b, Murgia et al. 1991).

Agriculture

Agriculture in the Yucatan has undergone many changes through the centuries. During the early Mayan civilization, people practiced subsistence farming and also produced food for the Mayan lords. Crops consisted mostly of corn, beans, squash, and chili peppers. In more modern time, henequin and chicle (forest crop) were very important crops to the world market and made many plantation owners rich. However, as synthetic substitutes were found, the crops quickly lost their competitive value. Today most of the people living in rural Yucatan still practice at least some form of subsistence-level farming (Andrews et al. 1998, Faust 1991).

Farming: As the human population in the preserve has grown, traditional family practices have changed. Families that once fished and practiced subsistence farming have begun to specialize. While some fish, others farm and sell the produce to those who no longer grow their own. In Ría Lagartos 12% of the economically active population practices farming (SPP 1989). In San Felipe 8.8% of the population farms (SPP 1989).

Cattle ranching: Land that has been cleared for lumber and is no longer productive for farming purposes is abandoned or put to use for cattle grazing. Cattle ranches compose most of the land south of the Ría Lagartos Preserve. In Ría Lagartos town, 4% of the economically active population practices cattle ranching (SPP 1989). In San Felipe, 2.7% of the population practices cattle ranching on 37% of the town's land (SPP 1989) and this is steadily increasing. As of 1995 an estimated 7,000 ha had been cleared for grazing (Andrews et al. 1998).

Other industries: Along with the growth in fishing, salt extraction, agriculture, and cattle ranching, the expected growth of service industries has occurred. A ship builder's yard, a sawmill, a gravel quarry, restaurants, stores and gas stations have all opened (Murgia 1989b).

Residential Growth

The urban growth within the preserve influences and is influenced by the preserve's economic growth. Early populations moved to the area in order to fish and extract salt. As these economic activities became increasingly profitable, more people moved to the preserve. The failure of economic industries inland contributed to the growing population on the coast. Statistics show that most of the people moving to the coastal towns come from interior areas within the state of Yucatan (SPP 1989). Most of these migrants are displaced workers who moved into a specific town for family reasons (i.e., marriage, or moving in with family members to find work) (SPP 1989).

Town development is not actively planned within the preserves. As in most of the Yucatan's coastal towns, development follows a functional "T" formation (Murgia 1989b). Early migrants, in an attempt to be as close as possible to the shore for fishing and salt operations, would create a road to the lagoon's shore and settle near it, the most desirable sites being on the shore and next to the main road. For a time development builds along the coastline in each direction, until the town limits are met, then begins to form along the central road perpendicular to the coast. In this way, the towns begin to form a "T" shape.

There is some provision for government control over development in the preserve. This is carried out by two planning documents, the "Esquema de Desarollo Urbana" (Plan for Urban Development) and the "Plan Director". The Plan for Urban Development is a technical advisory document. This is only a guideline and has no legal ramifications. This document is required for towns smaller than 2,500 persons. This can become a legal tool for town development if it is voted on and passed as such by leading town officials although there is rarely incentive to do so. The Plan Director is required for all towns with populations greater than 2,500 people. This is the legal administrative document and is a legally binding development guideline (from personal discussions with Alfredo Alonzo 1989).

At present SAHOP (Ministry of Human Settlements and Public Works) has drawn up Esquemas for San Felipe, Rio Lagartos, and El Cuyo in Ría Lagartos Biosphere Reserve and in Celestún, but they have been largely ignored (from personal discussions with Alfredo Alonzo 1989). There will be no need to develop Plan Directors for each town until they reach 2,500 inhabitants. At present much of the urban activity in the Ria Lagartos area is presided over by various "ejidos". The "ejido" structure is a rural communal unit similar to a small town (Uphoff 1985). These ejido units were a concept developed by revolutionaries such as Emiliano Zapata after the Mexican revolution (Chavarria 1988). The new government gave express rights to rural communities to use designated areas of public land (often areas taken back from Spanish colonists offspring) in order to live and farm. Areas of land were often granted to family units with the stipulation that they live on and work the land (similar to the Homestead Act in the United States).

Land Use and Tenancy at Ría Celestún

In 1985 the village occupied an area of 111.9 ha. Most of it was residential use. The land use in Celestún is related mostly to the construction of dwelling units with the activities of fishing and tourism/recreation.

The development trends of the urban core within Celestún are mostly toward the edges of the village, on both sides of the road, this being the "ejidos" area. These areas, mostly flooded, have been filled gradually by the immigrants, to whom the former municipal administration gave pieces of property, contributing to the increase of the value of more centric lots and other pieces of property along the beach.

The former administration of Celestún also filled land parcels and obtained credit from FONHAPO for the construction of houses for the fishermen, which as of 1996 are not occupied. On the other hand, and due to an increased demand of tourism uses in the north side, there is speculation with land lots that are now land reserves. An ecotourism complex was constructed here in 1996.

The Secretaria de Reforma Agraria reports that approximately 70% of the Wildlife Refuge area (36,000 ha) in the state of Yucatan is national (federal) property. The *ejido* Celestún has lots to the east of the reserve with a total area of 8,650 ha. The rest of the area is private property that is distributed in the coastal strip. There is no information about the portion that belongs to the state of Campeche (Biocenosis 1989).

The area of the Wildlife Refuge (59,130 ha) belongs to the municipalities of Celestún (Yucatan) and Calkini (Campeche). Nevertheless, Biocenosis (1989) says that in the cartographic reconstruction of the limits (borders), it seems that the southeast portion of the refuge is located in the municipality of Maxcanu, Yucatan.

Fishing is one of the most important economic activities. Fisheries currently employ about 90% of the population in Ría Celestún directly or indirectly. It has grown from 1,584 fishers with 391 small vessels and 3 large vessels in 1986 to 2,569 fishers with 584 small vessels and 11 large vessels in 1991 (Andrews et al. 1998). The main fishing products are white grunt, sea trout, mullet, sardine, anchovy, red snapper, and gray snapper, with secondary emphasis on grouper, octopus, *huachinango*, and shark. Most of the species are at maximum level of exploitation with an annual capture of 11,000 tons. Total fisheries production has remained the same, but economic impact per capita is decreasing (Andrews et al. 1998).

The salt industry is one of the oldest in the peninsula in terms of small-scale artisanal salt production. The coastal communities have been the most connected to the activity. Nevertheless today, although this activity could be contributing in the creation of jobs, there are problems with it regarding marketing and technology. There are saltpans in eight municipalities of the coast and the mine management has registered 83 estates and has the concession of 10,141 ha (Pare 1986). In 1994 ten Societies of Salt Employees (SSE) with a total of 190 members were engaged in production, but employment varies widely because of environmental conditions (Andrews et al. 1998).

Fish flour industry: This is no longer important for Celestún. Ría Celestún was in second place in fish production in Yucatan, and this was due to the volume of species that are destined to the fish flour industry.

Tourism: The tourism industry has recently expanded in Ría Celestún (see Figs. 8.11a and b). There are two types of tourists: in the first group are the tourists that come from within the state, mostly from Mérida, that usually own summer houses on the beach and the second group are the tourists that come to watch the flamingos and other birds, mostly national and international, and circulate around the lagoon and surrounding wetlands. A major issue affecting both biosphere reserves and the Mexican coast is the potential loss of natural values and environmental services derived from coastal mangrove forests when traded off for other land uses (Barbier 1993, Clark 1991, Ewel et al. 1998, Hernandez et al. 2001, Kaplowitz 1998, Twilley et al. 1992).

Fig. 8.11a Launching point for ecotourism boat tours from Celestún. Photo credit: Richard Smardon

Management of the Preserves

The Mexican government established the Ría Lagartos and Ría Celestún National Wildlife Refuges in 1979 (Andrews et al. 1998, Murgia 1989b) under article #4 of the Presidential Decree (Valdes 1988, Correa and Boege 1989). According to Mexican law, this implies that the main concern of the preserves are "to protect and preserve one or various plant and animal species" even if this may mean the restriction of human activities within the preserve (Anon 1989a).

Fig. 8.11b Boardwalk within Celestún mangrove constructed for ecotourism. Photo credit:
Richard Smardon

In 1982, the Ría Lagartos Preserve along with Ría Celestún and Punta
Cancun was designated federal ecological reserves based on the Environmental
Protection Law of 1982 (Vargas 1984). This listing was meant to address
problems of coastal ecosystem protection and restrict the impacts of human
activities in the preserves (Vargas 1984).

In 1985, Ría Lagartos was declared a biosphere reserve by the United
Nations Man and the Biosphere Program (Murgia 1989b, Anon 1989a, Correa
et al. 1989). This was to signify the preserve's importance to the world community and to better integrate the preserve's management of human activities with
the preserve's wildlife protection goals.

In 1986 the Ría Lagartos Wildlife Preserve became the only Mexican wetland
to be listed as a Ramsar site (Murgia 1989b, Anon 1989a, Correa et al. 1989).
This signifies it as a wetland of international importance. This also means
"Mexico is responsible to the world community to give high priority to nature
conservation at this site and to provide a high level of management compe-
tence" (Sprunt et al. 1989). In order to meet this commitment, the state of
Yucatan has formed a state-level coastal zone management program and a state
system of protected areas, which encompass the Ría Lagartos Preserve.

In 1992, the reserve came under the administration of the Secretariat of
Social Development (Secretaria de Desarrollo Social, SEDESOL) and in
December 1995 passed to the Secretariat of the Environment, Natural
Resources and Fisheries (Secretaria del Medio Ambiente, Recursos Naturales
y Pesca) SEMARNAP (Caillas et al. 1992). In May 1999, Ria Lagartos was

declared a biosphere reserve without the qualifier "Special" and in November 1999 SEMARNAP published the Ría Lagartos Biosphere Reserve Management Plan (Fraga 2006). There were lots of problems with how the management plan was developed and this is well documented by Fraga (2006). It took a long time to develop the plan and was essentially done by the research university from Monterrey, Mexico, which is not familiar with the specific area or the people living within the reserve.

A similar series of events affected the management plan for Ría Celestún Preserve. The author saw a management plan for this reserve in 1989. When he asked others as to whether this plan was recognized by those within the preserve, the answer was basically "nobody pays any attention to it" and "we were not asked to participate in its production". The plan was redone in the early 1990s again by a university or technical institute not from the area with World Bank funding – the same as Ría Lagartos. It is not known to the author whether there is an "official" management plan in place for Ría Celestún at the current time.

Actors Involved with Preserve Management

Mexico also has local NGOs, such as DUMAC (*Ducks Unlimited of Mexico, Asociación Civil*) and PRONATURA (*Programa para la Naturaleza*), that receive funds from both foreign governments and international NGOs, as well as from local businesses, periodic raffles, and other fundraising activities. These NGOs are involved in various projects of environmental education, training for members of local communities, and the setting aside of land in conservation trusts (*servidumbres de conservación*). Academic research, the programs of government agencies, and NGO activities made efforts to coordinate their research and share resources, with the encouragement of President Fox's administration. The *Comisión Nacional para el Conocimiento y Uso de la Biodiversidad* (CONABIO, National Commission for Knowledge and Use of Biodiversity, at http://www.conabio.gob.mx) and its parent ministry, the SEMARNAT (at http://www.semarnat.gob.mx), are also making efforts to facilitate the exchange of information among government agencies, research centers, universities, and public interest groups.

Institutions and Major Actors Involved in Mexico's Protected Natural Areas

One of the major institutional factors affecting Mexico's protected areas has been the radical change in personnel of all government agencies that routinely occurred every 6 years, with the election of a new national president. This was commonly associated with changes in agency mission, which has had costs in terms of continuity in programs, institutional memory, and the longitude of professional expertise (Mumme et al. 1988). Mexico does not have a professional civil service with permanently employed experts; those who are in government service for one 6-year period are frequently later to be found in academia or business, using their accumulated knowledge in other ways. In the latest case, the change of president has also involved the fall from power of one political party that had governed the country for over 70 years. Recognizing the danger of abrupt changes, President Fox created a transition team to help ease the process of change and has included some members of the previous government in his administration. The history of Mexico's experience with protected natural areas needs to be understood within this political and institutional setting.

Mexico's General Law for Environmental Protection (*La Ley General del Equilibrio Ecológico y la Protección al Ambiente*, LEEGEPA) was established only in 1988, but it reflects previous efforts dating back to Mexico's first national park, El Chico, established in 1812 in the state of Hidalgo, more than 60 years before the establishment of Yellowstone National Parks in the United States (http://semarnat.gob.mx, August 6, 2002) and another, Desierto de los Leones in the watershed of Mexico City, established in 1876 (http://conanp.gob.mx, August 8, 2002). By 1972, concerns over environmental contamination led to the establishment of an agency for environmental improvement (*Subsecretaría para el Mejoramiento del Ambiente*) within the Ministry of Health and Public Assistance (*Secretaría de Salubridad y Asistencia*).

The Ministry for Urban Development and Ecology (*Secretaría de Desarrollo Urbano y Ecología*, SEDUE) was formed in 1982 to implement new laws for environmental protection, including a new federal law, la *Ley Federal de Protección al Ambiente*.This was followed by the 1988 law, LEEGEPA, which instituted the Sistema *Nacional de Áreas Naturales Protegidas* (SINAP, the National System of Protected Natural Areas).

In 1992, the former duties of SEDUE were divided and assigned to other ministries and agencies. Most major environmental responsibilities were taken over by the newly created *Secretaría de Desarrollo Social* (SEDESOL, Ministry of Social Development), within which the *Instituto Nacional de Ecología* (INE, National Institute of Ecology) was formed as a semi-autonomous body, with regulation and control capabilities. However, the *Secretaría de Agricultura y Recursos Hidráulicos* (SARH, Mexico's Ministry of Agriculture and Water Resources) was expected once again to have responsibility for most parks, while SEDESOL managed the biosphere reserves. Under the new Forestry Law, non-governmental groups were permitted to manage federal protected areas, within the policies of SEDESOL and with the managerial oversight of SARH. The *Procuraduría Federal de Protección al Ambiente* (PROFEPA, a Federal Prosecutor for Protection of the Environment) was established, and the *Secretaría de Pesca* (Ministry of Fisheries) took over responsibility for the promotion, conservation, and development of the marine and freshwater flora and fauna as well as the establishment of breeding grounds, nurseries, and other aquatic reserves.

Two years later, with the election of President Zedillo in 1994, agencies were again renamed and responsibilities redistributed. For the first time an independent government ministry was established for the management of natural resources including forests, fisheries, biosphere reserves, and other protected areas. It was called the *Secretaría de Medio Ambiente, Recursos Naturales y Pesca* (SEMARNAP) and had a clear mandate to combine environmental protection with management of natural resource use, including forests and fisheries. For the 6-year presidential term, SEMARNAP was headed by a young female biologist, Julia Corrabias. It employed many enthusiastic young biologists in the drafting of legislation and in the establishment of management plans for protected areas and other programs for protecting biodiversity.

In 2001, newly elected President Fox slightly modified the name to SEMARNAT (Secretaría para el Manejo de Recursos Naturales), moving the oversight of fishing (and related activities) to the agricultural ministry. He also followed tradition in replacing the political appointees running the agency with those of his own administration. The SINAP formed in 1988 by the LEEGEPA lasted until 2000 when it was renamed and reconstituted by President Fox as the *Comisión Nacional de Áreas Naturales Protegidas* (CONANP, the National Commission for Protected Natural Areas), although the name SINAP continues to be used for the list of protected areas and a more restricted list called SINAP II appears to refer to those areas involved in a new program of the World Bank that involves both funding and supervision (Smardon and Faust 2006).

The CONANP continues SINAP's previous responsibilities for the supervision and integration of protected natural areas. Since the early 1990s, SINAP objectives included building capacity in each protected area for recreation, culture, research, and citizen involvement (Pérez-Gil and Jaramillo-Monroy 1992). With CONANP there is a clearer focus on the protection of these legally delimited areas while "priority regions" have been established for projects of "regional sustainable development"; these are to involve indigenous groups and other rural communities in the design, ownership, and operation of productive activities of a sustainable nature (http://conanp.gob.mx, April 24, 2002, page 2 of "Qué es Conanp?").

The nine categories originally used by the SINAP to classify protected areas have been transformed by CONANP into five, reflecting new international guidelines: Biosphere Reserves (31), National Parks (66), Natural Monuments (4), Areas for the Protection of Natural Resources (1), and Areas for the Protection of Flora and Fauna (23). Many of these protected areas were established in populated areas that have both cultural importance and longstanding histories of resource use, such as are the case in our study area of the Yucatan Peninsula.

The last two decades of the twentieth century were critical not only in the establishment of laws and government agencies dealing with environmental issues but also in the formation of private organizations and civil associations, which are increasingly referred to as NGOs. These groups began successfully promoting the establishment of protected areas and sometimes participated in efforts to manage them. Significant examples for Yucatan include

- Amigos de Sian Ka'an (Friends of Sian Ka'an) – works in the biosphere reserve of the same name in the state of Quintana Roo;
- Ducks Unlimited México, Asociación Civil (DUMAC) – maintaining and rehabilitating habitat for waterfowl in the coastal lagoons of Rías Lagartos and Celestún in the state of Yucatan (among others);
- PRONATURA (Programa para la Naturaleza, Program for Nature) – managing small private reserves such as the Rancho Limonar near the Reserva de Ría Lagartos and providing management support for reserves such as Calakmul, Ría Lagartos, and Ría Celestún (Andrews et al. 1998).

In 1992 The Global Environmental Facility (GEF 1992) approved a grant for Mexico that was predicated on Mexico's ability to support the indicated recurrent costs ($20 million for 17 protected areas over 3½ years, or approximately $33,600 per unit per year). Much of these funds were used to pay for the development of management plans for at least ten of these reserves. With that budget estimate, Mexico could only afford around seven protected areas at this level of recurrent costs. For the entire protected area estate to be funded at this rate, Mexico's park system would require an annual operating budget of over $20 million, nearly ten times the current budget estimate. Inflation, the debt crisis, and massive unemployment have created a difficult situation in Mexico, where protected area officials have had to struggle for resources to fulfill their mandate.

In Mexico, training needs have been cited as a principal factor limiting the effective management of protected areas. There is a shortage of research scientists and trained resource management specialists. No institution specializes in advanced training in conservation and management of resources, although there is a master of science program in human ecology in the Mérida Campus of CINVESTAV with a doctoral program planned, and another master of science in resource conservation and management at the Universidad Autónoma de Yucatán (UADY, Autonomous University of Yucatan), also in Mérida. El Colegio de la Frontera Sur (ECOSUR, The College of the Southern Border) also does research on resource management and conservation biology with local communities in the Yucatan Peninsula. El Centro de Investigación Científica de Yucatán (CICY) collaborates with the Institute of Ecología in Xalapa, Veracruz, to provide an inter-institutional doctorate in ecology. The Colegio de Posgrados de Chapingo has a branch in Mérida that provides training in agroforestry, while some research and training in conservation is also done by SEMARNAT within the agency's programs. In addition to the academic programs, some NGOs are offering short courses, but many of these are periodically curtailed due to lack of financing.

Other Mexican agencies of note for this case study are listed below:

SEPES – Ministry of Fisheries provides fisheries and fisherman cooperatives regulation, fisheries inventory, technical assistance, processing of fishing products, permits season regulations for capture, and also administers the budget for infrastructure and equipment (Valdes 1988). This agency's function is now absorbed into SEMARNAT.

SPP – Ministry of Budget and Programming approves the budgets for these agencies. These agencies all can undertake research projects singly or jointly in order "to gather data to help establish policies and objectives for protected coastal areas" (Valdes 1988).

SECTUR – the Ministry of Tourism regulates, promotes, and provides the financial support for tourism development (Valdes 1988).

Other federal agencies carrying out activities in the Rio Lagartos Preserve are as follows:

SAHOP – The Ministry of Human Settlements and Public Works
CFE – Federal Electric Commission
SCT – Ministry of Transport and Communications
SARH – Ministry of Agriculture
SEMIP – Ministry of Industry and Mining

Most of these government agencies and NGOs alike are interwoven into the management fabric of these two estuarine wetland complexes. To illustrate this the following section will focus on the major management issues for these two wetland areas and the respective roles played by the actors listed above.

Threats/Management Issues

Threats to the biosphere reserves include actions related to fisheries production due to various factors: impact of salt drying and production operations; impact of tourism activity; impacts of farming, ranching, gravel mining and logging; and residential development.

Impact on Fish Productivity

Of the eight northern coastal lagoons and surrounding waters of the Yucatan Peninsula, Ría Celestún is the largest and Ría Lagartos the second largest. Fishing is the most profitable legal industry within the preserve and benefits the widest number of local individuals (Andrews et al. 1998, Murgia 1989b, Correa et al. 1989). Recent unofficial accounts indicate that the size and amount of fish being caught by the fisherman at Ría Lagartos may be decreasing (Faust and Sinton 1991, Fraga 2006). Decrease in fishing productivity is a problem with two dimensions: (1) many families in the preserve depend on fishing as their source of income and more and more arrive intending to fish and (2) fish are a prime indicator of the ecological health of the lagoon and are a food source for many water birds. Two major contributing factors could include overfishing (Andrews et al. 1998, Fraga 2006) and changes in hydrology (Arellano-Torres et al. 2006, Vega-Cendejas et al. 1994) in the lagoon.

It is not clear that the fishermen mainly fish in the coastal waters with recent decreases in fish catch. However, lobster and octopus catch have recently increased. The use of the lagoons is mainly directed at crab, shrimp, and shellfish (comments from J. Andrews 1998and J. Frazer 1996). Better equipment and increased fishing activity by larger numbers of fishermen are putting increased pressure on fish populations in the area. Also considered a problem is the focus on only a few species of fish while others are ignored (Murgia 1989b). Season and catch limits along with protected area restrictions are widely ignored and difficult to enforce. There are some areas further up the lagoon to the east that is off limits to fishing.

As the lagoon is primarily a closed system, any change in water quality has effects on aquatic populations and water quality in the preserve is affected by land use in the watershed around the lagoon (Barbier and Strand 1998, Clark 1974, Flores-Verdugo et al. 1988, Reyes and Merino 1991, Sklar and Browder 1998). Changes in water temperature, turbidity, available oxygen, salinity, nutrient content, and confining layer in the water column will directly affect aquatic species (Young et al. 2005). Factors affecting water quality in the preserve are as follows:

1. Breaches in the Ría Lagartos barrier peninsula caused by hurricanes/tropical storms and stripping of vegetation and sand by towns, industry, and hotels – which allows water from the gulf to enter the lagoon directly in greater than normal quantities changing water temperature salinity, oxygen, and nutrient levels. Larger breaks may allow the physical action of tides and waves to affect the lagoon's aquatic and shore life plus allowing more sand to wash into the lagoon making it too shallow for boat use which in turn causes the need for dredging with its attendant impacts.
2. Upland vegetation clearing is significant, in that it causes a number of problems such as (a) removing a protective filter-like buffer which keeps excessive nutrients and particulate matter from entering the lagoon, (b) increasing erosion levels and allowing eroded material to more easily wash or blow into the lagoon, (c) removal of organic material vital to the nutrient cycle, and (d) decreasing shoreline stability and allowing for breaching. Vegetation clearing is caused primarily by the salt operation in some areas and urbanization near the towns.
3. Increased pollution over time from agricultural pesticide use can cause hazardous chemicals to build up and affect water quality (Lopez-Carrillo et al. 1996, Young et al. 2005) especially since the lagoon is a virtually closed system. Increased human populations in the Rías Lagartos and Celestún

Preserves have meant increased amounts of garbage and sewage. Specific concerns include organochloride pesticides such as DDT (Albert 1996), aldrin in nearby Terminos lagoon to the north (Albert 1996, endrine in the lagoons along the gulf coast plus heptachlor epoxide and endrine aldehyde (Albert 1996). Specific testing by Gold-Bouchot et al. (2006) found chlorobenzenes, HCHs, and PCBs in Ría Celestún lagoon sediment, HCHs and PCBs in Dzilam lagoon and the highest concentrations in Laguna de Terminos. Other concerns include excessive organic material and fecal coliforms, which have been reported in some coastal lagoons (Ortiz-Hernandez and Saenez-Moralez 1999) and along other points along the Mexico coast (Tran et al. 2002).

4. Decreasing inputs of freshwater affect the lagoon's saline/freshwater mix and may affect the confining layer in the water column and the coastal aquitard (Alonzo 2007, Batilori 1988and undated, Cable et al. 2002, Sklar and Browder 1998). The lagoon's aquatic life and vegetation have adapted themselves to the specific salinity regimes present in the lagoon, especially mangrove vegetation while salt tolerant, needs a freshwater influx in order to survive and grow.

5. Decreased water circulation slows nutrient cycling, lowers the oxygen content, and raises the salinity level in areas to the east of the obstruction which affect aquatic fish populations in the lagoon. Salt charcas (basins), impermeable roads, bridges, and land filling for house lots are the primary causes of decreased water circulation (Reyes and Marino 1991, Young et al. 2005).

6. Loss of habitat is linked to all the processes listed above, especially loss of mangrove vegetation which decreases valuable hiding and nursery areas for aquatic life. Changes in water quality from the alteration of circulation patterns, freshwater inputs, and salinity, nutrient, oxygen, pollution, and turbidity levels all decrease the success of aquatic species to adapt to different conditions. The ability to assess water quality conditions and loss of fish habitat is linked to the development of an integrated management plan for the preserve. Some of the critical research has been done by CINVESTAV. SEMARNAP is aware of the issues, but this is a tough problem that demands many agencies and NGOs to coordinate actions.

One remarkable effort, by the residents of San Felipe, was the creation of a marine reserve in 1988. The fishermen in the community manage this marine reserve, at the mouth of Ría Lagartos, without receiving official recognition from the state or the federal government (Fraga 2006). In April 1995, the directors of the fisherman's cooperative and the municipal government officials signed a document establishing the management rules for the marine reserve, and in December, the reserve was given a name "Actum Chuleb". This is a Mayan word meaning the "water where the birds drink" (Fraga 2006). The rules of the reserve, while imposed by the fishing cooperative, are mainly self-enforced by the members as they relate to the whole community's livelihood. After some initial problems of enforcement, this self-enforced system did work.

There was initial moratorium of fishing activity followed by restricted use of certain types of fishing gear. The sea grass beds improved and the fishery recovered to a great degree even though there was no official recognition for this self-declared marine reserve.

Salt Industry Operations

Given the above fisheries habitat and water quality issues, the following activity has contributed greatly to water quality and fisheries habitat impacts. The solar extraction process of salt production requires large areas of the lagoon to be dammed off into shallow pans of varying salinity gradients and a great deal of water pumping, both fresh and saline. This has created many impacts on the local ecology and has drawn criticism from area managers and fishermen. There are other reasons why the salt operation has sparked the most controversy of all the human activities in the preserve:

- Permission to obtain land and appropriate title was hastily granted just prior to the area's Wildlife Preserve designation in 1979 (Correa and Boege 1989).
- The salt company owns much of the most important wildlife habitat in the preserve and they intend to obtain more by buying out smaller landowners in the preserve. But with the impact of Hurricane Gilberto, the salt company lost all of its machinery; many of the charcas are still in disrepair, and the company is in fiscal difficulty (J. Andrews communication 1996, Ramsar 1989).
- The salt operation had carried out its activities without consideration of possible environmental side effects. This is less true today. There is a greater sense of the need for collaboration. This is because the reserve's manager is more effective, and family members with a better conservation ethic bought out other family members using bank loans.

For these reasons, the management of the Rio Lagartos Preserve believed in the past that the salt industry was the most destructive human activity operating in the preserve's ecology (from personal conversations with CINVESTAV, SEMARNAP, and PRONATURA 1989). Today there is more concern about the amount of land converted to cattle ranching (Andrews et al. 1998, Kaplowitz 1998).

The operation of a large-scale solar salt production facility in the Rio Lagartos Preserve has affected the hydrology of the preserve in many ways in the past, which is also critically linked to fisheries production including changes in water quality caused by breeches in the Ría Lagartos Peninsula and pumping of water into the charcas also affects water quality in the lagoon.

Breaches caused by vegetation destruction allow greater than normal circulation to the Gulf of Mexico. These flows combined with damming affects of the charcas raise water levels to the east of the charcas' narrowest point. These changes are exacerbated by periods of high storm-driven rainfall. The combination of artificially high water levels and early spring storms has devastated flamingo-nesting colonies for three nesting seasons (1989–1991) (personal communication with Jesus Garcia-Barron 1989). Note, there are differing opinions concerning this issue.

The salt company has also built saltpans in areas where prime nesting islands were located (by Mulsunilo), forcing the flamingos and other bird species (roseate spoonbill, egrets, and cormorants) to move to less desirable sites. There have also been observations that heavy equipment used to haul salt disturbs nesting and feeding birds, impeding nesting success (Murgia et al. 1989b). Most of the original breeding sites have been lost to the combined impacts caused by the salt company (Correa and Boege 1989, Hernandez and Baron 1989, and from personal conversations with Jesus Garcia Barron 1989).

In 1988 Hurricane Gilbert wiped out the salt operation. The owners of the salt operation wanted to greatly expand and were in the process of acquiring loans to rebuild the operation. Leaders of the fishing industry were worried about the affect of the expanded operation on fish productivity in the lagoon and preserve management was worried about long-term ecological impacts on the preserve as described above.

Relations between the biosphere reserve management, the salt company, and the fishermen have never been amicable. A confrontation between the three parties made this relationship more tenuous in 1990 (Faust and Sinton 1991). After the Hurricane Gilbert, the estuary system suffered some degradation as a result of the hurricane and reconstruction activities. Some of the "reconstruction activities" have in fact involved considerable unofficial expansion.

Biosphere reserve management believed that the factory had plans to expand the salt ponds in order to produce 1 million tons of salt by 1995. The Secretary for Ecology and Urban Development (then SEDUE) was attempting to stop the operation from unofficial expansion. In the summer of 1990, resistance by the salt operation owners to an injunction against expansion resulted in SEDUE (now SEMARNAP) locking up the pumps for the evaporation ponds. The operation then shut off the community's freshwater supply that runs through factory pipes. In the end, the community of Las Colorados interpreted SEDUE's action as a threat to their jobs and their domestic water supply. At that time, a group of academics working with the author, Professors Faust and Sinton plus students, were branded as "SEDUE spies" by some of the salt workers' union members, and it was decided to withdraw temporarily from the village. The salt works owners creating a conflict situation also pitted the salt workers' union against the fishermen's associations.

The two principal sources of revenue for workers were from work in the salt operation, with 148 employees, and from fishing there are 105 fisherman and perhaps 100 young men who help them. In contrast to the 105 fishermen in Las Colorados, there are 406 in San Felipe, 920 in Rio Lagartos, and 710 in El Cuyo (Faust and Sinton 1991). At that point both the fishing and salt operation jobs were endangered.

Members of the San Felipe and Rio Lagartos fishing cooperatives strongly supported SEDUE's injunction against salt operation expansion. The closing of the salt operation in July 1990 resulted not only in distrust of outsiders connected with SEDUE but also in general distrust between salt operations workers and fisheries. Salt factory workers accused fishermen of wanting to destroy their jobs, while fisherman grumbled that, if the federal government were not going to enforce its own laws and prevent ecological and hydrologic disturbances, there would be nothing left to do but dynamite the "salt factory" (Faust and Sinton 1991).

By the end of summer 1990, however, CINVESTAV staff facilitated communication between the two groups, and in September 1990 there were constructive meetings in Ría Lagartos among factory workers, owners, union leaders, fisherman, and preserve administrators. There has been an agreement to work together for mutual goals of sustainable development for all parties. The question remains whether "sustainable development" can include everyone's definition of that concept. It should also be pointed out that CINVESTAV has had the long-term working relationship with many of the parties and has built up a level of trust with villagers, salt workers, and fishermen so that they were in a good position to be the mediator for the dispute.

Although the salt company has expanded its operation, indeed, the two large charcas (basins) have not been rehabilitated since Gilberto. The company has constructed a pier to allow it to ship export salt. The company argued that without the pier, there would be unemployment and it could not survive. Export sales would absorb 250,000–500,000 tons produced annually (J. Andrews communication 1996).

Tourism and Ecotourism Impacts

Residents of Ría Lagartos and Ría Celestún would like to see increased ecotourism activity but not if it threatens the very resources people come to see (Andrews et al. 1998, Ría Lagartos, although it is a Ramsar wetland and is the place where the flamingos breed, is just far enough from both Cancun and Merida that it does not enjoy high number of ecotourists. Ría Celestún, however, is close to Mérida and is enjoying increased numbers of ecotourists from Mexico, Canada, the United States, and Europe in that order as well as large numbers of national Mexican tourists. It is also the area where two NGOs are playing major roles; PRONATURA (Peninsula de Yucatan) working with local people as well as GECE (Grupo Ecologica de Celestún) on ecotourism operations and DUMAC (Ducks Unlimited Mexico) in doing some of the basic research needed to manage the estuary and the species it hosts.

From the mid-1980s occasional tourist groups and CINVESTAV researchers have been paying local fisherman from Celestún to bring them out in the lagoon to see the flamingos (winter season), other water birds, petenes (freshwater upwellings), the mangroves, and petrified forest. The lagoon is very long so boat tours can include more features or less depending on what one wanted to spend. Tourists also go into the village of Celestún for lunch at several of the seafood restaurants. There was very little accommodation for overnight stays. Today this has changed. There is a major visitors' center at the bridge, large numbers of covered boats and guides plus increased accommodations within the village of Celestún.

What PRONATURA and SEMARNAP have done since 1988 is work to organize the local fishermen and others wanting to conduct boat tours into two associations. They sometimes fight over fares but generally get along. PRONATURA also helped them to get government loans to obtain tour boats with canopies for shade and adjustable outboard motors. PRONATURA has also worked with the associations to develop a code of conduct for the guides, e.g., standard fares, not to get too close to the flamingos, observe boat speeds in certain areas, provide ecological information to the tourist. PRONATURA has also worked with local village leadership such as Maria de Carmen, a hotel operator in Mérida and Celestún, to improve the amenities of the village, e.g., pick up the garbage and improve the beachfront facilities. In addition there have been several collaborative projects with the SUNY College of Environmental Science and Forestry to assist with ecotourism development in both Ría Celestún and Lagartos (Galicia and Baldassarre 1997, Moan 1992, Smardon 2006).

Disturbance of Wildlife

Wildlife provides one of the attractors necessary to ecotourism and as a food source for local people. However, impacts on the wildlife may result from even these non-consumptive uses:

- Tourist activity may disturb major species (such as the flamingos) limiting their feeding or breeding success or forcing the species to change their habits or location (Arengo and Baldassarre undated, Galicia and Baldassarre 1997, Yosef 2000). One example is the ecotourists inducing the guide to get the boat close to the flamingo flock while they are resting or feeding on brine shrimp on the lagoon shallows. They will spook and the whole flock moves to a different location. With the new guide association with code of conduct and quieter motors, this is less of an issue than before, but still occasionally happens.
- Increased tourism may degrade or destroy resources needed by a particular species, particularly activity on open beaches needed by sea turtles to lay eggs at night – or use of bright lights, which disorients sea turtles. This is somewhat of an issue at both Rías Lagartos and Celestún where the beaches are used by sea turtles. There are egg-gathering programs run by SEMARNAP in cooperation with CINVESTAV and PRONATURA in both places plus

educational programs to give emphasis to appropriate behavior so as not to disturb nesting sea turtles (Andrews et al. 1998, Frazier 2006).

- Promoting a few species for ecotourism may focus too much attention on a target species causing a market for live animals or products made with those species. This is not a problem for flamingos as hunting them has been outlawed. It is a continuing problem for sea turtles, even though hunting was prohibited. Poaching sea turtles and eggs is still a lucrative business for locals or others at these sites (Andrews et al. 1998, Frazier 2006).

- Impacts of catastrophic natural events on ecotourism and local livelihoods. Hurricane Gilbert literally "cleaned out" both Ría Lagartos and Ría Celestún in 1988 eliminating traditional food sources and nesting sites for flamingos and other species causing them to look for alternate sites along the coast for a few years. The hurricane also resuspended the old lead shot used from hunting ducks on the bottom of Ría Celestún. Researchers at DUMAC started documenting that this resuspension of lead was poisoning flamingos at Celestún. DUMAC, PRONATURA, and (then SEDUE) SEMARNAP collaborated in the research on lead poisoning, which led to banning the use of lead shot on the coast (Andrew communication 1996). Other concerns related to climate change include increased storm activity and flooding of coastal areas (Mulholland et al. 1998, Day et al. 1995) plus increased vulnerability of coastal villages in general (Nicholls 1994).

Increased resource demands: Peak use of Mexican vacation houses is during the 2-week period for Easter break, April 1, and during the summer months of July and August (Mallen 1989). If foreign and domestic tourist seasons coincide, they could stress resources beyond capacity, particularly potable water. This is not a problem for Ría Celestún as the flamingo-viewing season is the winter months, whereas there is more likelihood of conflict at Ría Lagartos where the flamingo-viewing season is the summer months which is the same time that Mexicans use their vacation homes (Alonzo 2007).

Impacts due to an increase in vehicular use: There is increased bus and auto traffic to the boat launch point just across the bridge to Celestún. The use of ATVs may also become a problem causing erosion, scaring wildlife, and destroying vegetation. This does not appear to be a major issue yet, but some beach and dune areas at Ría Lagartos could be susceptible to ATV-induced erosion. The major issue is that the tour boat's keels and outboard motors stir up the sediment in shallow lagoon areas. The use of adjustable outboard motors has partially solved this problem, but there is still sediment disturbance. A secondary issue is the contamination of oil and gasoline in the lagoon waters as well as litter from the boats. The latter does not appear to be a major issue at Ría Celestún or Ría Lagartos.

Increased infrastructure development: In Ría Celestún there are plans for new ecolodges. One would be in town and the other would be east of town. Many of the restaurants are sprucing up their facilities and there is not as much garbage as there used to be because of the efforts of local citizens organized with GECE and PRONATURA. The fishing village of Celestún is growing, with an influx of new comers from the interior of Mexico looking to make a living fishing. Many of the town's residents are indifferent to ecotourism development, but if there were development they would like it to be in town or close to town to benefit local merchants, not just restaurant owners and tour guides.

The four communities in Ría Lagartos Preserve are growing from small fishing villages into medium-sized towns. Restaurants, curio shops, and sleeping accommodations are being built in order to meet and increase tourism activity in the preserve. There was planned a major eco-resort to the south of El Cuyo. Most of the present tourism development in the preserve is of low frequency and often operate from people's homes, with the exception of the Hotel Nefertiti (now closed) in Rio Lagartos and cabins in El Cuyo. Impacts on local resources such as clean food and water are critical constraints. There is a new hotel in San Felipe that is currently expanding. Increased coastal development from tourism is a major issue on the Caribbean side of the Yucatan Peninsula and is also increasing on either side of Progresso between the two coastal biosphere reserves (Andrews et al. 1998, Meyer-Arendt 1991, 2001, Reyes 1986).

Impacts of Agricultural Activities and Other Land Uses

Land clearing for agricultural purposes has destroyed large quantities of the preserve's densely forested wildlife habitat (Barbier and Strand 1993, 1997, Ewel et al. 1998, Hernandez et al. 2001, Kaplowitz 1998). This is forcing more of the preserve's wildlife to compete for less suitable habitat. Long-term reduction of diversity of flora and fauna results. Associated impacts include reduction of microclimate moisture, erosion and runoff affecting water quality, fertilizer and pesticides additions to runoff and water quality, and changing the preserve's vegetation type from medium and low forest to savanna and agriculture.

Impacts of Development

Bridge development: There are two bridges in the Rio Lagartos Preserve, which enable the citizens of Los Colorados and El Cuyo to move from the mainland onto the Ría Lagartos Peninsula. The causeway to El Cuyo has caused some major problems for the flamingo populations in an area of the preserve. The causeway approaches sit too low over the water and during heavy rains it backs up the water in the eastern end of the lagoon, as a dam would. Unfortunately, the heavy rains also seem to coincide with the peak flamingo-nesting season during July and August. Floods caused by the backed-up water have devastated islands full of nesting flamingos (Correa et al. 1989). This was first noticed in1983 (Correa and Boege 1989). There is one bridge in the Ría Celestún Preserve connecting to the coastal town of Celestún and was recently rebuilt.

Development on the barrier islands: The towns of Los Colorados and El Cuyo are built on the Ría Lagartos Peninsula, which forms the reserve's northern limit. Construction and continued activity on the island have destroyed vast stretches of the island's dunes and vegetation. This has compromised the dunes' ability to withstand wind and wave erosion, lowering the protection value of the barrier island from storms and tides and increasing the chances of breaching (Clark 1991). There does not appear to be as much of similar modification near Ría Celestún.

This location also places the inhabitants of these towns at extreme risk to tropical storms and hurricanes. There is no protection between the Ria Lagartos Peninsula and the gulf. The full force of potential storms could hit the towns causing flooding, destroying houses, and wiping out the bridges – the only escape route. In 1988, during Hurricane Gilbert, Las Colorados and El Cuyo suffered the most damage (Ramsar 1989, Sprunt et al. 1988). Ría Celestún was also hit further up the peninsula. Studies by Jauregui and Cruz (1980) show that 25 hurricanes have struck the northeastern coast of the Yucatan Peninsula and that 47 have passed within 250 km of the coast since 1886.

Increased amounts of waste– particularly garbage and sewage are a problem. Many Towns people still throw their trash out their back door (traditional Mayan way) and it is not as organic because of the packaging and plastics, so it is not decomposing. This buildup of garbage allows places for mosquitoes to breed, can help spread disease throughout the community, and attracts vermin and scavengers (Andrews et al. 1998, Smardon 1991). The primary species, which act as scavengers in this situation (i.e., raccoon, turkey, vulture, and rats), are some of the same species, which prey on flamingo eggs and young. Supplying food for these scavengers increases their populations, further threatening flamingo-nesting success. This increased waste also increases the opportunity for bacteria such as *Salmonella* spp. and *E. coli* to grow. These bacteria have been implicated in a number of flamingo deaths (Hernandez and Barron 1989, Anon 1989b), although some biologists dispute this. During a period of 3 years from 1993 to 1996, there have been cases of cholera in Celestún (J. Andrews communication 1996).

Increasing amount of sewage is becoming a problem in the Rio Lagartos Preserve. Disposal of sewage is taken care of primarily by outdoor latrines. Many of these latrines empty into stagnant pools of standing water creating prime disease vector conditions. The waste quickly enters the porous limestone subsurface polluting sources of freshwater and adding large amounts of organic material to the lagoon and raising BOD (biological oxygen demand) levels (Batilori undated and Young et al. 2005).

According to the 1990 census, less than 30% of the houses of Celestún were registered having septic systems. Although the septic systems might work properly, the soil is not suitable for that use because of its permeability; thus it becomes direct contamination of the water bodies. There are still wells with fresh spring water within the village to which people go when there is no potable water supply, which is piped overland into the village (whenever there are failures in the system such as leaks in the pipes or pump problems). The phreatic level of the water is less than 2 m. The water has a salty flavor and it is necessary to boil it. Gastrointestinal diseases and dehydration are very common in infants. The direct inflow into the lagoon is a serious problem because its length of stay in the system is very high (about 50 days, Batilori, 1988). Thus, the pollutants stay for long periods of time propitiating eutrophication, especially in the northeast area of the lagoon.

Increased water use: With an increase in population comes an increased need for potable water. As the population in the Ría Lagartos Preserve has risen, it has added to the increased drain on cenotes and petenes caused by increasing economic needs. Overdrawing from these freshwater sources located in the preserve is causing saltwater intrusion which makes them unfit for drinking (Andrews et al. 1998, Correa and Boege 1989).

Introduction of exotic species: A pressing problem for both the Rías Lagartos and Celestún Preserves is the introduction of domestic animal species. Many of the townspeople have pigs, chickens, and dogs, which run loose throughout the towns. These animals could become potential competition for native wildlife. Pigs and goats will eat a wide variety of vegetation and could cause problems for vegetation management in the preserve. In the future, the preserve management wished to introduce peccaries or tapirs (*Tapis terrestris*) to the preserve; they may have to compete with the escaped domestic animals. Similarly, reintroduction of the white-tailed deer, jaguar, and other mammal species may be jeopardized by packs of wild dogs.

Many of these community development problems are within the domain and control of the preserve towns themselves. There are federal mandates for planning and land use control, but realistically these have little effect. For townspeople in the two biosphere reserves, their only meaningful resources and influences have been the NGOs. So, for instance, CINVESTAV has been doing community education in towns such as Los Colorados for years – educating both children and townspeople about ecologically sound practices addressing waste management, water management, ecotourism development, and counter messages to combat wildlife poaching of sea turtles, deer, and birds. For Celestún, PRONATURA has been working with town leadership to develop methods of addressing waste collection and recycling, community development, counter poaching messages, and even English classes for children and grownups. Such strategies are gradually working to build trust and "action infrastructure" within the communities themselves rather than "top-down" management plans and decrees from the Mexican government agencies, which had been the norm before.

Institutional Analysis

If we look at both Ría Lagartos and Ría Celestún Preserves collectively, there are some major issues, which permeate all the management issues we have reviewed thus far. These issues affect government agencies and NGO roles as well. They are

- lack of resources and decision-making power;
- lack of interagency cooperation;
- lack of data and plans;
- lack of public participation in planning and management.

Lack of Resources and Decision-Making Power

"Funds, personnel, and training are all minimum standards for countries with the economic status of Mexico" (Sprunt et al. 1989). These are hard economic times and the Mexican government believes it cannot afford to spend much on the environment (the author admits this is a gross simplification). This means that there is little government funding for projects and studies within the preserve. Scientists complain that they must spend as much time applying for aid and grants as doing research (from personal conversation with members of CINVESTAV and SEMARNAP 1989). On the other hand there are specific programs like the North American Waterfowl Treaty grant program that funds this type of research activity, and administrators of the program cite the lack of Mexican/American/Canadian proposals and those that are approved are scarce. There are also the World Bank funds for management plan preparation. So the issue is not so much lack of funds, but lack of easily accessible government funds for direct preserve management activities. The NGOs such as PRONATURA, using funds from TNC, have even supplemented the salaries of SEDUE preserve management staff (Andrews et al. 1998). Recently though many regional NGOs are receiving less funding for administrative support from multinational NGOs because of recent shifts in policy.

Lack of Inter-agency Cooperation

Many agencies involved with the Rio Lagartos Preserve have formed partnerships and are cooperating in order to gather data and complete a management plan. Unfortunately, some agencies are operating on a narrow agenda and have not taken environmental issues into consideration (Correa and Boege 1989). For a more detailed discussion of instructional factors involved with natural area management in Mexico, please see Smardon and Faust (2006) and Mumme et al. (1988).

Mexico's agencies are responsible for the development of complete projects within the project without contacting SEMARNAP, the agency in charge of natural areas, for permission and do not consider the environmental consequences of their actions. On the other hand, regional, national, and international NGOs can become a resource for focusing attention on significant projects or actions that can cause harm to the environment and are not respecting SEMARNAP's authority. The salt processing operation in Rio Lagartos is such an example. NGOs also have been accused of having their own organizational objectives – which has to do with forming or maintaining an image (see Frazier 2006).

Lack of Site Data and Plans

Although there are many studies going on in the preserves, little systematic information is available and much needs to be done. Without funding or full government support, long-term projects and research cannot be done, scientists cannot be paid, equipment cannot be bought, and local opposition threatens projects. All these factors cause instability and uncertainty, slow the gathering of data, and planning of projects vital for the ecological health of the preserves. CINVESTAV has done much of the biological research for the preserve supplemented by DUMAC's work on flamingo and waterfowl habitat and PRONATURA's work and support for sustainable economic-related projects. In fact the university and NGO contribution to site data accumulation is sometimes a major benefit. The issue here is that certain types of systematic studies are needed for both preserves such as hydrologic and trophic-level analyses and models that address critical management problems presented in this case study. This is changing as scientists and agencies develop frameworks to address such issues (Clark 1991, Comtreras-Espimosa and Warner 2004, Euan-Avilia and Witter 2002, Rivera-Monroy et al. 2004, Yanez-Arancibia et al. 2004).

Comprehensive management plans although completed for the two reserves often did not include local knowledge as part of the process (Fraga 2006). All the preceding factors, lack of funding, lack of interagency support, and the resulting lack of information, have delayed completion of the preserve's management plans. Actually there was a management plan done for the Ría Celestún Preserve but there was so little participation by other agencies and the Celestún town's people that it was ignored and not accepted. When the World Bank funds were finally made available to produce management plans for both preserves – a key concern was how much participation and involvement of local population, community organizations, and NGOs was allowed or encouraged. This is also changing in southern Mexico and the Caribbean as documented by Fraga et al. (2006) and Mazzotti et al. (2005) have proposed an ecological model for the Sian Ka'an Biosphere Reserve.

Mexico's system of land ownership makes it difficult to protect natural areas. Town expansion and population growth are two of the largest problems in the Ría Lagartos Preserve and significant issues for Ría Celestún. In Mexico, one management tool for urban planning is the Urban Development Plan (Esquema de Desarollo Urbana). This document acts as a technical advisory plan for town development and growth. As such, it is only a guideline and has no legal ramifications. If the town's leading figures ratify this document it becomes a "Plan Director". The Plan Director, with some reservations, then becomes a legal tool for development. This sets up enforceable guidelines and restrictions for town growth.

At this point no town has voluntarily ratified its Esquema de Desarollo Urbana, and it does not seem like any are interested in doing so (from personal discussion with Alfredo Alonzo 1989). It seems a more successful strategy would be to directly address some of the undesirable side affects of town growth, i.e., waste generation, contaminated water supply, disease vector habitat, feral dog populations, and economic development alternatives like ecotourism activities; by working with NGO-sponsored projects like those of PRONATURA.

Lack of Public Participation in Planning and Management

Traditionally in Mexico there are few options for the average citizen to affect or participate in planning and management issues. Decisions concerning land development are typically imposed from federal or state agencies to the local official or ejido leader. Unfortunately, these local leaders often use their position for their own personal gain and do not concern themselves with the needs of their constituents (Uphoff 1985). This often creates situations where the local population is angered by and ignores the dictates of the local powers that be. When these dictates involve changes in land use or protected areas and resources locals may continue prohibited activities. There are also issues of working with indigenous or traditional communities such as the Maya throughout the Yucatan Peninsula. Some of these issues are presented by Faust (1991) and Smardon and Faust (2006) and it is also notable that the Ramsar Bureau has a separate handbook for wetland management with indigenous communities (Ramsar 2000).

Together, these problems make it difficult for the management of both preserves to carry out its functions properly. The key, which is beginning to happen in Celestún, is to involve local community members as stakeholders. This is seen in PRONATURA's efforts to organize the boat guide associations for organized tours of the estuary and the organization of town's people for garbage pick-up and other community development purposes. Specific projects that result in direct benefits to local residents, either economic or quality of life, create foundations for taking on more difficult issues. This can also be seen in Fraga's (2006) long-term work with the villages of San Felipe and Rio Lagartos – to develop a place for local knowledge as part of resource decision making. Management of fisheries and other aquatic resources has, in some places, moved toward co-management or common pool resource model (Begossi and Brown 2003) as exemplified by the local marine reserve creation in San Felipe.

So, in summary, on the surface it would appear that CINVESTAV and PRONATURA are playing subsidiary support roles for the government agencies such as SEMARNAP. There are divisive issues of whose interests are being served. Some maintain (Frazier 2006) that the NGOs are merely maintaining an image to further future funding. Others are maintaining that these same NGOs are fighting to survive, given the funding shifts by major international NGOs and granting foundations (Andrews 2006).

In reality, these NGOs are getting results and involvement of preserve residents in projects that directly benefit them. Even more important is the level of trust local residents and government agencies alike place in the NGOs. This is illustrated by CINVESTAV's role as a mediator in the dispute between the fishermen, preserve management, and salt workers in Ría Lagartos and PRONATURA's sponsored projects in Ría Celestún, which got local residents involved in activities that they previously would not. These same NGOs lend a sense of continuity for either community or preserve management, because as everyone in Mexico knows, elections and resultant restructuring of government agencies create constant change and shifts in program direction and goals.

Acronyms

CICY: Center for Scientific Investigations of the Yucatan
CINVESTAV: Centre de Investigacions Y Estudios Avanzados
CONABIO: National Commission for Knowledge and Use of Biodiversity
CONANP: National Commission on Protected Areas
CPE: Federal Electric Commission
CZM: Coastal Zone Management
DUMAC: Ducks Unlimited of Mexico
ECOSUR: The College of the Southern Border
FONHAPO: Trust Fund of National People's Rooms
GEF: Global Environmental Facility
INE: National Institute of Ecology
IUCN: International Union for the Conservation of Nature
ISYSA: Yucatan's Industrial Salt Society
LEEGEPA: Mexico's General Law for Environmental Protection
NAWT: North American Waterfowl Treaty
PROFERA: Federal Prosecutor for Protection of the Environment
PRONATURA: Program for Nature

SAHOP: Ministry of Human Settlements and Public Works
SARAH: Ministry of Agriculture and Natural Resources
SECTUR: Ministry of Tourism
SEDESOL: Ministry of Social Development
SEDUE: Ministry of Urban Development and Ecology
SEMARNAP: Secretariat of the Environment, Natural Resources and Fisheries
SEMARNAT: Secretariat for the Management of Natural Resources
SEMID: Ministry of Industry and Mining
SEPES: Ministry of Fisheries
SINAP: National System of Protected Area Management
SOT: Ministry of Transportation and Communication
SPP: Ministry of Budget and Programming
TNC: The Nature Conservancy
UADY: Autonomous University of Yucatan
USDI: US Department of the Interior

References

Albert, L. A. 1996. Persistent pesticides in Mexico. *Reviews of Environmental Contamination & Toxicology*, 147: 1–44.

Alonzo, G. M. 2007. Science and NGO's: Collaboration for the conservation of groundwater resources in the Yucatan Peninsula. In NAS, *Sustainable Management of Groundwater in Mexico: Proceedings of a Workshop*, pp. 97–101. Washington, DC: NAS, Science and Technology for Sustainability Program.

Andrews, J. M. 2006. Shifts of strategies and focus of the conservation efforts of PRONATURA on the Yucatan Peninsula: A personal history. *Landscape and Urban Planning*, 74: 193–203.

Andrews, J. M., R. M. Von Bertrab, S. Rojas, A. S. Mendez, and D. A. Rose. 1998. Mexico: Ria Celestun and Ria Lagartos Special Biosphere Reserves. In K. Brandon, K. H. Redford, and S. E. Sanderson (eds.) *Parks in Peril – People, Politics and Protected Areas*, pp. 78–105. Washington, DC: The Nature Conservancy and Island Press.

Anon. 1989a. *Ecological Evaluation of the Rio Lagartos Wildlife Preserve, Yucatan, Mexico*. Merida, YUC: CINVESTAV-IPN Working Paper.

Anon. 1989b. *Phoenicoterus Ruber Ruber in the Yucatan*. Merida, YUC: CINVESTAV-IPN Working Paper.

Arellano-Torres, A., R. Perez-Castaneda, and O. Defo. 2006. Effects of a fishing gear on an Artesanal multispecific penaeid fishery in a coastal lagoon of Mexico: Mesh size, selectivity and management implications. *Fisheries Management & Ecology*, 13(5): 309–317.

Arengo, F. and G. Baldassarre. undated. *American Flamingos and Ecotourism on the Yucatan Peninsula, Mexico*. SUNY College of Environmental Science and Forestry, Syracuse, NY, 11 pp.

Barbier, E. B. 1993. Sustainable use of wetlands. Valuing tropical wetlands benefits: Economic methodologies and applications. *The Geographic Journal*, 159(1): 22–32.

Barbier, E. B. and I. Strand. 1998. Valuing mangrove-fishery linkages – A case study of Campeche, Mexico. *Environment and Resource Economics*, 12(2): 151–166.

Barron, J. and J. Correa. 1990. *Los Flamencos en Mexico*. Merida, YUC: CINVESTAV-IPN Working Paper, 35 pp (in Spanish).

Batilori, S. E. 1988 *Productividad Seconaria en el Estero de Celestun.* Tesis Grad Maestria. CINVESTAV – IPN, Unidad, Merida (in Spanish).

Batilori, S. E. 1990. *Caracterizacion Ecologica del Reguio Faunistico "Ria de Celestun" al Noroeste de la Peninsula de Yucatan.* Merida, YUC: A CINVESTAV Working Paper, Merida, 83 pp (in Spanish).

Batilori, E. C. undated. *Algunos Apsetos de la Hidrologia de Rio Lagartos.* Merida, YUC: A CINVESTAV Working Paper, 21 pp (in Spanish).

Beard, J. S. 1944. Climax vegetation in tropical America. *Ecology*, 25(1944): 125–158.

Begossi, A. and D. Brown. 2003. Experiences with fisheries co-management in Latin America and the Caribbean. In D. C. Wilson, J. R. Nielsen, and P. Degnbd (eds.) *Co-Management Experience Accomplishments, Challenges and Prospects*, pp. 135–152. Dordrecht/Boston/London: Kluwer.

Bianchi, T. S., J. R. Pennock, and R. W. Twilley. 1999. *Biochemistry of Gulf of Mexico Estuaries.* New York: John Wiley & Sons, 448 pp.

Biocenosis, A. C. 1989. Estudio para del Sureste v su Aprovechamiento. Ed. *IMERNARAC2*: 215–271, Mexico (in Spanish).

Boo, E. 1990. *Ecotourism: The Potentials and Pitfalls*, Vols. I and II. Washington, DC: World Wildlife Fund.

Bourdelle, E. 1956. *"Essai d Unification de la Nomenclature en Matiere de Protection de la Nature, In Chavarria*, 1988, p. 8.

Brooks, D. M., R. E., Bodner, and S. Matola. 1997. *Tapirs: Status Survey and Conservation Action Plan.*IUCH/SSC Tapir Specialist Group. Gland, Switzerland and Cambridge, UK: IUCN, viii – 164 pp. and online at http://www.tapirback.com/tapirgal/iucn-ssc/tsg/action97cover.htm

Cable, J. E., D. R. Corbett and M. W. Walsh. 2002. Phosphate uptake in coastal limestone aquifers: A fresh look at wastewater management. *Limnology and Oceanography Bulletin*, 11(2): 29–32.

Caillas, C. A. et al. 1992. *Programa Conceptual de Manejo de la Reserva Especial de la Biosphera Ria Lagartos.* ITESM campus CECAREMA Unidad de Infromacion Biografica 206 pp, plus appendices (in Spanish).

Cesar-Dachary, A. and S. M. Arnaiz. 1984. *Estudios Economicos Preliminares de Quintana Roo. El Territorio Y la Populacion (1982 1983).* Cancun, QR: CIQRO, 294 pp (in Spanish).

Chavarria, E. 1988. *Coastal Protected Areas in Mexico: A Management Assessment.* Unpublished Masters Thesis, Oregon State University, Corvallis.

Clark, J. R. 1974. *Coastal Ecosystems; Ecological Considerations for Management of the Coastal Zone.* Washington, DC: The Conservation Foundation.

Clark, J. R. 1991. Management of coastal barrier biosphere reserves. *Bioscience*, 41(5): 331–336.

Cole, L. J. 1910. The caverns and people of the Northern Yucatan. *Bulletin American Geographic Society*, 42: 321–335.

Conroy, M. 1998. *Sustainable Use of Natural Resources in the Yucatan Peninsula, Mexico.* Washington, DC: World Bank/WBLS CBNRM Initiative, 7 pp.

Comtreras-Espimosa, F. and B. G. Warner. 2004. Ecosystem characteristics and management considerations of coastal wetlands in Mexico. *Hydrobiologia*, 511(1–3): 233–245.

Correa, J. and E. Boege. 1989. *Problemas que Afectan al Refugio Faunistico de Rio Lagartos.* Merida, YUC: CINVESTAV-IPN Working Paper (in Spanish).

Correa, J. and J. Garcia. 1991. *Estado actual de la polacion de flamingos (Phoenicoterus ruber ruber) en la Peninsula de Yucatan.* Merida, YUC: CINVESTAV-IPN Working Paper (in Spanish).

Correa, J., E. Batllori, and G. de la Cruz. 1989. *Rio Lagartos Ecosystem: A Case Study of Coastal Resource Management.* Merida, YUC: CINVESTAV-IPN Working Paper.

Day, J. W., D. Pont, P. F. Hensel, and C. Ibanez. 1995. Impacts of sea-level rise on deltas in the Gulf of Mexico and the Mediterranean: The importance of pulsing events to sustainability. *Estuaries*, 18(4): 636–647.

Espejel, I. 1984. La Vegetacion de las Dunas Costeras de la Peninsula de Yucatan. I. Analisis Floristico del Estado de Yucatan. *Biotica*, 9(2): 183–210 (in Spanish).

Espino-Barrios, R. and G. A. Baldassarre. 1989a. Activity and habitat patterns of Breeding Caribbean flamingos in Yucatan, Mexico. *Condor*, 91: 585–591.

Espino-Barrios, R. and G. A. Baldassarre. 1989b. Numbers, migration chronology, and activity patterns of non-breeding Caribbean flamingos in Yucatan, Mexico. *Condor*, 91: 592–597.

Euan-Avilia, J. I. and S. G. Witter. 2002. Promoting integrated coastal management in the Yucatan Peninsula, Mexico. *Journal of Policy Studies*, No. 12.

Ewel, K. C., R. R. Twilley, and J. E. Ong. 1998. Different kinds of mangrove forests provide different goods and services. *Global Ecology and Biography Letters*, 7(1): 83–94.

Faust, B. B. 1991. Maya culture and Maya participation in the International Ecotourism and Resource Conservation Project. In J. Kusler (compiler). *Ecotourism and Resource Conservation*, pp. 224–226. Berne, NY: Association of Wetland Managers.

Faust, B. B. and J. Sinton. 1991. Let's Dynamite the Salt Factory; Communication, coalitions, and sustainable use among users of a Biosphere Reserve. In J. Kusler (compiler). *Ecotourism and Resource Conservation*, pp. 602–624. Berne, NY: Association of Wetland Managers.

Flores-Verdugo, F. J., J. W. Day, L. Mee, and R. Briseno-Duenas. 1988. Phytoplankton production and seasonal biomass variation of seagrass, *Ruppia maritima*, in a Mexican lagoon with an ephemeral inlet. *Estuaries*, 11(1): 51–56.

Fraga, J. 2006. Local perspectives in conservation politics; the case of the Ria Lagartos Biosphere Reserve, Yucatan, Mexico. *Landscape and Urban Planning*, 74(3–4): 285–295.

Fraga, J., Y. Arias, and J. Angulo. 2006. Chapter 4: Communities and stakeholders in marine protected areas of Mexico, Dominican Republic and Cuba. In Y. Breton, D. Brown, B. Davy, M. Haughton, and L. Ovares. *Coastal Resource Management in the Wider Caribbean: Resilience, Adaptation, and Community Diversity*. Toronto: IDRC Publications.

Frazier, J. 2006. Biosphere reserves and the "Yucatan" syndrome; another look at the role of NGO's. *Landscape and Urban Planning*, 74(3–4): 313–333.

Galicia, E. and G. A. Baldassarre. 1997. Effects of motorized tour boats on the behavior of non-breeding American Flamingos in Yucatan, Mexico. *Conservation Biology*, 11(5): 1159–1165.

Garret, W. E. 1989. La Ruta Maya. *National Geographic*, Oct. 1989: 424–479.

Gold-Bouchot, G., V. Ceja-Moreno, J. P. Rodas-Ortiz, J. Dominguez-Maldonado, D. Espinola-Panti, P. Ku-Chan, and M. Yarto. 2006. Organochloride pesticides and PCBs in sediments in the Southern Gulf of Mexico and the Yucatan Peninsula. *Organohalogen Compounds*, 68: 2113–2116.

Heimo, M., A. H. Siemens, and R. Hebda. 2004. Pre-Hispanic changes in wetland topography and their implications to past and future wetland agriculture at Lagunda Mandinga, Veracruz, Mexico. *Agriculture and Human Values*, 21(4): 313–327.

Hernandez, M. A. and J. G. Barron. 1989. Estudio del Flamingo en la Peninsula de Yucatan. *Bosques Y Fauna*, 13: 3–13 (in Spanish).

Herrera-Silveira, J. 1987. *Productividad Primera Fitoplanctonica en la Laguna de Celestun, Yucatan*. Tesis Profesional, U.A.G. Mexico (in Spanish).

Hurley, B-C. 2005. "Crocodylus, moreletii" (On-line). Animal Diversity Web. http://animal-diversity.ummz.umich.edu/site/accounts/information/crocodylus_moreletii.html.

IUCN, Conservation International and Nature Serve. 2000. *Global Amphibian Assessment*. Washington, DC: IUCH/SSC-CI/CABS Biodiversity Assessment Unit.

Kaplowitz, M. D. 1998. Conflicting Agendas for Mangrove Wetlands in Yucatan, Mexico. Working Paper for 1990 Latin American Studies Association, Chicago, IL, p. 43.

Jauregui, E., J. Vidal, and F. Cruz. 1980. Los Ciclones Y Tormentas Tropicala en Quintana Roo Durante el Periodo 1871–1978. In CIQRO (ed.) *Memorios del Simposio Quintana Roo: Problematicos Y Perspectivaa*, pp. 47–63. Cancun, QR: CIQRO (in Spanish).

Lara-Dominguez, A. L., J. W. Day, G. V. Zapata, R. W. Twilley, H. A. Guillen, and A. Yanez-Arancibia. 2005. Structure of unique inland mangrove forest assemblage in fossil lagoons on the Caribbean coast of Mexico. *Wetlands Ecology and Management,* 13(2): 111–122.

Lee, S. Y. 1995. Mangrove outwelling: A review. *Hydrobiologia,* 295(1–3): 203–212.

Lopez-Carrillo, L., L. Torres-Arreola, L. Torres-Sanchez, F. Espinosa-Torres, C. Jimenez, M. Cebrian, S. Waliszewski, and O. Saldate. 1996. Is DDT use a public health problem in Mexico? *Environmental Health Perspectives,* 104(6): 584–588.

Lugo, A. E., S. Brown, and M. W. Brinson. 1988. Forested wetlands in freshwater and salt-water environments. *Limnology and Oceanography,* 33(4): 894–909.

Mallen, C. 1989. *Guide to the Yucatan Peninsula, including Belize,* 2nd, ed. Moon Publications.

Mazzotti, F. J., H. E. Fling, G. Merediz, M. Lazcano, C. Lasch, and T. Barnes. 2005. Conceptual ecological model of the Sian Ka'an Biosphere Reserve, Quintana Roo, Mexico. *Wetlands,* 25(4): 980–997.

Meyer-Arendt, K. J. 1991. Tourism development on the north Yucatan coast: Human response to shoreline erosion and hurricanes. *GeoJournal,* 23(4): 327–336.

Meyer-Arendt, K. J. 2001. Recreational development and shoreline modification along the north coast of Yucatan, Mexico. *Tourism Geographies,* 3(1): 87–104.

Miranda, F. 1958. La Vegetacion de la Peninsula Yucaeca. En Beltran, E. Los Recursos Naturales de Sureste y su Aprovechamiento. Ed. *IMERNARAC,* 2: 215–271, Mexico (in Spanish).

Moan, S. 1992. Ecotourism in the Yucatan Peninsula; Ecotourism potentials for the Ria Lagartos Wildlife Reserve. Unpublished Masters Project, SUNY College of Environmental Science and Forestry, Syracuse, NY, 280 pp.

Montel, S., A. Estrada, and P. Leon. 2006. Bat assemblages in a naturally fragmented ecosystem in the Yucatan Peninsula, Mexico: species richness, diversity and spatio-temporal dynamics. *Journal of Tropical Ecology,* 22: 267–276.

Mulholland, D. J., G. R. Best, C. C. Coutant, G. M. Hornberger, J. L. Myer, D. J. Robinson, J. R. Sternberg, R. E. Turner, F. Vera-Herrera, and R. G. Wetzel. 1998. Effects of climate change on freshwater ecosystems of the Southeastern United States and the Gulf Coast of Mexico. *Hydrological Processes,* 11(8): 949–970.

Mumme, S., C. R. Bath, and V. J. Assetto. 1988. Political development and environmental policy in Mexico. *Latin Research Review,* 23(1): 7–35.

Murgia, R. R. 1989a. *Notes About an Ecotourism Program on the North Coast of the Yucatan Peninsula.* Merida, YUC: CINVESTAV-IPN Working Paper (in Spanish).

Murgia, R. R. 1989b. *The Eco-Archeological Tourism: An Alternative for the Preservation and Management of the Complex Archeological Site-Tropical Jungle.* Merida, YUC: CINVES-TAV-IPN Working Paper (in Spanish).

Murgia, R. R., R. C. Smardon, and S. Moan. 1991. Developing principals of natural and human ecological carrying capacity, and natural disaster risk vulnerability for application to ecotourism development in the Yucatan Peninsula. In J. Kusler (compiler). *Ecotourism and Resource Conservation.* Berne, NY: Assoc of Wetland Managers, pp. 740–751.

Navarrete, A. J. and J. J. Olivia-Rivera. 2002. Litter Production of Rhizophora Mangle at Bacalar Chico, Southern Quintana Roo, Mexico. *Universidad y Ciencia,* 18(36): 79–86.

Nicholls, R. J. 1994. Synthesis of Vulnerability Analysis Studies. In *Proceedings of World Coast '93,* Rijkswaterstaat, The Netherlands Coastal Zone Management Centre, 41 pp.

Ortiz-Hernandez, M. C. and R. Saenz-Morales. 1999. Effects of organic material and distribution of fecal coliforms in Chetumal Bay, Quintana Roo, Mexico. *Earth and Environmental Science,* 55(3): 423–434.

Pare, M. L. 1986. *Regionalizacion Demografica y Socioeconomica del Estado de Yucatan.* Merida, YUC: CONAPO-CINVESTAV-IPN (in Spanish).

Pérez-Gil, R. and F. Jaramillo-Monroy. 1992. Natural Resources in Mexico: A report to IUCN and the Interamerica Development Bank.

ParksWatch. 2002. *Park Profile – Mexico Ria Celestun Biosphere Reserve*. ParksWatch at
    http://www.parkswatch.org

Perry, E. C. 1991. *Hydrologic, Hydrogeologic, and Geochemical Study of Rio Lagartos Lagoon,
    Yucatan, Mexico*. An unpublished research proposal, University of Illinois, Urbana.

Platt, S. G. and J. B. Thorbjarnason. 2000. Population status and conservation of Morelets
    Crocodile. *Biological Conservation*, 96(1): 21–29.

Ramo, C. and B. Busto. 1993. Resource use by herons in a Yucatan wetland during the
    breeding season. *Wilson Bulletin*, 105(4): 573–586.

Ramos-Miranda, J., L. Quinicu, D. Flores-Hernandez, T. Do Chi, L. Ayala-Perez, and A.
    Sosa-Lopez. 2005. Spatial and temporal changes in the nekton of the terminus lagoon,
    Campeche, Mexico. *Journal of Fish Biology*, 66(2): 513–530.

Ramsar Convention. 1989. *Ramsar Advisory Missions: Report No. 12. Ria Lagartos, Mexico
    (1989)*. Gland, Switzerland: Ramsar Bureau.

Ramsar Convention. 2000. *Handbook 5: Establishing and Strengthening Local Communities
    and Indigenous People's Participation in the Management of Wetlands: Annex: Case Studies
    on Local and Indigenous People's Involvement in Wetland Management*. Gland, Switzer-
    land: Ramsar Convention Bureau.

Rejmankova, E., K. O. Pope, M. D. Pohl, and J. M. Rey-Benayas. 1995. Freshwater wetland
    plant communities in Northern Belize: Implications for paleaecological studies of Maya
    wetland agriculture. *Biotropica*, 27(1): 28–36.

Rejmankova, E., K. O. Pope, R. Post, and E. Maltby. 2007. Herbaceous wetlands of the
    Yucatan peninsula: communities at extreme ends of environmental gradients. *Hydrobio-
    logia and Hydrographie*, 81(2): 223–252.

Reyes, P. J. 1986. *Las Technologicas de Ecodesarrollo como Alternativa a los desequilibrios del
    Proceso de Urbanizacion en la Peninsula de Yucatan. Proquesta para el uso del Sitema
    Integral de Reciclamiento de Desechos Organicos en Celestun, Yucatan*. Tesis Profesional,
    Facultad de Arquitectura, U. A. N. L., Mexico (in Spanish)

Reyes, E. and M. Merino. 1991. Diel dissolved oxygen dynamics and eutrophication in a
    shallow well-mixed tropical lagoon (Cancun, Mexico). *Estuaries*, 14(4): 372–381.

Rivera-Monroy, V. H., C. J. Madden, J. W. Day, R. R. Twilley, F. Vera-Herrera, and H.
    Alvarez-Guillen. 1998. Seasonal coupling of a tropical mangrove forest and an estuarine
    water column: enhancement of aquatic primary productivity. *Hydrobiologia* 379(1–3): 41–53.

Rivera-Monroy, V. H. and R. R. Twilley. 1996. The relative role of denitrification and
    immobilization in the fate of inorganic nitrogen in mangrove sediments. (Terminos
    Lagoon, Mexico). *Limnology and Oceanography*, 41(2): 284–296.

Rivera-Monroy, V. H., R. R., Twilley, D. Bone, D. L. Childers, C. Coronado-Molina, I. C.
    Feller, J. Herrera-Silveira, R. Jaffe, E. Mancera, E. Rejmankova, J. E. Salisbury, and E.
    Weil. 2004. A conceptual framework to developing long-term ecological research & mgmt.
    objectives in the wider Caribbean region. *BioScience*, 54(9): 843–856.

SAHOP 1982. (Secretaria de Asentamientos Humanos Y Obras Publicas) *Ley General de Bienes
    Nacionales, Dario Oficial, Viernes 8 de Enero de 1982*, Mexico, pp. 14–33 (in Spanish).

Sauer, J. 1967. *Geographic Reconnaissance of Vegetation Along the Mexican Gulf Coast*. Baton
    Rouge; LSU Press.

Schmitz, R. A. and G. A. Baldassarre. 1992a. Contest asymmetry and multiple bird conflicts
    during foraging among non-breeding American flamingos in Yucatan, Mexico. *Condor*,
    94: 254–259.

Schmitz, R. A. and G. A. Baldassarre. 1992b. Correlates of flock size and behavior of foraging
    American flamingos following Hurricane Gilbert in Yucatan Mexico. *Condor*, 94: 260–264.

Silva, M. and I. Delvestre. 1986. Marine and coastal protected areas in Latin America: A
    preliminary assessment. *Coastal Management*, 15(1): 311–345.

Sklar, F. H. and J. A. Browder. 1998. Coastal environmental impacts brought about by
    alterations to freshwater flow in the Gulf of Mexico. *Environmental Management*, 22(4):
    547–562.

Smardon, R. C. 1991. Ecotourism and landscape planning, design, and management. In J. Kusler (compiler). *Ecotourism and Resource Conservation*, pp. 704–709. Berne, NY: Assoc of Wetland Managers.

Smardon, R. C. 2006. Heritage values and functions of wetlands in Southern Mexico. *Landscape and Urban Planning*, 74(3–4): 296–312.

Smardon, R. C. and B. B. Faust. 2006. Introduction: International policy in the biosphere reserves of Mexico's Yucatan Peninsula. *Landscape and Urban Planning*, 74(3–4): 160–192.

SPP. 1989 *Esquema de Dessarrollo Urbano' Rio Lagartos en San Felipe.* Working documents for town development (in Spanish).

Sprunt, A. IV, S. C. Snedaker, and J. R. Clark. 1988. *The Ecological Status of Rio Lagartos following Hurricane Gilbert. Report of a Visiting team of Experts, July 10–17, 1988.* A report filed for CINVESTAV presented at the 1989 Ecotourism Conference held in Merida, Mexico.

Thompson, J. D. and G. A. Baldassarre. 1990. Carcass composition of non-breeding Blue-Winged Teal and Northern Pintails in Yucatan, Mexico. *The Condor*, 92(4): 1057–1065.

Tran, K. C., D. Valdes, J. Euan-Avila, E. Real, and E. Gil. 2002. Status of water quality at Holbox Island, Quintana Roo State, Mexico. *Aquatic Ecosystem Health and Management*, 5(2): 173–189.

Trejo, A. 1986. *Estudio de la Vegetacion de la Zona Costera Inundable Perteneciente a l;os Bordes de la Laguna de Celestun, Yucatan, Los Manglares.* Reporte de Servicio Social, U.A.M.-Iztapalapa, Mexico (in Spanish).

Twilley, R. R., R. H. Chen, and T. Hargis. 1992. Carbon sinks in mangroves and their implications to carbon budget of tropical coastal ecosystems. *Water, Air & Soil Pollution*, 64(1): 265–288.

Uphoff, N. 1985. Fitting projects to people. In M. Cernea (ed.) *Putting People First*, pp. 359–395. Oxford University Press.

Valdes, C. 1988. *Regional Level Coastal Management in Mexico: A Proposal for Quintana Roo.* Unpublished Masters Thesis, Oregon State University, Corvallis.

Vargas, F. 1984. *Parques Nationales de Mexico Y Reservas Equivalentes.* Mexico: Instituto de Investigations Economicas, UNAM, 110 pp (in Spanish).

Vazquez, J. A. C., H. C. Perez, and J. J. S. Soto. 2005. Composition and spatio-temporal variation of the fish community in the Chacmochuch Lagoon system, Quintana Roo, Mexico. *Hydrobiologia*, 15(2): 215–225.

Vega-Cendejas, E. and H. DeSantilliana. 2004. Fish community structure and dynamics in a coastal hyper saline lagoon: Rio Lagartos, Yucatan, Mexico. Estuarine, *Coastal and Shelf Science*, 60(2): 285–299.

Vega-Cendejas, M. E. and F. Arregun-Sanchez. 2001. Energy fluxes in a mangrove ecosystem from a coastal lagoon in Yucatan Peninsula, Mexico. *Ecological Modeling*, 137: 119–133.

Vega-Cendejas, M. E., U. Ordomez, and M. Hernandez. 1994. Day-night variation of fish population in a Mexican tropical mangrove coastal lagoon. In R. G. Wetzel, A. G. van deValk, R.E, Turner, W. J. Mitsch, and B. Gopol (eds.) *Recent Studies on Ecology and Management of Wetlands.* New Delhi, India: Vedams eBooks.

Wilson, E. M. 1980. Physical Geography of the Yucatan Peninsula. In E. H. Moseley and E. B. Terry (eds.) *Yucatan: A World Apart.* University of Alabama Press.

Wilson, E. M. and A. William, Jr. 1987. A coastal ecosystem in Northwestern Yucatan. In *Conference of Latin American Geographers Yearbook*.

Withers, K. 2002. Shorebird use of coastal wetland and barrier island habitat in the Gulf of Mexico. *The Scientific World Journal*, 2: 514–536.

Woodin, M. C. 2004. Use of saltwater and freshwater by wintering Redheads in southern Texas. *Hydrobiologia*, 279–280(1): 279–287.

Yanez-Arancibia, A., A. L. Lara-Dominguez, J. L. Rojas-Galaviz, P. Sanchez-Gil, J. W. Day, and C. J. Madden. 1988. Seasonal biomass and diversity of estuarine fishes coupled with tropical habitat heterogeneity (Southern Gulf of Mexico). *Journal of Fish Biology*, 33: 191–200.

Yanez-Arancibia, A., A. L. Lara-Dominguez, J. L. Rojas-Galaviz, D. J. Zarate Lomeli, G. J., Villalobos, and P. Sanchez-Gil. 2004. Integrating science and management on coastal marine protected areas in the Southern Gulf of Mexico. *Ocean and Coastal Management*, 42(2): 319–344.

Yanez-Arancibia, A., A. L. Lara-Dominguez, and J. W. Day. 1993. Interactions between mangrove and seagrass habitats mediated by estuarine nekton assemblages: coupling primary and secondary production. *Hydrobiologia*, 264: 1–12.

Yosef, R. 2000. Individual distances among Greater Flamingos as indicators of tourism pressure. *Waterbirds: The International J. of Waterbird Biology*, 23(1): 26–31.

Chapter 9
The Mankote Mangrove: Microcosm of the Caribbean

Introduction and Caribbean and Latin American Wetland Policy Context

Wetlands in Latin America and the Caribbean have sustained human activity since pre-Columbian times (Davidson and Gauthier 1993, Lugo 2002, Smardon 2006), but it is only recently that wetland protection policies have been addressed (Davidson and Gauthier 1993, Castro 1995). In 1985 the International Waterfowl and Wetlands Research Bureau (IWRB), with support from other agencies, coordinated the first comprehensive survey of Neotropical wetlands of international importance. This work resulted in the publication of the *Directory of Neotropical Wetlands* (Scott and Carbonell 1986), which lists and describes wetlands in each Latin American and Caribbean country. This directory was used as a baseline for identifying wetlands in Central America, with emphasis on those wetlands of importance to humans as well as high biological diversity.

An IWRB Workshop was held in Florida in November 1992, which provided an opportunity for organizations from the United States, Europe, and Mexico to meet and comment on a new wetland strategy report. This report, "Wetland Conservation in Central America" (Davidson and Gauthier 1993) was the first region-specific assessment of wetlands in Latin America and the Caribbean. In late 1993, Wetlands International (WI) (formerly wetlands of the Americas) approached USAID for funding to compile and publish the first comprehensive ecological assessment and policy review directed at setting a conservation agenda for South America's diverse wetlands (Castro 1995). This effort became known as "The Wetlands of South America: An Agenda for the Conservation of Biodiversity and for Policy Development". Later efforts by the World Wildlife Fund focused on biodiversity conservation for terrestrial ecosystems for the region.

A workshop was held in Santa Cruz, Bolivia, in the fall of 1995 to coincide with Wetland International's review of its assessment of wetlands in the Latin American Caribbean region (LAC). Difficulties of working through such priorities were noted by Castro (1995) as well as several emerging trends affecting conservation efforts:

R.C. Smardon, *Sustaining the World's Wetlands,*
DOI 10.1007/978-0-387-49429-6_9, © Springer Science+Business Media, LLC 2009

- Emergence of civil governments plus proliferation of NGOs
- Economic reforms causing strengthening of LAC economies
- Continued high rate of urbanization throughout the LAC region

These trends caused major actors such as WWF to move away from being omnipresent – working broadly and widely at a small scale in many areas and projects – to playing a deeper catalytic role in the field and at policy levels by

- identifying conservation priorities for large-scale funders;
- starting projects which can serve as demonstration and then scaling up, or replicating by other funders;
- building up the conservation infrastructure that can take advantage of this increased funding;
- filling gaps in international support.

The other result from the 1995 to 1996 deliberations was a policy document produced by the Inter-America Development Bank entitled "Freshwater Ecosystem Conservation: Toward a Comprehensive Water Resources Management Strategy" (Bucher et al. 1997). Thus from the 1980s to the mid-1990s there have been major shifts, with attention paid to conservation of wetlands systems as part of regional development decision making in the Latin American Caribbean region.

The major actors in the Latin American/Caribbean region include

- International Financial Institutions such as the World Bank, the Inter-American Development Bank (IDB), and the Organization of American States (OAS);
- bilateral aid agencies in North America with major programs affecting water resources include the US Agency for International development (USAID), Canada International Development Agency (CIDA), and the Organization for Economic Cooperation and Development (OECD) which has finalized "Guidelines for Aid Agencies for Improved Conservation and Sustainable Use of Tropical and Subtropical Wetlands";
- non-governmental conservation organizations that have major freshwater programs in the Latin American Caribbean region include World Wide Fund for Nature (World Wildlife Fund), Wetlands International, and the International Union for the Conservation of Nature (IUCN), as well as Conservation International and the Nature Conservancy.

Specific regional Caribbean programs include

- Caribbean Coastal Marine Productivity Program (CARICOMP), which supports long-term, region-wide comparative studies of biodiversity and productivity of Caribbean coastal ecosystems;
- The Society for the Conservation and Study of Caribbean Birds – a regional organization committed to the conservation of wild birds and their habitats in the Greater Caribbean region;
- Wildlife Without Borders – Latin America and the Caribbean – includes specific projects in the Caribbean with matching funding from the US Fish and Wildlife service and other leveraged funding;

- World Bank/WBI-National Strategies for Sustainable Development of Caribbean Small Island States from vision to action includes grants to assist countries in the Eastern Caribbean to develop National Strategies for Sustainable Development (NSSDs) and provide better integration of government plans and programs in relation to the environment. Such support was utilized by the government of St. Lucia (country of the following case study) to develop a national strategy for sustainability development plan.

The Wetland Resource

Most Caribbean countries have saltwater wetlands such as seagrass beds and mangrove and some Caribbean countries have freshwater marshes, forested wetlands, and freshwater aquatic wetlands (Bossi and Cintron 1990, Cintron and Schaefer 1992, Delgado and Stedman, Lugo 1990, West 1977). In terms of value and productivity, seagrass beds and mangrove are critical to fisheries production in the Caribbean (Delgado and Stedman, Faunce and Serafy 2006, Pauly and Yanez 1994, Yanez 1994). Mangroves also produce both market and non-market values through both wood and nonwood products (Ewel et al. 1998, FAO 1994) such as charcoal production and subsistence food harvesting but receive the least attention from conservation donors and agencies (Dinerstein et al. 1995, Lugo 2002). According to Lugo (2002) much ecological research (Lugo and Snedaker 1974) has been done on mangroves in both Latin America and the Caribbean, but he maintains that "studies of the dynamics of mangrove ecosystems lag behind the need for new information for conserving the ecosystem" (Lugo 2002, p. 6).

The mangroves of Latin America and the Caribbean (Fig. 9.1) cover between 4.1 million hectares (Lacerda et al. 1993) and 5.8 million hectares (FAO 1994) or about 30–35% of the worlds total mangrove area. According to Thom (1982), Lugo (2002), and Cintron and Schaeffer (1992), mangroves grow in the following eight environmental settings:

- Low tidal range with ample sediment input
- High tidal range and sediment input
- High wave energy and low sediment input
- High wave energy and high river discharge
- Drowned river valley
- Low-energy carbonate platforms
- Coral rampart or protective soil barrier
- Low-energy embayments without protective barriers

Examples of specific mangrove flora and fauna are described within this case study and the previous case study (Chapter 8) as well as in Cardona and Botero (1998), Chapman (1976), Jimenez (1992), Lugo (1990), and Lugo et al. (1981), Yanez and Lara (1999).

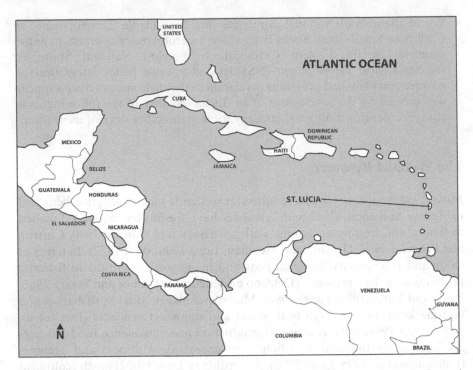

Fig. 9.1 Location of St. Lucia within the Greater Caribbean Region. Map drawn by Samuel Gordon adapted from CIA Web site http://www.cia.gov/cia/publications/factbook/reference.maps/central.america.htm

Specific ecological and human consumptive functions include (see Fig. 9.2)

- fish and crustacean habitat (Delgado and Stedman, Faunce and Serafy 2000, Pauly and Yanez 1994, Ramsunda 2005, Sheridan and Hayes 2005, Verweij et al. 2006, Yanez et al. 1994);
- avian migratory habitat (Frederick et al. 1997, John 2004, Lefebvre et al. 1994, Wunderle and Waide 1994);
- food web connections to microbes, fish, and animals living outside the mangal (Farnsworth 1998, Lopez et al. 1988, Odum 1982, Twilley et al. 1992, Yanez et al. 1993, 1999, Pauly and Yanez 1994, Yanez et al. 1983, FAO 1994);
- food web and nutrient connections between mangals, estuarine waters, seagrass communities, coral reefs, mangrove lagoons, other marine ecosystems, floodplains, and montane communities (Chen and Twilley 1999, Ellison 2002, Odum 1982, Lugo 1986, Lopez et al. 1988, Twilley et al. 1992, Yanez et al. 1993, 1994, Jimenez 1994a);
- absorb upland nitrogen inputs to protect seagrass beds from coastal eutrophication (Chen and Twilley 1999, Feller 1996, Feller et al. 1999, Valiela and Cole 2002);

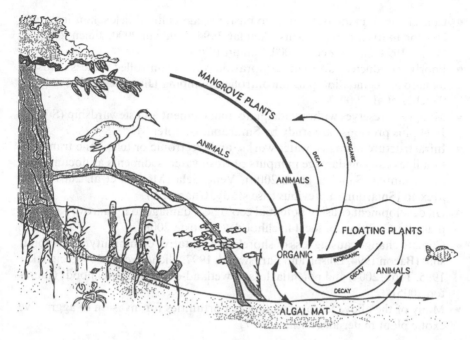

Fig. 9.2 The mangrove-nutrient exchange system. Drawn by Samuel Gordon and adapted from Scott Moan, 1992. Ecotourism in the Yucatan Peninsula; Ecotourism potentials for the Ria Lagartos Wildlife Reserve. Unpublished master's project, SUNY College of Environmental Science and Forestry, Syracuse, NY, p. 112

- use for subsistence food and fiber and local livelihoods (Bacon 1993, Baptiste 2008, Emerton 2005, Ewel et al. 1998, FAO 1994, Kovacs 1999) as well as aesthetic and cultural values (Baptiste 2008, Sanoja 1992, Smardon 2006, Turbey 2004);
- as carbon sinks (Twilley et al. 1992).

Causes of Mangrove Wetland Stress and Degradation

There are two books that address resource management and the impacts of development in the Caribbean (Barker and McGregor 1995, Goodbody and Thomas-Hope 2002) as well as a number of journal articles specifically addressing impacts on mangrove wetlands (Corredor et al. 1999, Ellison and Farnsworth 1996, Kovacs 2000, Lugo 1996) as well as restoration issues (Bacon 1999, Ellison 2000, Imbat et al. 2000). The following is a partial list of existing and potential causes of stress and degradation to mangroves in Latin America and the Caribbean derived from Lugo (2002) and other sources:

- Conversion of mangroves to intensive aquaculture, such as in coastal Ecuador, with attendant loss of habitat ecological services, indigenous peoples livelihoods plus water quality impacts (Olsen and Arriaga 1989, Twilley 1989)

- Conversion of mangrove forests to pastures, agricultural fields, housing, marinas, or tourism developments (Aguilar 1994, Baldwin 2000, Jimenez 1994a, Osorio 1994, Sanchez et al. 2000, Suman 1994)
- Poorly conducted artesian exploration or uncontrolled access causing mangrove degradation plus uncontrolled dumping (Jimenez 1992, 1994a, b, Sanchez et al. 2000)
- Mangrove reserves without adequate management and stewardship (Suman 1994 plus previous case study by Smardon, Chapter 8)
- Infrastructure of roads, water works, and electronic or telephone transmission lines causing damage in inputs of freshwater, sediments as documented in Columbia (Sanchez et al. 2000), Venezuela (Medina et al. 2001), and Mexico (Smardon in previous case study, Chapter 8)
- Oil development causing potential ecological damage to mangrove wetlands plus attendant loss of local livelihoods (Baptiste 2008)
- Climate change causing loss of shoreline, vegetative community, and habitat shift (Bacon 1994, Ellison and Farnsworth 1997, Nicholls et al. 1999, Snedaker 1995, Trotz 2004) and potential loss of wetland-dependent livelihoods (Emerton 2005)
- Many of the above factors creating opportunities for invasion of aggressive exotic plant materials (Bernier 2007)

More productive or sustainable uses (Pons and Fiselier 1991) of mangrove wetlands in the Caribbean include

- use for nature tourism illustrated by Caroni Swamp in Trinidad (Bacon 1970, Blohm and Pannier 1989) and in other potential Caribbean locations (SEDU 2002, Weaver 2004);
- sustainable harvesting for wood or other renewable resource use such as in San Juan River, Venezuela (Lugo 1986, 1996, Hamilton and Snedaker 1984), plus this case study in St. Lucia;
- commercial shrimp and other fisheries that are mangrove dependent, artisanal fishing in mangroves and fish patch production (Delgado and Stedman, Dinerstein et al. 1995, Windevoxhel 1994).

What follows is a specific case study of Mankote mangrove that illustrates many of the wetland mangrove issues throughout Latin America and the Caribbean as well as other parts of the world that have mangrove wetlands.

Introduction to Mankote Mangrove Case Study

Mangrove wetlands are under pressure throughout the Caribbean (Lugo et al. 1981). So much so that management includes the cutting of mangroves for tanning and fuel wood, draining of swamps and reclamation of the land for marinas, and aquaculture or strict protection of the area. There have been efforts to halt degradation of mangroves by community-based (CBO) and non-government organizations (NGOs) in Trinidad and Tobago, St. Lucia, and Barbados (Homer et al. 1991). It is St. Lucia that draws our attention because of the roles of both a community-based association and an NGO – the Caribbean Natural Resources Institute (CANARI) – and the case study does serve as instructional for both problems and opportunities with sustainable use of mangrove swamps in the Caribbean.

Geographical Context

St. Lucia is located among the Windward Islands on the southern part of the Antillean archipelago. Its 238 sq. miles (616 km) hosts a population of approximately 150,000, which is largely concentrated in coastal areas. The interior of the island is mostly mountainous – a testimony to its recent volcanic origin – with numerous valleys and steep ravines cutting through its slopes to tumble down to the clear waters of the Caribbean Sea on the west side and the Atlantic Ocean on the east side. St. Lucia's climate is largely influenced by its broken topography and by weather systems of the Atlantic Ocean, especially in the seasonal passage of hurricanes. The highest part of the island receives the most rainfall, while coastal areas are much drier. Biological life zones occur accordingly, with a succession from the rain forests at higher elevations to xerophytic formations in drier parts, cactus and scrub under more arid conditions, and typical mangrove, beach, and cliff formations on the shoreline.

The government of St. Lucia has been heavily involved with integrated coastal zone planning (Gov. St. Lucia 2004a, 2001b, Walker undated), sustainability planning (Gov. of St. Lucia 2001a), and fisheries management planning (Gov. of St. Lucia 2001b) since 2000. How these planning activities relate to mangrove wetland management will be covered in the summary section at the end of this chapter.

St. Lucia (Fig. 9.3) has about 200 ha of mangrove, and the 60 ha Mankote mangrove on the southeast coast is the largest of the 14 principal mangrove areas (Portecop and Benito-Espinal 1985) on the island and only one of two basin mangroves on the island. Mankote is a basin mangrove cut off from the sea for much of the year (Fig. 9.4) and floods at least once a year. It contains the four most common salt-tolerant mangrove species found in the region, white mangrove (*Laguncularia racemosa*), red mangrove (*Rhizophora mangle*), black mangrove (*Avicennia germinans*), and buttonwood (*Conocarpus erecta*) (see Figs. 9.6 and 9.7).

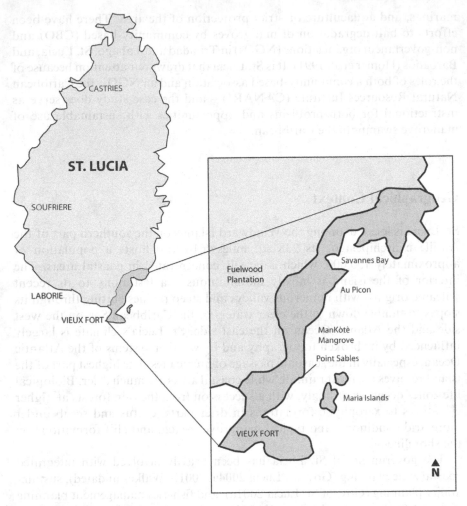

Fig. 9.3 Location map for Mankote mangrove, St Lucia. Map drawn by Samuel Gordon adapted from CIA Web site http://www.cia.gov/cia/publications/factbook/reference.maps/central.america.htm

Wetland Ecology

The vegetation on the beach includes *Cocos nucifera*, *Sophora tomentosa*, and *Sporobolus virginicus*. This beach vegetation grades into salt flat vegetation behind the sand ridge containing *Sesuvium portulacastrum*, *Fimbristylis spathacea*, and *Spartina patens*. Behind both the salt marsh and beach vegetation are alternating bands of red mangrove and mixed black and white mangrove. The mangrove ecology is typical of what we would find in fringe/basin coastal wetlands throughout the Caribbean (see Lugo 1990, Lugo et al. 1988, Lugo and Snedaker, 1974, West 1977).

Fig. 9.4 Southeastern coast of St. Lucia. Photo credit: Richard Smardon

Fig. 9.5 Overview of Mankote mangrove and surrounding area. Photo credit: Richard Smardon

This particular mangrove has a limited value for fisheries but is quite good for wildlife. Portecop and Benito-Espinal (1985) note that it is quite rich in bird species (see Table 9.1), though all are found in small numbers. The most numerous species is the carib grackle (*Quiscalus lugubris*), but the little blue heron (*Egretta caerulea*) is found in large numbers. Migratory species such as

Fig. 9.6 Typical shore/beach vegetation in Mankote mangrove. Photo credit: Richard Smardon

Fig. 9.7 Red mangrove with deeper water levels in the Mankote mangrove. Photo credit: Richard Smardon

the spotted sandpiper (*Actitis macularia*), the northern water thrush (*Seirus noveboracensis*), and the great blue heron (*Ardea herodias*) were also noted on the site (see also John 2004).

These mangle systems serve the following functions: maintaining coastal stability, limited fish breeding and nursery, avifauna habitat, silt trap, water

Table 9.1 List of birds using Mankote mangrove

Local species Scientific name	Common name
Bubulcus ibis	Cattle egret
Butorides virescens	Green heron
Coereba flaveola	Bananaquit
Dendroica adelaide	Adelaides warbler
Elaenia martinica	Caribbean elaenia
Eulampis holosericeus	Green-throated carib
Icterus laudabilis	St. Lucia oriole
Loxigilla noctis	Lesser Antillean bullfinch
Orthorhyncus cristatus	Antillean-crested hummingbird
Quiscalus lugubris	Carib grackle
Saltator albicoloris	Lesser Antillean saltator
Vireo altiloquus	Black-whiskered Vireo
Migratory species **Scientific name**	**Common name**
Anas americana	American widgeon
Anas discors	Blue-winged teal
Ardea alba	Greater egret
Ardea herodias	Greater blue heron
Arenaria interpres	Ruddy turnstones
Actitis macularia	Spotted sandpiper
Aythya affinis	Lesser scaup
Calidris alba	Sanderling
Calidris fuscicollis	White rumped sandpiper
Calidris himantopus	Silt sandpiper
Calidris mauri	Western sandpiper
Calidris melanotos	Pectoral sandpiper
Calidris minutilla	Least sandpiper
Calidris pusilla	Semipalmated sandpiper
Catoptrophorus semipalmatus	Willet
Ceryle alcyon	Belted kingfisher
Charadrius semipalmatus	Semipalmated plover
Circus cyaneus	Northern harrier
Dendrocygna autumnalis	Black-bellied whistling duck
Egretta gularis	Western reef heron
Egretta thula	Snowy egret
Egretta tricolor	Tricolor heron
Falco columbarius	Merlin
Falco peregrinus	Peregrine falcon
Fulica caribaea	Caribbean coot
Limnodromus griseus	Short-billed dowitcher
Limosa haemastica	Hudsonian godwit
Numenius phaeopus	Whimbrel
Pandion haliaetus	Osprey
Pluvialis squatarola	Black-bellied plover
Porphyrula martinica	Purple gallinule
Porzana carolina	Sora

Table 9.1 (continued)

Protonotaria citrea	Prothonotary warbler
Seirus motacilla	Louisiana water thrush
Seirus noveboracensis	Northern water thrush
Tringa flavipes	Lesser yellowlegs
Tringa melanoleuca	Greater yellowlegs
Tringa solitaria	Solitary sandpiper

Source: De Beauville-Scott, 2000.

quality maintenance, and nutrient uptake (Bacon and Alleng 1992, Chen and Twilley 1998, Ellison 2002, Ewel et al. 1998, Faunce and Serafy 2006, Farnsworth 1998, Sheridan and Hays 2005, Twilley et al. 1992, Verweij et al. 2006). They contribute to biological productivity by recycling nutrients from leaf decomposition (Chen and Twilley 1999, Feller 1996, Feller et al. 1999, 2003). The diversity of this habitat type in St. Lucia ranges from a few scattered patches to more diverse riverine and fringing mangal systems. Mangroves account for about 179.3 ha or 0.29% of St. Lucia's biomass.

The Mankote mangrove has been cited as a case study in a number of books and other studies as a case history of successful wetland management by local community-based organizations (see Bustos et al. 2004, Hudson 1998, Novelli and Burns 2006, OECS 2004, Polunin and Curme 1997, Barker and McGregor 1995, World Resource Institute 2000). Specific sources relied upon for the following Mankote mangrove case history include studies by Geoghegan and Smith (1998), Homer et al. (1991), Hudson (1998), Renard (1994), Romulus (1987), Samuel and Smith (2002), and Smith and Berkes (1991, 1993) as well as interviews with actual mangrove cutters plus Matius Burt, Yves Renard, and Allen Smith in January 1996.

Local Land Use/Cultural History

The cultural history (from Walters and Burt 1991a) is revealing in terms of the different uses and misuses of the Mankote mangrove. Before 1939 the Mankote mangrove was part of the Bellevue Sugar Estate. Wood from the mangrove was cut for fuel consumption and export to Barbados.

Early Problems

From 1941 to 1946 Mankote was used by the US military as a site to camouflage aircraft and dump garbage. You can still see the pits that they used to push the planes into during the day and hang camouflage-netting overhead. Access was restricted and no cutting took place.

Post 1946 local charcoal producers to supply fuel wood to nearby towns, especially Vieux Fort, used the Mankote. The mangrove continued to be used as a local dumping area for domestic garbage and industrial waste, as well as a site for cattle grazing and pig rearing – it was the time of benign neglect.

1979–1981: A youth agriculture project is undertaken at Aupicon adjacent to the future Aupicon fuel wood plantation site. While the project disbands before the Mankote-Aupcion project commences, it serves as an inspiration to the local charcoal producers to enter into agricultural production later on.

1981: The Ministry of Health, at the request of the nearby Halcyon Days Hotel, initiates a mosquito eradication program involving extensive spraying and some clearing of the mangroves. At the same time, the Eastern Caribbean Natural Area Management Program (ECNAMP), later to become CANARI, undertakes an extensive survey of the Lesser Antilles and identifies the southeast coast region of Saint Lucia, including the Mankote mangrove, as a priority site for conservation (see Fig. 9.2). ECNAMP with Yves Renard as a principal consultant is enlisted by the National Trust and government of St. Lucia to study the conservation and development requirements of the southeast coast region. It is at this time a group of Vieux Fort Senior Secondary (1991) students undertook a survey of charcoal producers using the Mankote mangrove.

Charcoal Production

At that time (1981) the students found out that there were nine groups of charcoal producers, seven of which lived in nearby Pierrot. One of the groups was a whole family, which would later become more of a tradition involving the children as well. For the production of charcoal, only two species were used – the white mangrove and the buttonwood. The red and black mangrove wood was not as good for charcoal production and the red mangrove sites usually had standing water. The larger branches are cut up and stumps are left to sprout. The branches are assembled into a pit, which is covered with leaves and earth and set fire. Each pit can yield an average of bags of charcoal. Each producer has several pits for a maximum production of 2,700 bags of charcoal per year. A selling price of $20/bag yielded a total production of $54,000 per year for the whole mangrove. For 1995, the average production was 162 bags per month, with the price per bag ranging from EC $25–35.

Key problems included the difficulty of working in the rainy season, cutting of trees reducing the amount of wood available, mosquito eradication, and the effect of pesticides on workers. Finally since the land belongs to the government – created uncertainty about the availability of land access to work the area in the future.

Studies undertaken since 1981 document that local people are using the area extensively for a variety of potentially sustainable purposes. Unlike much of the adjacent public lands, use of the mangrove appeared to be regulated to some extent by the community of users, particularly the charcoal producers (Walters and Burt 1991a,b).

Charcoal making, undertaken by small-scale producers, is an important cottage industry in St. Lucia and throughout the Caribbean. Charcoal makers in Mankote work individually or in small groups, helping one another on a reciprocal basis. Each producer uses one named cutting area per season (two seasons per year, before and after the rains), and rotate cutting areas, returning to a cutover area after about 2 years when new mangrove has regenerated. Regeneration occurs primarily through coppicing – where new stems regenerate from stumps. Charcoal producers actually leave a few larger seed trees to mark location of cutover areas and uncut areas.

Charcoal producers cut selectively in strips or patches for 10–20 m, zigzagging to access clusters of suitable stems (see Figs. 9.8 and 9.9). Cutting area of

Fig. 9.8 Cutting a swath through the black, white, and buttonbush mangrove. Photo credit: Richard Smardon

each working area is generally known to the others in a given season, but with use of larger seed/sprouting trees as markers helps to avoid conflicts. Related individuals often cut in adjacent areas to facilitate exchange of help. Cut stems are placed in rectangular pits dug in the forest floor, about 4–6 m long, partially covered with grass and leaves and then with soil, and then fired for about 3 days (see Figs. 9.10 and 9.11). The charcoal is then bagged in old flour sacks, each sack holding about 22 kg and selling for about EC $30 (US $11 in 1992). Charcoal is retailed in small lots in town markets and rural areas.

Fig. 9.9 Piling up the recently cut stems. Photo credit: Richard Smardon

The three or four producers that started in the early 1960s have increased to 15–20 in the early 1990s. From a loose group, the producers with CANARI's assistance have organized themselves into an informal cooperative. Their cutting rights, tenuous at best until recent years, have been recognized as customary rights, although as of 1992, they still lacked formal rights to use of public land nor had legal authority to manage resources.

For many St. Lucians, charcoal remains the cooking fuel of choice because it is slow burning, easy to transport, imparts a pleasant taste to food, can be purchased in small amounts at low cost, and produces less smoke than fuel wood. The Mankote mangrove has been the main supply of charcoal to some 1500 residents of Vieux Fort and the surrounding communities in the southeast of the island. The alternative cooking fuel is bottled gas, but for most households charcoal is at least as important, particularly for longer cooking tasks. Since charcoal production was a major consumptive use of the mangrove, and the activities of producers posed a visible and immediate threat to the remaining forest, conservation planning was directed primarily at charcoal producers.

Fig. 9.10 Trimming the recently cut mangrove stems. Photo credit: Richard Smardon

Era of Increased Wetland Management

As was mentioned previously, the Mankote mangrove had already been identified in 1981 as a priority area for conservation by the Eastern Caribbean Natural Area Management Program (ECNAMP, later renamed CANARI). A descriptive survey was carried out in 1985 and a monitoring program was started in 1986. The initial goals were to describe and monitor the status of the Mankote mangrove and the level of use and to assess the practices of the mangrove users. Based on this information, the ultimate objective was to ensure the conservation of the mangrove, while providing the resource users with the social and economic benefits from the sustainable use of the mangrove and alternative resources.

The project entailed two major components – first to improve the existing uses of Mankote mangrove by means of community participation and co-management and second to reduce the pressure on the mangrove. A plantation of alternative fuel wood for charcoal making was started in 1983, using

Leucaena leucocephala. This plantation did not live up to initial expectations after three plantings and development effort was broadened from 1987 onward to include a community vegetable garden. In the meantime, the other major impacts on the Mankote mangrove were

- animals roaming along the outer parts of the mangrove and overgrazing;
- indiscriminate dumping of garbage;
- a mosquito eradication program;
- dumping of industrial waste;
- cleared areas of mangrove and decrease in tree diameter from 6–8″ to 1–3″;
- footpaths cleared in the mangrove.

1986: The charcoal producers meet to discuss the concept of forming a producers' cooperative. The by-laws for the group are drafted and the Aupcion Charcoal and Agricultural Producers Group (ACADG) are formed with 14 initial members. Biweekly meetings are convened. ECNAMP and ACADG members participated in a regional meeting of Leucaena project coordinators. Progress on the woodlot is mixed as ACADG formulates a request to the National Development Corporation (NDC) for agricultural land adjacent to the woodlot, but there is no reply. Leucaena in the woodlot are measured by the Forestry Division to evaluate growth rates but annual woodlot planting is not done. Trial marketing of charcoal in supermarkets is undertaken. A decision is made by the ACADG to build a dam to supply water for the proposed agricultural garden at Aupcion. The Mankote mangrove is designated a Marine Reserve under the Fisheries Act, as are all mangroves in St. Lucia.

1987: A major fire destroys 5 acres of the plantation seedlings, but the agricultural component of the Aupcion project assumes prominence and several charcoal producers plant vegetables in a new garden site adjacent to the woodlot at Aupcion. The Jamaican anthropologist Charles Carnegie evaluates the institutional/organizational alternatives available for the charcoal producers group and University of the West Indies student Giles Romulus (1987) examines the Mankote-Aupcion project as a case study of community-based conservation and development. A workshop is held involving the Ministry for Community Development, the National Research and Development Foundation (NRDF), and representatives from the Aupicon/Pierrot/Cacao/Morne Caillandre areas to explore wide community development initiatives.

1988: A delegation led by the Ministry of Agriculture visits the Aupcion project and discusses the relevance and potential of community-based resources management initiatives elsewhere in St. Lucia. The first dam is constructed but much of the agricultural produce from the first planting spoils. The Forestry Division carries out some maintenance work of the woodlot and plants gmelina and cordia species. ACADG members provide paid labor for planting as well as some voluntary labor. The project officer leaves the project and internal conflicts lead to demobilization and temporary disbanding of ACADG.

1989: A formal partnership for project coordination is established between ENCAMP and the National Research and Development Foundation (NRDF).

Inter-American Foundation Funding is obtained and a new project officer, Matius Burt, begins coordination work. Weekly ACADG meetings are convened and formal agreements are signed with all members. A small on-site tree nursery is established at Aupcion. Local varieties of seedlings are planted and approximately 1,200 seedlings are produced. Goats destroy most of the seedlings in the nursery, but the Forestry Division plants 3,000 Leucaena seedlings. An irrigation system for the garden is purchased and a project storage shed is constructed. Koudmen (traditional system of shared labor) involving the charcoal producers and several local groups are used to replant acres of mangrove in Mankote and nearby Savannes Bay. A group of students from the University of Puerto Rico visit the Mankote-Aupcion project and the National Youth Council and several local school groups visit the project.

1990: The producers' group, with assistance from the Forestry Division, makes the first harvest of charcoal from the fuel wood plantation. Serious problems are encountered using a metal kiln provided by the Forestry Division and much of the charcoal is lost because of incorrect use of the kiln. The producers' group purchases two power tillers for the agricultural project and the producers construct a gate to keep the goats off the Aupcion site. A Koudmen involving several ACADG members is used to reconstruct the dam in order to ensure adequate water supply to the agricultural garden. Five Aupcion group members cultivate crops and three agricultural plots realize production of melon, cucumber, corn, and cantaloupe. The Sunshine Harvest Cooperative assists the ACADG to market their produce. The Forestry Division plants 300 leucaena, 1,000 casuarina, and 250 gmelina seedlings at the Aupcion site.

American anthropologist Stephen Koester evaluates the Aupcion project on behalf of the Inter-American Foundation and a fish species inventory of the mangrove is initiated. A community forestry workshop is held involving the Aupcion group, CANARI (formerly ECNAMP), and Forestry Division and NRDF. The Aupcion site is used as a demonstration model. A dozen community groups to the south of St. Lucia attend the workshop. Meetings with all the above parties and additional departments are held and decisions are made to develop a co-management arrangement for the mangrove involving the ACADG. A formal request to the cabinet is made for the vesting of the Mankote mangrove with the National Trust.

In 1990, CANARI is approached by the local Pierrot Youth Organization (PYO) to initiate a community wide tree-planting program. The PYO and Aupicon Development Committee participate in a voluntary tree planting of the local gliricidia species in the Aupcion woodlot. Bellevue Farmers Cooperative began negotiations with the ACADG for land on the project to start chicken and fish farms.

News is revealed of a preliminary plan for a major motel and golf course development that would destroy most of the Mankote mangrove. Cabinet sends a formal letter of refusal for the request to vest the Mankote mangrove in the National Trust.

Fig. 9.11 Pile of mangrove stems about to be put into the pit. Photo credit: Richard Smardon

1991: A workshop involving CANARI, the Forestry Department, and representatives from several community groups is held to discuss the Aupcion Project and three community faculty initiatives. The ACADG refuse the offer of the Bellevue Farmers Cooperative to develop agriculture and livestock at Aupcion. CANARI discusses with the Forestry Department the experimental pruning of Leucaena coppices. Members of ACADG begin clearing land again for the spring vegetable planting. CANARI, with ACADG, hosts a field trip of the Mankote mangrove to sensitize teachers about the values of the mangrove as well as evaluates the potential of ACADG members to lead guided tours of the mangrove. About 60 people attend, including most of the ACADG, several government officials, and more than 30 teachers.

After that workshop in 1991, there have been guided tours of the mangrove by some of the producers mostly for school groups. In addition there has been a trail linked to an observation tower (see Fig. 9.12) constructed by the ACADG. Around the observation tower, which overlooks the mangrove, there are several interpretive signs (see Figs. 9.13–9.15) explaining charcoal making, mangrove vegetation and wildlife, and history of the Mankote mangrove. Some of the charcoal producers also go fishing for crabs and tilapia in the wetter portions of the red mangrove areas.

Current Management of the Mankote Mangrove

Since 1991, CANARI has done an internal evaluation of the project (Walters and Burt 1991aand 1991b), and the CANARI scientists have started to evaluate whether the utilization of the mangrove for charcoal production is sustainable

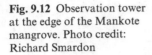

Fig. 9.12 Observation tower at the edge of the Mankote mangrove. Photo credit: Richard Smardon

(Smith and Berkes 1991). CANARI director Renard has also summarized the lessons learned from a participatory and group organization perspective.

Bird hunting was eliminated soon after the Mankote mangrove was designated a Marine Reserve in 1986. Waste dumping, which was a major degrading use, has almost stopped due to enforcement actions such as forcing people to clean up their own dumping. The producers usually report dumping and enforcement is by the St. Lucia Department of Fisheries. The same mechanism is used to enforce illegal cutting by nonmembers. Occasional grazing still occurs at the mangrove edges but is not perceived as a threat. The use of the mangrove for scientific and educational purposes started in the 1980s and the visitor tours in the early 1990s. According to Pantin et al. (undated), the major stress on mangrove systems in St. Lucia is illegal mangrove cutting and dumping leading to habitat destruction, degradation of the resource base, and aesthetic impacts affecting tourism.

Fig. 9.13 Interpretative entry sign for the Mankote mangrove. Photo credit: Richard Smardon

Fig. 9.14 Habitat interpretation sign along the Mankote mangrove trail. Photo credit: Richard Smardon

Fig. 9.15 Interpretative sign for Mankote mangrove vegetation. Photo credit: Richard Smardon

Policy and Government Stakeholders

The major national stakeholders include the Department of Fisheries, which is responsible for the management of marine reserves; the Forestry Department, which is responsible for forest and wildlife management of government lands; the St. Lucia National Trust (SLNT), the country's lead organization of natural and cultural heritage; and the National Development Corporation (NDC), the agency responsible for government lands and legal owner of Mankote (Geoghegan and Smith 1996).

The need for legal provision of cutting rights for the existing subsistence-level charcoal producers was first noted in 1981 and began to be generally accepted around 1990, but did not actually happen until 1996, and then only in a form of a letter from the Deputy Chief Fisheries Officer.

The main legal instruments governing forest use and management are the following:

- The Forest, Soil and Water Conservation Ordinance of 1946, amended in 1956 and 1983. It stipulates the conditions for timber harvesting, makes provisions for control of squatting, and defines other offenses.
- The Wildlife Protection Act of 1980 places authority for wildlife legislation in the hands of the Minister of Agriculture and makes provision for the conservation and management of wildlife through the listing of species, the establishment of reserves, and the setting of fines for offenses.

- The Crown Lands Ordinance of 1946 establishes the position of commissioner of crown lands and sets the conditions for the management of Crown Lands.
- The Land Conservation and Improvement Act of 1992 establishes a Land Conservation Board and gives it a broad mandate with respect to the management of land and water resources.

The government is also part of international conventions, which provide additional support to national policies governing natural resource management. The major convention here is the Biodiversity Convention and specific pertinent articles include

Article 6: General measures for conservation and sustainable use
Article 7: Identification and monitoring
Article 8: In situ conservation
Article 10: Sustainable use of components of biological diversity and
Article 12: Research and training. It should be noted that St. Lucia submitted a report on the "Benefit sharing arrangements in the Mankote mangrove" as part of the Conference of the Parties to the Convention on Biological Diversity Sixth Meeting at the Hague in 2002 (UNEP 2002).

Other pertinent international conventions include

- The International Convention on the Trade of Endangered Species;
- The World Heritage Convention;
- The Convention on the Protection and Management of the Coastal and Marine Environment of the Caribbean (Cartagena Convention).

It should also be noted that the government of St. Lucia is engaged in developing a national framework for sustainable development as part of the Barbados Programme of Action (Gov. of St. Lucia 2004b). This strategy reflects many of the issues that the Mankote mangrove illustrates on a smaller scale. Also as part of ongoing planning efforts the St. Lucia government has developed an integrated coastal zone plan for St. Lucia (Gov. of St. Lucia 2004a, 2001b).

The Role of Scientific Assessment

In terms of the mangrove health itself, Smith and Berkes (1993) have taken measurements since 1986 and there has been a significant increase in the mean stem density from 1989 to 1992. There are current plans to continue detailed biomass accounting as well as aerial photography via kite to show area-wide community changes. The increased regeneration in 1992 stems was particularly important as it followed a year of relatively high charcoal production in 1991. Smith and Berkes (1993) feel that improved regeneration is the result of the change in cutting practices, which contrast to prior clear cutting and indiscriminate slashing of earlier years.

The Mankote practice is simply based on going to a location which has good-sized stems of white mangrove and buttonbush and cutting in zigzagging strips before relocating to a new area in the next season. There is no regularized rotation by producers, no formal rules of allocation (e.g., by lottery), but simply constant communication, first-comer's rights within the group of users, and mutual respect for one another's cutting areas.

The net affect of these cutting practices, however, is that a cover of larger trees cannot be restored at the present rate of cutting. The cutting pressure has to be reduced and that was the reason for planting the Leucaena fuel wood plantation outside the mangrove – to reduce pressure on the mangrove and to lengthen the cutting cycle. Together with other alternative rural development measures, such as agricultural crops, it may be possible to lengthen the cycle permanently and to allow trees in some areas to mature and restore the mangrove forest. In the mean time, the major issue is getting everyone to leave the core red mangrove areas alone which has been pretty much a de facto management agreement.

One of the charcoal producers compiles figures for the number of bags produced by each member and submits these to CANARI each month.

Management and Social Structure

What are the socio-economic and organizational conditions and change in outlook affecting management practices? The more secure resource use rights of the charcoal producers precipitated a change in behavior and attitude. Instead of cutting wood indiscriminately, the security of tenure makes it possible to cut with more care and conserve for the medium or even long term. Smith and Berkes (1993) also maintain that integrated conservation-development projects have good potential to be effective if they can lead to the avoidance of open access conditions and the specification of property rights. This still allows preservation of the core area and resources, which are the year-round open-water red mangrove areas. This approach provides a social context by which the local community of resource users has certain rights and responsibilities. So the users enforce the rules that avoid open access conditions. Not because the users necessarily believe in conservation (although most users now have a conservation ethic), but because avoiding open access is also in their best interests.

Co-management principles, notwithstanding, there have been and continue to be some communication and organizational problems within and outside the producers' association. Rogue members who do not follow the rules or do not come to meetings are shut out. Some long-time charcoal producers are caught, if one of their family members breaks the rules or they have not attended meetings; their counsel is discounted or ignored. Some members have taken the opportunities of agricultural development; nature tour guiding, and other rural development options and some have not. Most members have an attitude that, in the long term, things will work out.

This history of participation and working things out eventually led to the formulation of a plan of co-management of the Mankote mangrove. After the area was declared a Marine Reserve, the project (CANARI and Producers) developed a management agreement among the Ministry of Agriculture (Forestry Division) and the National Development Corporation, and the community of harvesters, making them all responsible for the area's management. While this agreement has not been officially adopted, it is de facto in force.

On the institutional front, the project has demonstrated the need for a diversity of groups to assume part of the area's management responsibility. However, its principal weakness has been its inability to formalize the management agreement among the various government agencies and the community. This is due in part (according to Renard 1994), to bureaucratic delays, but perhaps more importantly to two sets of negative attitudes toward the co-management approach. On the one hand, the country's political directorate does not appear ready to commit itself to the protection of such an area and remains prepared to entertain development proposals – including construction of a golf course – which would significantly alter the area and its resources. On the other hand, some of the resource management and conservation agencies remain opposed to the concept of sustainable harvesting and would favor an end to all extractive uses.

Summary/Roles of CBOs and NGOs

Some commentary is in order, given the last points raised. First, the issue was raised whether the management agreement really needs to be signed if the management of the mangrove is de facto working. No answer was given, but it appears the National Development Authority or most powerful government agency wants to keep its options open. Second, even though the Forestry Department and the National Trust might favor reduction of mangrove cutting for charcoal production, it is the presence and self-policing action of the producers that guard against both overcutting and non-desirable uses of the mangrove such as illegal dumping and hunting.

While these questions remain and are unresolved, it seems useful to comment on the role of ENCAMP/CANARI historically and at present. The most critical role early on was helping the mangrove charcoal producers' association organize themselves and gain legitimacy in the negotiations and discussions about mangrove management. CANARI staff admits themselves that meaningful participation processes involved a learning curve for CANARI staff as well, e.g., early participation was more like consultation for the growers' association, which then moved to local empowerment, and finally progressed to co-management and shared decision making. Without such support the growers' association would not exist and develop.

The second role of CANARI was that of research by Smith and Berkes (1993). The major function of research, in this case, was to find out what was working or not working – sustainable or not in regard to the natural functions and reproduction of the mangrove vegetation. This research served to externally validate the program, e.g., the mangrove grows back after 2 years and stem size is increasing. It also provides internal feedback to the mangrove charcoal producers, e.g., we know we are not overcutting with use of certain practices and this ensures a future supply of mangrove wood.

The third role of CANARI is externalization of the lessons learned from the mangrove wetland. Can sustainable mangrove production be replicated elsewhere in St. Lucia, in the Caribbean, or elsewhere in the tropical world where mangrove grows? Toward this goal of extension and training, the project undertook three important training activities that confirmed the Mankote mangrove's value for demonstration and extension. First, a case study was produced that provided a useful account of the project and identified the lessons learned in its implementation (Walters and Burt 1991a,b).

Second, a regional workshop on people's participation in development and natural resources was hosted by the project. The Caribbean Association, Coordination in Development (CODEL), and the Government of St. Lucia sponsored the workshop. Third, the project served as a case study for a regional workshop on coastal zone management held in St. Lucia in July 1985 under the auspices of the Commonwealth Science Council. Both of these events made the project known to the Caribbean region.

Another workshop was held in August 1998 with funding from the World Wildlife Fund (Ramsar 2003, WWF 2003). This project used the Mankote mangrove site in Saint Lucia as a model for the preparation and testing of training modules (Brown et al. 2000). This project has also been used as an OECS case study for protected areas and associated alternative livelihoods (OECS 2004).

The externalization of the Mankote mangrove story to the Caribbean community has been done by CANARI, but there is greater potential for this project for extension and training. There is a long-term commitment by CANARI to continue monitoring as well as assistance to the producers' association through extension and research. Although this is a relatively small mangrove wetland, the case study is significant because it pertains to the conservation of diversity of fuel wood resources, which are under pressure in various parts of the world such as India and Africa.

There are a number of questions that remain concerning the sustainability of mangrove utilization for such uses in the face of local need for fuel and other wood products (Ewel et al. 1998, Pons and Fiselier 1991). There have been questions raised in other evaluations (De Beauville-Scott 2000, Hudson 1998) about the socio-economic sustainability of such an operation. The other related issue is the pressure for conversion on the mangrove for another land use related to tourism or port development, which is a microcosm of situations throughout the Caribbean (Baldwin 2000, Novelli and Burns 2006, Pons and Fiselier 1991).

There are other questions of mangrove maintenance and restoration throughout the Caribbean region (Bacon 1999, Cardona and Botero 1998, Corredor et al. 1999, Ellison 2000, Ellison and Farnsworth 1996, Imbat et al. 2000). In this regard, one of the major future concerns are effects of global warming, regional climate change and coastal sea level changes, and resultant effect on coastal ecosystems and mangroves in particular (Bacon 1994, Ellison and Farnsworth 1997, Nicholls et al. 1999, Schleupner Undated, Snedaker 1995).

Then the resultant issue is the capacity of the government of St. Lucia, other NGOs, and stakeholders to protect and manage the remaining mangroves, like the Mankote mangrove, from the various pressures listed above. It looks like there are serious efforts underway for sustainability planning (Gov. of St. Lucia 2001a, 2004b, Creary 2003, Rosenberg and Thomas 2005), natural resource-dependent livelihood analysis (OECS 2004, Pantin et al. undated), biodiversity conservation (UNEP 2002, World Resource Institute 2000), and integrated coastal zone planning (Gov. of St. Lucia 2001b, 2004a, Walker undated). The other critical capacity issue is the underpinning environmental science needed for resource decision making, and there seems to be movement for development of research frameworks toward such ends from a biophysical perspective (Rivera-Monroy et al. 2004) and a socio-economic wetland valuation perspective (Brander et al. 2006, Bustos et al. 2004, OECS 2004).

The study is an example of an integrated development project (ICDP) and has a number of relatively unusual characteristics. First, it is based on strengthening the organization of local users and their resource use rights, rather than trying to eliminate them, and building a community-based management system to provide user incentives to conserve, rather than relying on conservation by government control. Second, it employs integrated rural development approaches to diversify the economic base of the community which may otherwise have no alternative but destroy its own resource base.

Acronyms

ACAGG: Aupcion Charcoal and Agricultural Producers Group
CANARI: Caribbean Natural Resources Institute
CARICOMP: Caribbean Coastal Marine Productivity Program
CIDA: Canadian International Development Agency
CODEL: Caribbean Association for Coordination in Development
ECNAMP: Eastern Caribbean Natural Resources management Program
IDB: Inter-American Development Bank
IDP: Integrated Development Project
IWRB: International Waterfowl and Wetlands Research Bureau
NDC: National Development Corporation

NRDF: National Research and Development Foundation
OAS: Organization of American States
OECS: Organization of Eastern Caribbean States
PYU: Pierrot Youth Organization
SLNT: St. Lucia National Trust
WBI: World Bank International
WI: Wetlands International
USAID: US International Development Agency
WWF: Worldwide Fund for Nature-World Wildlife Fund

References

Aguilar, R. X. 1994. CODDEFFAGOLF: los defensores de los manglares del Golfo de Fonseca, Honduras. *Revista Forestal Centroamericana*, 3(9): 27–32.

Bacon, P. R. 1970. *The Ecology of the Caroni Swamp*. Trinidad: Central Statistical Office, 68 pp.

Bacon, P. R. 1993. Mangroves in the Lesser Antilles, Jamaica and Trinidad and Tobago. In L. D. Lacerda (cord.) *Conservation and Sustainable Utilization of Mangrove Forests In Latin America and African Regions*, pp. 155–209. Yokohama, Japan: International Society for Mangrove Ecosystems and International Tropical Timber Organization Project PD114/90(F).

Bacon, P. R. 1994. Template for evaluation of impacts of sea level rise on Caribbean coastal wetlands. *Ecological Engineering*, 3(2): 171–186.

Bacon, P. R. 1999. Wetland rehabilitation in the Caribbean. In. W. Streever and B. Streever (eds.) *An International Perspective on Wetland Rehabilitation*, New York: Springer-Verlag.

Bacon, P. R. and G. P. Alleng. 1992. The management of insular Caribbean mangroves in relation to site location and community type. *Hydrobiologia*, 247(1–3): 235–241.

Baldwin, J. 2000. Tourism development, wetland degradation and beach erosion in Antigua, West Indies. *Tourism Geographies*, 2(2): 193–218.

Baptiste, A. K. 2008. Evaluating Environmental Awareness: A Case Study of the Nariva Swamp Trinidad. Unpublished Dissertation, SUNY College of Environmental Science and Forestry, Syracuse, NY 204 pp.

Barker, D. B. and D. F. M. McGregor. 1995. *Environment and Development in the Caribbean: Geographical Perspectives*. University of West Indies Press, 320 pp.

Bernier, L. P. 2007. Invasion of Caribbean Island Wetlands by Melaleuca quinquenervia: Ecology and management of Alien Wetlands. In *Fort Lauderdale*, Florida: USDA Agricultural Research Service Archives 1PLR.

Blohm, C. and F. Pannier. 1989. Manglares. Editorial *Ex Libris*, Caracas.

Bossi, R. and G. Cintron. 1990. *Mangroves of the Wider Caribbean*. Nairobi, Kenya: Caribbean Conservation Association, The Pano Institute and United Nations Environment Programme, 33 pp.

Brander, L. M., J. G. M. Florax, and J. E. Mermaat. 2006. The empirics of wetland valuation: a comprehensive summary and a meta-analysis of the literature. *Environmental and Resource Economics*, 33(2): 223–250.

Brown, N. A., Y. Renard, and A. Smith. 2000. *A Guide to Teaching Participatory and Collaborative Approaches to Natural Resource Management*.Vieux Fort, St. Lucia: CANARI Technical Report No. 267.

Bucher, E., G. Castro, and V. Floris. 1997. *Freshwater Ecosystem Conservation: Towards a Comprehensive Water Resources Management Strategy*, Washington, DC: Inter-American Development Bank Paper no. ENV-114, 42 pp.

Bustos, B., N. Borregaard, and M. Stilwell. 2004. *The Use of Economic Instruments in Environmental Policy: Opportunities and Challenges*, UN Environmental Program, Division of Technology, Industry and Economics, Economics and Trade Unit, UNEP Program/ Earthprint, pp. 104–106.

Burt, M., interview in Vieux -Fort, St. Lucia, West Indies, January 27, 1996.

De Beauville-Scott, S. 2000. A preliminary assessment of the Basin of the Mankote Mangrove, Saint Lucia, West Indies. Cave Hill, Barbados: Natural Resource management Program, Department of Natural Resource management, Faculty of Science and Technology (unpublished Draft).

Cardona, P. and L. Botero. 1998. Soil characteristics and vegetative structure in a heavily deteriorated mangrove forest in the Caribbean coast of Colombia. *Biotropica* 30(1): 24–34.

Castro, G. 1995. A Freshwater Initiative for Latin America and the Caribbean. Unpublished Report, World Wildlife Fund, 23 pp.

Chapman, V. J. 1976. *Mangrove Vegetation*. Valduz, Crammer.

Chen, R. and R. R. Twilley. 1998. Gap dynamic model of mangrove forest development along gradients of soil salinity and nutrient resources. *Journal of Ecology*, 86(1): 37–51.

Chen, R. and R. R. Twilley. 1999. A simulation model organic matter and nutrient accumulation in mangrove wetland soils. *Biogeochemistry*, 44(1): 93–118.

Cintron, M. G. and Y. N. Schaefer. 1992. Ecology and management of the world's mangroves. In U. Seeliger (ed.) *Coastal Plant Communities of Latin America*, pp. 233–258. San Diego, CA: Academic Press.

Corredor, J. E., R. W. Havarth, R. R. Twilley, and J. M. Morell. 1999. Nitrogen cycling and anthropogenic impact in the tropical inter-American seas. *Biogeochemistry*, 46 (1–3): 163–178.

Creary, M. 2003. *Methodologies, Tools and Best Practices for Managing information for Decision-Making on Sustainable Development in the Caribbean SIDS*. Kingston, Jamaica: University of the West Indies, Centre for Environment and Development (UWICED), 27pp.

Davidson, I. and M. Gauthier. 1993. *Wetland Conservation in Central America*. Report Number 93-3, Ottawa, ON: North American Wetland Conservation Council (Canada), 87pp.

Delgado, P. and S.-M. Stedman. Undated. *The US Caribbean Region: Wetlands and Fish; A Vital Connection*. Silver Spring, MD: NOAA Fisheries, Office of Habitat Conservation.

Dinerstein, E, D. M. Olson, D. J. Graham, A. L. Webster, S. A. Primm, M. P. Bookbinder, and G. Ledec. 1995. *A Conservation Assessment of the Terrestrial Eco-regions of Latin America and the Caribbean*. Washington, DC: World Wildlife Fund and World Bank, 129pp plus maps.

Ellison, A. M. 2000. Mangrove restoration: Do we know enough? *Restoration Ecology*, 8(3): 219–229.

Ellison, A. M. 2002. Macroecology of mangroves: Large-scale patterns and processes in tropical coastal forests. *Trees – Structure and Function*, 16(2–3): 181–194.

Ellison, A. M. and E. J. Farnsworth. 1996. Anthropogenic disturbance of Caribbean mangrove ecosystems: past impacts, present trends, and future predictions. *Biotropica*, 28(4A): 549–565.

Ellison, A. M. and E. J. Farnsworth. 1997. Simulated sea level change alters anatomy, physiology, growth, and reproduction of red mangrove (*Rhizophora mangle L.*). *Oecologia*, 112(4): 435–446.

Emerton. L. (ed.). 2005. *Values and Rewards: Counting and Capturing Ecosystem Water Services for Sustainable Development*. Sri Lanka, Colombo: IUCN Water, Nature and Economics Technical Paper no. 1, The World Conservation Union, Ecosystems and Livelihoods Group Asia.

Ewel, K. C., R. R. Twilley and J. E. Org. 1998. Different kinds of mangrove forests provide different goods and services. *Global Ecology and Biogeography Letters*, 7(1): 83–94.

FAO (Food and Agriculture Organization). 1994. *Mangrove Forest Management Guidelines*. Rome Italy, FAO Forestry Paper 117, 319 pp.

Farnsworth, E. J. 1998. Issues of spatial, taxonomic and temporal scale in delineating links between mangrove diversity and ecosystem function. *Global Ecology and Biogeography Letters*, 7(1): 15–25.

Faunce, C. H. and J. E. Serafy. 2006. Mangroves as fish habitat: 50 years of field studies. *Marine Ecology Progress Series*, 318: 1–18.

Feller, L. C. 1996. Effects of nutrient enrichment on leaf anatomy of Dwarf Rhizophora Mangle L. (Red Mangrove). *Biotropica*, 28(1): 13–22.

Feller, L. C., K. L. McKee, D. F. Whigham, and J. P. O'Neill. 2003. Nitrogen vs. phosphorous limitation across an ecoternal gradient in a mangrove forest. *Biogeochemistry*, 62(2): 145–175.

Feller, I. C., D. F. Whigham, J. P. O'Neill, and K. L. McKee. 1999. Effects of nutrient enrichment on within-stand cycling in a mangrove forest. *Ecology*, 80(7): 2193–2205.

Frederick, P. C., J. Correa, C. Luthin, and M. Spaulding. 1997. The importance of the Caribbean coastal wetlands to Nicaragua and the Honduras to Central American populations of water birds and Jabiru storks (*Jabiru mycteria*). *Journal of Field Ornithology*, 68(2): 287–295.

Geoghegan, T. and A. H. Smith. 1998. *Conservation and sustainable livelihoods; Collaborative management of the Mankote Mangrove, St. Lucia: Community Participation in Forest management.* Vieux Fort, St. Lucia: CANARI.

Goodbody, I. and E. M. Thomas-Hope (eds.). 2002. *Natural Resource Management for Sustainable Development in the Caribbean.* Canoe Press, 416 pp.

Government of St. Lucia. 2004a. *Coastal Zone Management in St. Lucia: Policy Guidelines and Selected Projects.* Vieux-Fort: Government of St. Lucia.

Government of St. Lucia. 2004b. Draft Final Report. *To Review the Implementation of the Barbados Programme of Action (BPOA).* Vieux-Fort: Ministry of Physical Development, Environment and Housing, Government of St. Lucia, Executive Summary, pp. 1–12.

Government of St. Lucia. 2001a. *Plan for Managing the Fisheries of St. Lucia.* Vieux-Fort: Government of St. Lucia.

Government of St. Lucia. 2001b. Chapter 3. Current Coastal Area Management Issues. In *Integrating the Management of Watersheds and Coastal Areas in St. Lucia.* Vieux-Fort: Government of St. Lucia, pp. 50–81.

Hamilton, L. S. and S. C. Snedaker (eds.). 1984. *Handbook for Mangrove Area Management.* Honolulu, HI: United Nations Environment Programme and East-West Center, Environment and Policy Institute, 123 pp.

Homer, F., Y. St. Hill, and Y. Renard. 1991. *Management of coastal wetlands: The role of community based on nongovernmental organizations.* Vieux Fort, St. Lucia, WI: CANARI Communication no. 16, Caribbean Natural Resources Institute (CANARI), 9pp.

Hudson, B. 1998. A Socio-Economic Study of the Community Based management of Mangrove Resources in St. Lucia. In *"Crossing Boundaries". 7th Annual Conference of Int'l. Assoc. for the Study of Common Property.* Vancouver, British Columbia, 28pp.

Imbat, D., A. Rousteau, and D. Scherrer. 2000. Ecology of mangrove growth and recovery in the Lesser Antilles: State of knowledge and basis for restoration projects. *Restoration Ecology*, 8(3): 230–236.

Jimenez, J. A. 1992. Mangrove forests of the Pacific Coast of Central America. In U. Seeliger (ed.) *Coastal Plant Communities of Latin America*, pp. 259–267. San Diego, CA: Academic Press.

Jimenez, J. A. 1994a. *Los Manglares del Pacifico Centroamericana.* Heredia, Costa Rica: Editorial Fundacion UNA, 336 pp.

Jimenez, J. A. 1994b. Bosques de manglares en la Costa Pacifica de America Central. *Revista Forestal Centroamericana*, 3(9): 13–17.

John, C. L. 2004. *Migrant Bird Records for Saint Lucia, West Indies.* Vieux, Fort, St. Lucia: CANARI.

Kovacs, J. M. 1999. Assessing mangrove use at the local scale. *Landscape and Urban Planning*, 43: 201–208.

Kovacs, J. M. 2000. Perceptions of environmental change in tropical coastal wetland. *Land Degradation and Development*, 11: 209–220.

Lacerda, L. D. (Coord.) 1993. *Conservation and Sustainable Utilization of Mangrove Forests in Latin America and Africa Regions*. Yokohama, Japan: International Society for mangrove Ecosystems and International Tropical Timber Organization Project PD114/90(F), 272 pp.

Lefebvre, G., B. Poulin, and R. McNeil. 1994. Temporal dynamics of Mangrove bird communities in Venezuela with special reference to migrant warblers. *The Auk*, 111: 405–415.

Lopez, J. M., A. W. Stoner, J. R. Garcia, and I. M. Garcia. 1988. Marine food webs associated with Caribbean island mangrove wetlands. *Acta Cientifica*, 2: 94–123.

Lugo, A. E. 1986. *Water and the Ecosystems of the Luquillo Experimental Forest*. New Orleans, LA: USDA Forest Service, Southern Forest Experiment Station General technical Report SO-63, 17 pp.

Lugo, A. E. 1990. Fringe wetlands. In *Forested Wetlands, Ecosystems of the World, 15*. New York: Elsevier, pp. 143–169.

Lugo, A. E. 1996. Caribbean island landscapes: indicators of the effects of economic growth on the region. *Environment and Development Economics* 1(1): 128–136.

Lugo, A. E. 2002. Conserving Latin American and Caribbean mangroves: issues and challenges. *Madera y Bosques Numero Especial*, 2002: 5–25.

Lugo, A. E., S. Brown, and M. M. Brinson. 1988. Forested wetlands in freshwater and saltwater environments. *Limnology and Oceanography*, 33(4): 894–909.

Lugo, A. E., R. Schmidt, and S. Brown. 1981. Tropical forests in the Caribbean. *Ambio*, 10(6): 318–324.

Lugo, A. E. and S. C. Snedaker. 1974. The ecology of mangroves. *Annual Review of Ecology and Systematics*, 5: 39–64.

Mangrove cutters, interviews in Vieux-Fort, St. Lucia, West Indies, January 26–27, 1996.

Medina, E., H. Fonseca, F. Barboza, and M. Francisco. 2001. Natural and man-induced changes in a tidal channel mangrove system under tropical semiarid climate at the entrance of the Maracaibo lake (Western Venezuela). *Wetlands Ecology and Management*, 9: 233–243.

Nicholls, R., F. M. J. Hoozemans, and M. Marchand. 1999. Increasing flood risk and wetland losses due to global sea-level rise: Regional and global analyses. *Global Change*, 9: 569–587.

Novelli, M. and P. M. Burns. 2006. *Tourism and Social Identities*. New York: Elsevier.

OECS. 2004. *Additional Annex 14: Social Assessment and Public Participation OECS Countries: OECS Protected Areas and Associated Alternative Livelihood*. Castries, Saint Lucia: Organization of Eastern Caribbean States (OECS), pp. 8–15

Odum, W. E. 1982. *The Ecology of the Mangroves of South Florida: A Community Profile*. Washington, DC: US Department of the Interior, Fish and Wildlife Service, Biological Services Program, FWS/OBS-81/24, 144pp.

Olsen, S. and L. Arriga (eds.). 1989. *A Sustainable Shrimp Mariculture Industry for Ecuador*. Technical Report Series TR-E-6, Narragansett, RI: International Coastal Resources Management Project, 276pp.

Osorio, O. 1994. Proyecto INRENARE/OIMIT al rescate de los mangles de Panama. *Revista Forestal Centroamericana*, 3(9): 33–37.

Pantin, D., D. Brown, M. Mycoo, C. Topin-Allahor, J. Gobin, W. Rennie, and J. Hancock. Undated. *Feasibility of Alternative Sustainable Coastal Resource – Based Enhanced Livelihood Strategies*. Trinidad and Tobago: Sustainable Economic Development Unit, University of West Indies, St. Augustine Campus, pp. 43–57.

Pantin, D., D. Brown, M. Mycoo, C. Toppin-Allahar, J. Gobin, W. Rennie, and J. Hancock. Undated. *Scientific Annex; Feasibility of Alternative, Sustainable Coastal Resource–Based Livelihood Strategies (R8135)*. Trinidad and Tobago: Sustainable Economic Development Unit (SEDU), University of West Indies, St. Augustine Campus, pp. 41–46, St. Lucia Case Study.

Pauly, D. and A. A. Yanez. 1994. Fisheries in coastal lagoons. In B. J. Kjerfve (ed.) *Coastal Lagoon Processes*, The Netherlands, Amsterdam: Elsevier, pp. 377–399.

Polunin, N. and L. M. Curme. 1997. *World Who is Who and Does What in Environment and Conservation*. James and James/Earthscan.

Pons, L. J. and J. L. Fiselier. 1991. Sustainable development of mangroves. *Landscape and Urban Planning*, 20: 103–109.

Portecop, J. and E. Benito-Espinal. 1985. *The Mangroves of St. Lucia: A Preliminary Survey*. Vieux-Fort, St. Lucia, WI: ECNAMP, pp. 43–454.

Ramsar. 2003. *CANARI Completes WFF Project on Participatory Management Training*. Gland, Switzerland: Ramsar Convention Secretariat.

Ramsunda, R. H. 2005. The distribution and abundance of wetland icthyofauna and exploitation of fisheries of the Godineau Swamp, Trinidad-Case Study. *Revista de Biología Tropical*, 53(1): 1–13.

Renard, Y., interview in Vieux-Fort, St. Lucia, West Indies, January 27, 1996.

Renard, Y. 1994. Community Participation in St. Lucia. *Number 2 in Community and the Environment; Lessons from the Caribbean*. Fort-Vieux, St. Lucia: A series by the Panos Institute and the Caribbean Natural Resources Institute (CANARI).

Rivera-Monroy, V. H., R. R. Twilley, D. Bone, D. L. Childers, C. Coronado-Molina, I. C. Feller, J. Herrera-Silveira, R. Jaffe, E. Mancera, E. Rejmankova, J. E. Salisbury, and E. Weil. 2004. A conceptual framework to develop long-term ecological research and management objectives in the wider Caribbean region. *BioScience*, 54(9): 843–856.

Romulus, G. 1987. *Micro-Study of Charcoal Production in the Mankote Mangrove with an Evaluation of a Conservation Strategy for Sustainable Development*. Barbados: Centre for Resource Management and Environmental Studies, University of West Indies, Cave Hill Campus.

Rosenberg, J. and L. L. Thomas. 2005. Participating or just talking: Sustainable development councils and the implementation of Agenda 21. *Global Environmental Politics*, 5(2): 61–87.

Samuel, N. and A. Smith. 2000. *Popular Knowledge and Science: Using the Information that Counts in Managing Use of a Mangrove in St. Lucia, West Indies*. Paper Presented at: Quebec 2000 Millennium Wetland Event, Quebec, 6–12 August 2000. Fort Vieux, St. Lucia: CANARI Communication no. 278, 5 pp.

Sánchez-Páez, H., R. Alvarez-León, O. A. Guevara–Mancera, and G. A. Ulloa-Delgado. 2000. *Hacia la Recuperacion de los Manglares del Carbe de Columbia*. Santa Fe de Bogotá, Columbia: Impresos Panamericana, 294 pp.

Sanoja, M. 1992. Wetland Ecosystems and the management of cultural heritage. In A. E. Lugo and B. Bayle (eds.) *Wetland Management in the Caribbean and the Role of Forestry in the Economy*. New Orleans, LA: USDA Forest Service, Southern Forest Station, pp. 66–73.

Scott, D. A. and M. Carbonell. 1986. *Directory of Neotropical Wetlands*. Cambridge, UK: International Waterfowl and Wetlands Research Bureau, Slimbridge and IUCN.

SEDU, Dept. of Economics, University of West Indies. 2002. "Synopsis of the Study of the Insertion of Environmental Management in Sartorial Polices; The Tourism Case in the Caribbean". Produced by SEDU, Department of Economics, University of West Indies, Trinidad and Tobago for Inter-American Development Bank Regional Policy Dialogue, Washington, DC.

Sheridan, P. and C. Hayes. 2005. Are mangroves nursery habitat for transient fishes and decapods? *Wetlands*, 23(2): 449–458.

Smardon, R. C. 2006. Heritage values and functions of wetlands in Southern Mexico. *Landscape and Urban Planning*, 74(3–4): 296–312.

Smith, A., interview in Vieux -Fort, St. Lucia, West Indies, January 26, 1996.

Smith, A. H. and F. Berkes. 1993. Community based use of mangrove resources in St. Lucia. *International Journal of Environmental Studies*, 43: 123–131.

Smith, A. H. and F. Berkes. 1991. Solutions to the Tragedy of the Commons: Sea urchin management in St. Lucia, West Indies. *Environmental Conservation*, 18(2): 131–135.

Snedaker, S. C. 1995. Mangroves and climate change in the Florida and Caribbean region: Scenarios and hypotheses. *Hydrobiologia*, 295(1–3): 43–49.

Suman, D. 1994. Legislacion y administracion de los manglares de America Central. *Revista Forestal Centroamericana*, 3(9): 6–12.

Thom, B. G. 1982. Mangrove ecology – a geomorphologic perspective. In B. F. Clough (ed.) *Mangrove Ecosystems in Australia: Structure, Function and Management*. Canberra, Australia: Australian Institute of Marine Science and Australian National University Press, pp. 3–17.

Trotz, U. 2004. "Developing Country Dialogue on Future International Actions to Address Global Climate Change". Presentation given in Mexico City, Mexico Nov. 16, 2004.

Turbey, S. 2004. Folklore and popular conceptions regarding the fauna of a wetland area on the Caribbean coast of Columbia. *Agriculture and Human Values*, 21(2–3): 105–110.

Twilley, R. R. 1989. Impacts of shrimp mariculture practices on the ecology of coastal ecosystems in Ecuador. In S. Olsen and L. Arriga (eds.) *A Sustainable Shrimp Mariculture Industry for Ecuador*. Technical Report Series TR-E-6, Narragansett, RI: International Coastal resources Management Project, The University of Rhode Island Coastal Resource Center, pp. 91–120.

Twilley, R. R., R. H. Chen, and T. Hargis. 1992. Carbon sinks in mangroves and their implications to carbon budget of tropical coastal ecosystems. *Water, Air and Soil Pollution*, 64(1–2): 265–288.

UNEP. 2002. *Conference of the Parties to the Convention on Biological Diversity, Sixth Meeting; Incentive measures*. UNEP/CBD/COP/6/12/Add.3, pp. 1–5.

Valiela, I. and M. L. Cole 2002. Comparative evidence that salt marshes and mangroves may protect seagrass meadows form land derived nitrogen loads. *Ecosystems*, 5: 92–102.

Verweij, M. C., I. Nagelkerken, S. L. J. Wartenbergh, I. R. Pen, and G. van derVelde. 2006. Caribbean mangroves and seagrass beds as daytime feeding habitats for juvenile French grunts. *Haemulon Flavolineatum. Marine Biology*, 149(6): 1291–1299.

Vieux-Fort Senior Secondary School. 1991. *Preliminary Report on a Survey of Charcoal Production in Mankote (Vieux-Fort)*. St. Lucia, West Indies: Vieux-Fort Secondary School.

Walker, L. A. Undated. *Towards the Development of a Coastal Zone Management Strategy and Action Plan for Saint Lucia*. Vieux Fort: Government of Saint Lucia, 118 pp.

Walters, B. B. and M. Burt. 1991a. *Community-based Management of Mangrove and Fuel wood Resources: A Case Study of the Mankote-Aupicon Project, St. Lucia, West Indies*. Vieux-Fort, St. Lucia: Caribbean Natural Resources Institute (CANARI), 40 pp.

Walters, B. B. and M. Burt. 1991b. *Integrated Management of Common Property; Fuel wood Resources from Natural and Plantation Forests in St. Lucia*. Paper prepared for the IDRC Workshop on Common Property Resources, Winnipeg, Canada. Vieux-Fort, St. Lucia, WI: CANARI, 21 pp.

Weaver, D. B. 2004. Ecotourism in the small island Caribbean. *GeoJournal*, 31(4): 457–465.

West, R. C. 1977. Tidal salt marsh and mangal formations of Middle and South America. In V. J. Chapman (ed.) *Wet Coastal Ecosystems: Ecosystems of the World*, pp. 193–213. New York: Elsevier.

Windevoxhel. N. 1994. Valoracion econnomica de los manglares: demonstrando la rentabilidasd sostenible, Caso heroes y martires de Veracruz, Nicaragua. *Revista Forestal Centroamericana*, 3(9): 18–26.

WWF. 2003. CANARI competes WWF project on participatory management training. WFF at http://www.ramsar.org/wff/wff_rpts_stlucia_canari.htm

World Resource Institute. 2000. Managing Mankote Mangrove. *World Resources Institute 2000-2001: People and Ecosystems; The Fraying Web of Life*, pp. 176–177. Washington, DC: World Resource Institute.

Wunderle, J. M. Jr. and R. B. Waide. 1994. Future prospects for Nearartic migrants wintering in Caribbean forests. *Bird Conservation International*, 4: 191–207.

Yanez, A. A. and A. L. Lara D. (eds.). 1999. *Mangrove Ecosystems in Tropical America*. Instituto de Ecologia, A. C. Xalapa. Mexico, IUCN/ORAMA, Costa Rica and NOAA-NMFS, Silver Springs MD, 380 pp.

Yanez, A. A., A. L. Lara Dominguez, and J. W. Day Jr. 1983. Interactions between mangrove and seagrass habitats mediated by estuarine nekton assemblages; Coupling of primary and secondary production. *Hydrobiologia*, 264: 1–12.

Yanez, A. A., A. L. Lara Dominguez, and D. Pauly. 1994. Coastal lagoons as fish habitats. In B. J. Kjerfve (ed.) *Coastal Lagoon Processes*, pp. 363–376. The Netherlands, Amsterdam: Elsevier.

Chapter 10
Review of Wetland Management Roles, Functions, and Innovations

Introduction

Faced with a loss of 50% of the world's wetlands plus increasing stress on the remaining wetlands (Maltby 1986, Millennium Ecosystem Assessment 2005a), it is time to review the eight previous case studies for lessons learned and any wetland management innovations that could be applicable to other wetland systems and other regions of the world. The focus of this final chapter will be to address these issues as well as to offer any guidance for sustainable wetland management from an international perspective.

From Table 10.1 we can see a number of case studies ranging from coastal estuarine wetlands to riverine systems to lake-related wetland systems. Some are very large scale wetland systems such as the Wadden Sea (Chapter 2) and the Great Lakes coastal wetlands (Chapter 7) to relatively small wetland systems such as the Mankote mangrove (Chapter 9) and Kolkata wetlands (Chapter 5).

Stakeholder Roles

In terms of government stakeholders, we see a range of scale of government policies from the elaborate Tripartite Wadden Sea Agreement and the US/Canada/Mexico Great Lakes wetland programs such as the North American Wetlands Treaty to almost no wetland policy such as Vietnam. Vietnam has one or two Ramsar wetlands but almost no wetland management policy. In almost all instances we can see the value of the guidance provided by the Ramsar Convention and Bureau as well as by the IUCN biosphere reserve program.

International NGOs include World Wildlife Foundation that is very active in Africa, Europe, Asia, and the Americas. Some international NGOs are quasi-governmental such as the Ramsar Convention Bureau and the IUCN. Others are more focused such as the International Crane Foundation that highly values crane habitat maintenance and the World Bank which occasionally provides funding for the Kolkata wetlands fisheries project or biosphere reserve management plans for Mexico.

R.C. Smardon, *Sustaining the World's Wetlands*,
DOI 10.1007/978-0-387-49429-6_10, © Springer Science+Business Media, LLC 2009

Table 10.1 Wetland studies – stakeholders, uses, and innovations

Case study locations	Stakeholder's gov. agencies	INGOs	NGOs/CBOs	Management innovations
Wadden Sea wetlands	Denmark Ministry of Energy & Env. Germany Ministry Env. Nat. Cons. & Nuclear Netherlands Ministry Agri, Fisheries & Nature Mgmt	WWF IUCN Ramsar	Denmark Wadden Group WWF Germany Friends of the Wadden Sea	Trilateral mgmt agreement & conference + shared mgmt. principles
Axios River Delta Greece	Ministry Env. & Physical Plann Ministry Agri. Ministry Industry Energy & Tech	Ramsar MedWet IUCN	WWF/Greece	Thessaloniki Com on wetlands Joint Ministerial Decision (JMD) Red Alert Prog
Kafue Flats Zambia	Dept. Nature & Parks Zambia Wild. Auth ZAMA	WWF Ramsar IUCN	WWF/Zambia U. of Zambia	local Chiefs dist Councils Formal partnerships – sugar industry/energy/ ZAMA/chiefs/Tourism firms
Brace Bridge Nature Park Kolkata wetlands	Calcutta Port Trust Calcutta Mun.Corp W. Bengal Fisheries Dept. India Env. Dept.	World Bank Ramsar	Fish Farmers Devel. Agency	MFCS Multiuse urban wetlands for PUBLIC WQ treatment, fisheries, agri. nature park
Tram Chim Nat & Wildlife Preserve Mekong Delta Vietnam		Intern. Crane Foundation IUCN Brehm Fund	WWWG	Dong Thop Province Leadership negotiated mgmt agreement + hydrologic Restoration

Table 10.1 (continued)

Case study locations	Stakeholder's gov. agencies	INGOs	NGOs/CBOs	Management innovations
Great Lakes Coastal Wetlands US/Canada	USEPA, COE USF&WS, NRCS 8 US states Env Canada Wildlife Service Ontario Ministry Natural Resources	Ducks Unlimited GL Wetlands Policy Consortium (GLWPC)		NAW Treaty vision statement +agenda GLW Conser-vation Action Plan
Ria Celestun Ria Lagartos Biosphere Reserves Mexico	SEMARNAT CONAP, SECTUR SAHOP	Nature Cons. World bank Ramsar IUCN	PRONATURA CINVESTAV Town of San Felipe Town of Celestun Local fisherman associations	Biosphere Reserve Mgmt Plans San Felipe Marine Res. Ecotourism associations
Mankote Mangrove St. Lucia	St. Lucia Fisheries Dept. Port Authority	CANARI> WWF (funding)	ACADG ECNAMP	Sustainable charcoal other uses

National or regional NGOs provide key support roles such as PRONA-TURA for biosphere reserve management and ecotourism development in Mexico or WWF Greece's support of the Red Alert Project in the Axios River Delta wetlands. There are often issues of who controls the funds and project management especially between big international NGOs and national/regional NGOs. For instance in Latin America and the Caribbean, WWF has shifted their strategic role from direct project management to encourage others to take on project management and/or funding.

Another key stakeholder is academia or national or state universities. The role is that of researcher or developer of management technology, which is then utilized by the government, NGO, or CBO. A good illustration of this is the early work by the WWWG from Hanoi University on the Saurus Crane habitat in Tram Chin Nature Reserve in Vietnam or the University of Zambia's early and continuing work on the ecology of Kafue Flats in Zambia.

At the local level, we have community-based organizations, which, to this author, are the real story with wetland management in many cases. Excellent examples are the Mudialy Fisherman's Cooperative Society that oversees aquaculture production in the Kolkata lagoon wetlands to the Mangrove Producers Association in the Mankote mangrove in Saint Lucia, West Indies. Sometimes local initiatives are less formal, such as the leadership shown by the DongThop Province for restoring parts of the Mekong Delta in Vietnam and the fisherman/residents creating the San Felipe Marine Reserve within the Ria Lagartos Biosphere Reserve in Mexico.

This brings us to the role of local residents and management of protected areas. A brief history is presented below but a much more elaborate account of the international biosphere reserve system management can be seen in Batisse (1996), McNeeley (1999), UNESCO (1984, 1995), and Smardon and Faust (2006).

Biosphere Reserves and Stakeholder Management History

In the autumn of 1983, the First International Congress on Biosphere Reserves was held in Minsk, Belarus. It provided a major opportunity for taking a critical and constructive look and eventually led to the formation and adoption in 1984 of the "Action Plan for Biosphere Reserves" (UNESCO 1984), which was formally endorsed by UNESCO, the United Nations Environment Program (UNEP), and the World Conservation Union (IUCN). While the action plan itself was a rather complex (and not always clear) document, it constituted a new starting point for the development of an information network and for the refinement of the biosphere reserve concept (Batisse 1996). What had previously been considered as a rather "loose label" became a much more formal recognition, in particular due to the criteria defined by the scientific advisory panel.

The Fourth World Congress on National Parks and Protected Areas was held in Caracas, Venezuela, in February 1992. Many of the world's protected area planners and managers gathered and, as a body, approved a resolution in support of biosphere reserves. They also produced formal policy statements concerning community involvement and international collaboration that are now essential aspects of these reserves. The emphasis on and clarification of policy concerning collaboration with local communities built on the initial arguments made in 1962 at the First World Congress on National Parks by Mexican conservationist Enrique Beltra'n (1964).

Since the 1992 meeting in Caracas, there have been further innovations in the management of biosphere reserves. New methodologies have been developed for involving local residents in decision-making processes and the resolution of conflicts (see McNeeley 1992, Oviedo and Brown 1999, Jeanrenaud 1999, Erickson 2006). Increased attention has been given to the need to use regional approaches (see Dyer and Vinogrado 1990, Batisse 1996 for examples; Stolton and Dudley 1999 for a general discussion). New kinds of biosphere reserves such as cluster and transboundary reserves have been devised. Many biosphere reserves that began with a primary focus on conservation evolved into greater integration of local uses, as initial conflicts led to improved communication strategies and increased cooperation among local residents, reserve managers, NGOs, and scientific researchers – often referred to jointly as "stakeholders" (UNESCO 1995). This was a move toward the position originally taken by Mexican conservationists such as Beltra'n (1949) that conservation should be combined with long-term development strategies and includes participation by and benefits to local communities.

Recent biosphere reserve management has also focused on economic management policy and the role of incentives, economic valuation of protected areas plus ecotourism potential, and funding mechanisms for such (Munasinghe and McNeely 1994). Likewise, recent Ramsar, IUCN, and WWF publications (Emerton 2005, Schuyt and Brander 2004) have moved toward economic valuation of wetland services such as nutrient flows, flood abatement, and water supply. These publications have drawn heavily from the ecological economics literature (Barbier 1993, Barbier et al. 1997, Bardecki 1998, Brander et al. 2003, Costanza et al. 1997, Turner et al. 2000, Woodward and Wui 2001). There is a good deal of controversy over how environmental services of wetlands are valued (see Brander et al. 2003). Put simply, the issue is that wetlands provide a myriad of services and functions (see Table 10.1) that cannot always be appropriately quantified or valued. This author and others (Millennium Ecosystems Assessment 2005, Emerton 2005, Ratner et al. 2004, Schuyt and Brander 2004) are most concerned with subsistence wetland utilization and wetland-dependent livelihoods within rural developing country contexts.

Even if we have good participation of stakeholders in sustainable wetland management, we sometimes have conflicts over power, credit, and money between government agencies, international NGOs, national/regional NGOs, and CBOs (see Frazier 2006, Terborgh 1999). The author has already alluded to the conflict in project control and funds between BINGOs and local NGOs. This seems to be spreading throughout the world where the BINGO has large donors that are very much concerned with regional or even international habitat/species protection issues. A regional or local NGO, in many cases, must balance habitat/protection goals with local sustainability issues of a local human resident population. Hence, there is a potential management conflict. This is well illustrated by the International Crane Foundations work with Tram Chim Wildlife Preserve in the Mekong Delta, Vietnam, in Chapter 6. ICF was very much interested in crane habitat restoration while the local residents were interested in fisheries, rice, and *Melaleuca* forestry restoration. It was because of the painstaking negotiations for development of the management plan that all functions were allowed to be realized.

The other major issue is sustainable or "wise" use of wetlands for food, fiber, and other human renewable resource use vs. wetland habitat preservation/restoration (Smart and Kanters 1991). In North America we are much interested in the control of exotic vegetation and species that do not (ecologically) belong in a particular wetland community type. In many places in the world, substantially altered wetlands are heavily utilized for food and/or fiber production or may be permanently developed. Examples include the Kolkata wetlands in Chapter 5, which are used for sewage treatment plus pisciculture plus agriculture or otherwise lost to land development; or the Mankote mangrove in Chapter 9, which is heavily utilized for charcoal production when under constant threat of development of other uses by the local port authority. So the operative phrase may be "use it or lose it" for many areas of the world where there is this constant conversion or development threats.

This type of pressure even happens for Ramsar wetlands. The author was amazed at the general ecosystem health threats to the Axios River Delta wetlands in Greece in Chapter 4 as well as the Ria Lagartos lagoon in San Felipe, Yucatan, Mexico, in Chapter 8. Even combining aspects of biosphere reserve management, Ramsar protection, and other international and national environmental protection measures may not be enough to protect wetlands. Maybe there should be an international classification system that recognizes a range of sustainable uses, stakeholders, and ecosystem integrity of the wetland systems (see Groot 1992, Turner et al. 2000).

Wetland System Management Innovations

At the larger wetland system scale, certainly the Tripartite Wadden Sea agreement between Netherlands, Germany, and Denmark stands out as an outstanding example in Chapter 2. This example illustrates how the environmental agencies and NGOs of the three countries can work together to achieve mutually beneficial ends. Certainly the North American Waterfowl treaty between the United States, Canada, and Mexico is equally impressive in terms of cooperative process of government agencies and international NGOs working together as well as in terms of results of waterfowl increase in numbers along the North Atlantic flyway.

For a national scale innovation we have to tip our hat to Canada. Even though the Great Lakes Wetlands Policy Consortium was a bi-national effort, the Canadians turned this initiative into an implemented program with the Great Lakes Wetlands Conservation Action Plan (Environment Canada 2006). Also St. Lucia for a small island country is moving toward sustainable development planning, incorporating "wise use" of their wetlands (Government of St. Lucia 2001a,b, 2004a,b). On the other hand, both Greece and Mexico are struggling to operationalize meaningful wetland protection policies.

From a regional perspective we have to applaud the World Wildlife Fund for their innovation of utilizing partnerships with key actors in the Kafue Flats in Zambia in Chapter 3. This has been a difficult management situation because of the many stresses on the altered floodplain wetlands, but involving the sugar producers, hydroelectric company, Zambia government, local chiefdoms, and tourist companies is a unique combination of partnerships. The role of the NGO PRONATURA and the university researchers with CINVESTAV with support for wetland management and ecotourism development in the Yucatan should also be recognized as a regional model of cooperation.

From a local or community perspective, we should give credit to three of the case studies in Chapters 4, 5, and 9 – the Kolkata wetlands, the Tram Trim Nature Preserve in Vietnam, and the Mankote wetlands in St. Lucia. For the Kolkata wetlands the roles of local CBOs, other NGOs, and local universities are all key to maintaining a viable multiple use of a created lagoon wetland system. Similarly the International Crane Foundation with local leadership plus Vietnam University was key to development of the multiple use management plan for the Trim Tram Nature Preserve. Finally, the roles of CANARI and local organizations in managing the Mankote mangrove for sustainable charcoal production and outreach communication are laudable.

International Wetland Management Principles

International wetland management principles can be found in the various RAMSAR documents (IUCN 1980, 1985, IUCN/IWRB 1980, 1984, Maltby 1991, Navid 1988) and are succinctly stated by Dugan (1988, 1990). These are also summarized in Chapter 1 of this book. In addition the Millennium Ecosystem Assessment Project (2005) presents the latest international wetland management strategies in Chapter 1. The issue the author would like to address is the nexus of participatory management of wetlands by multiple parties for sustainable or "wise" use from a Ramsar perspective.

Much of the recent North American wetland management literature focuses on biophysical management of basic wetland functions with little reference to traditional or heritage wetland uses or functions. It has long been this author's view (Smardon 1983, 2003, 2006) that traditional/heritage wetland use/management is a key perspective that should not be lost as we struggle with such issues. The major push to recognize human use values of wetlands was the Leiden Netherlands meeting in 1990 (see Marchand and de Haes 1990, 1991). This was the first international Ramsar meeting to focus on human use values of wetland systems and contained a wide range of presentations and papers.

Add to the previous cited literature the need for participatory management of protected areas by indigenous or traditional populations that live and depend on these areas for their livelihoods (see McNeeley 1992, Wilcox and Duin 1995, Oviedo and Brown 1999, Jeanrenaud 1999, Erickson 2006, Ramsar 2000) as well as the notion that the sustainable use of such areas constitute use of "natural capitol" (Bustos et al. 2004, Costanza et al. 1997, Gotz et al. 1999, Inamdar et al. 1999, Turner et al. 2000) by these same populations.

So the following principles for wetland management are offered given the emphasis offered above as well as innovations found from the case studies contained in the previous chapters:

- As part of the basic wetland inventory/assessment, traditional heritage uses of wetlands as well as current usage patterns should be documented (Baptiste 2008, Smardon 2003, 2006).
- Livelihood analysis should be done to document socio-economic benefits of such uses (Millennium Ecosystem Assessment 2005a, Whitten and Bennett 2005).
- Participatory processes such as partnering or multiparty negotiations should be used to develop wetland management plans for specific wetland areas or larger systems (Drijver 1991, Faust and Smardon 2001, Solton and Dudley 1999).
- Parties to such negotiations could be international NGOs, national or regional NGOs, government agencies, local government, CBOs, and local companies (World Wildlife Fund 1992).
- Implementation of single use or multiple use wetland management should involve local community-based organizations as key actors for implementation and monitoring with resources coming from other organizations listed above.
- Implementation should involve cross-training with other CBOs and/or government agencies with similar wetland systems in other locations to share their experiences.
- International agencies such as RAMSAR/IUCN and/or international NGOs such as Wetlands International should give recognition to sustainable and/or innovative wetland management efforts (Emerton 2005, Hails 1997, Ramsar 2000) so they can be shared with others.

When the author started this effort in 1990, he had no way of knowing if some of the case studies would result in irresolvable problems and dead ends. Successful or sustainable wetland management practices could become unsustainable or worse. The author does not think any of the case studies, ended at the time this book was finished, was truly negative with no redeeming characteristics. Even the direst of situations later resulted in organizations taking different paths and trying new approaches. This is a very strong message.

Combinations of NGOs, CBOs, government, and industry can work together toward innovative wetland management partnerships.

Acronyms

BINGO: Big International NGO
CANARI: Caribbean Natural Resources Institute
CBO: community-based organization
CINVESTAV: Centre de Investigacions Y Estudios Avanzados
ICF: International Crane Foundation
IUCN: Union for the Conservation of Nature
NGO: non-government organization
PRONATURA: Program for Nature
UNESCO: United Nations Program for Environment, Society and Culture
UNEP: United Nations Environment Program
WWF: World Wildlife Fund
WWWG: Wetlands Working Group – University of Vietnam

References

Baptiste, A. K. 2008. Evaluating Environmental Awareness: A Case Study of the Nariva Swamp Trinidad. Unpublished Dissertation, SUNY College of Environmental Science and Forestry, Syracuse, NY, 204 pp.

Barbier, E. B. 1993. Sustainable use of wetlands: Valuing tropical wetland benefits. *The Geographical Journal*, 159: 22–32.

Barbier, E. B., M. Acreman, and D. Knowler 1997. *Economic Valuation of Wetlands: A Guide for Policy Makers and Planners*. Gland, Switzerland: Ramsar Convention Bureau.

Bardecki, M. J. 1998. *Wetlands and Economics: An Annotated Review of the Literature, 1988-1998*. Ontario: Environment Canada.

Batisse, M. 1996. Biosphere reserves and regional planning: A prospective vision. *Natural Resources*, 32(3): 20–30.

Beltra'n, E. 1949. *La proteccio'n de la natueraleza: Principias y problemas*. Mexico: Secretari'a de Educacion Publica.

Brander, L. M., R. J. G. M. Florax, and J. E. Verman. 2003. *The Empirics of Wetland Valuation: A Comprehensive Summary and Meta-Analysis of the Literature. Institute for Environmental Studies*. Amsterdam: Vrije University.

Bustos, B., N. Borregaard, and M. Stilwell. 2004. The Use of Economic Instruments in Environmental Policy: Opportunities and Challenges, UNEP, Division of Technology, Industry and Economics, Economics and Trade Unit, UNEP/Earthprint, pp. 104–106.

Costanza, R., R. d'Arge, R. de Groot, S. Farber, M. Grasso, B. Hannon, K. Limburg, S. Naeem, R. V. O'Neill, J. Paruelo, R. G. Roskin, P. Sutton, and M. van den Belt. 1997. The value of the world's ecosystem services and natural capitol. *Nature*, 387: 253–260.

Drijver, C. A. 1991. People's participation in environmental projects in developing countries. *Landscape and Urban Planning*, 35: 7–23.

Dudley, N., B. Gujja, B. Jackson, J.-P. Jeanrenaud, G. Oviedo, A. Philips, P. Rosabel, S. Stolton, and S. Wells. 1999. Challenges for protected areas in the 21st century. In S. Solton and N. Dudley (eds.) *Partnerships for Protection: New Strategies for Planning and Management of Protected Areas*, pp. 3–12. London: IUCN and Earthscan.

Dugan, P. J. 1988. The importance of rural communities in wetlands conservation and development. In D. D. Cook et al. (eds.) *The Ecology and management of Wetlands Volume 2; Management Use and value of Wetlands*, pp. 3–11. Portland, OR: Timber Press.

Dugan, P. J. 1990. *Wetland Conservation: A Review of Current Issues and required Action*. Gland, Switzerland: IUCN The World Conservation Union.

Dyer, M., and B.V. Vinogrado. 1990. The role of Biosphere reserves in landscape and ecosystem studies. *Natural Resources*, 36(1) 19–26.

Emerton, L. (ed.). 2005. *Values and Rewards: Counting and Capturing Ecosystem Water Services for Sustainable Development*. Colombo, Sri Lanka: IUCN Water, Nature and Economics Technical Paper no. 1, IUCN – The World Conservation Union, Ecosystems and Livelihood Group Asia, 93 pp.

Environment Canada 2006. *Great Lakes Wetlands Conservation Action Plan Highlights Report 2003–2005*. Toronto, Ontario: Environment Canada, 24 pp and at http://www/on.ec.gc.ca/wildlife/publications-e.html

Erickson, J. 2006. A participation approach to conservation in the Calakmul Biosphere Reserve Campeche, Mexico. *Landscape Urban Planning*, 74(2006): 242–266.

Faust, B. B. and R. C. Smardon. 2001. Introduction and overview: environmental knowledge, rights and ethics: Co-managing with communities. *Environmental Science Policy*, 4(4/5): 147–151.

Frazier, J. 2006. Biosphere reserves and the "Yucatan" syndrome: Another look at the role of NGO's. *Landscape Urban Planning*, 24(2006): 313–333.

Gotz, W. N., L. Forthman, D. Cumming, J. di Felt, J. Hilty, R. Martin, M. Murphee, N. Owen Smith, A. M. Starfield, and M. I. Westphal. 1999. Sustaining natural and human capitol; villagers and scientists. *Science*, 283: 1855–1856.

Government of St. Lucia. 2001a. *Plan for Managing the Fisheries of St. Lucia*. Vieux-Fort: Government of St. Lucia.

Government of St. Lucia. 2001b. Chapter 3. Current Coastal Area Management Issues. In *Integrating the Management of Watersheds and Coastal Areas in St. Lucia*, pp. 50–81. Vieux-Fort: Government of St. Lucia.

Government of St. Lucia. 2004a. *Coastal Zone Management in St. Lucia: Policy Guidelines and Selected Projects*. Vieux-Fort: Government of St. Lucia.

Government of St. Lucia. 2004b. Draft Final Report. *To Review the Implementation of the Barbados Programme of Action (BPOA)*. Vieux-Fort, St. Lucia: Ministry of Physical Development, Environment and Housing, Government of St. Lucia, Executive Summary, pp. 1–12.

de Groot, R. S. 1992. *Functions of Nature: Evolution of Nature in Environmental Planning, Management and Decision-Making*. Groningen, The Netherlands: Wolters Noordhoff.

Hails, A. J. (ed.). 1997. *Wetlands, Biodiversity and the Ramsar Convention: The Role of the Convention on Wetlands in the Conservation and Wise Use of Biodiversity*. Gland, Switzerland: Ramsar Convention Bureau, 71 pp.

Inamdar, A., H. de Jode, K. Lindsey, and S. Cobb. 1999. Capitalizing on nature: Protected area management. *Science*, 233: 1856–1857.

IUCN. 1985. *Wetlands Conservation Programme*. Gland, Switzerland: IUCN.

IUCN. 1980. *The Ramsar Convention: A Legal Review, Conference on the Convention of Wetlands for International Importance Especially as Waterfowl Habitat*. Conf./5, Cagliari, Italy, Nov. 24–29, 1980. Gland, Switzerland: IUCN.

IUCN/IWRB. 1980. *The Ramsar Convention: A Technical Review: Conference on the Convention of Wetlands of International Importance Especially as Waterfowl Habitat*. Conf. 4, Cagliari, Italy, Nov. 24–29, 1980. Gland, Switzerland: IUCN.

IUCN/IWRB. 1984. *Overview of National Reports Submitted by Contracting Parties and Review of Developments Since the First Conference of the Parties*. IUCN/IWRB Doc. C2. 6, Groningen, Netherlands, May 7–12, 1984.

Jeanrenaud, S. 1999. People-oriented conservation progress to date. In S. Solton and N. Dudley (eds.) *Partnerships for Protection: New Strategies for Planning and Management of Protected Areas*, pp. 126–134. London: IUCN and Earthscan.

Maltby, E. 1991. Wetland management goals, wise use and conservation. *Landscape and Urban Planning*, 20(1–3): 9–18.

Maltby, E. 1986. *Waterlogged Wealth: Why Waste the World's Wet Places*. London, UK: Earthscan.

Marchand, M. and H. A. Udo de Haes (eds.). 1990. *The Peoples Role in Wetland Management*. Leiden, The Netherlands: Center for Environmental Studies, Leiden University.

Marchand, M. and H. A. Udo de Haes (eds.). 1991. The Peoples Role in Wetland Management: Wetlands Special Issue. *Landscape and Urban Planning*, 20(1–3): 1–276.

McNeeley, J. A. 1992. Nature and culture: conservation needs them both. *Natural Resources*, 28(3): 37–43.

McNeeley, J. A. 1999. Protected area institutions. In S. Stolton and N. Dudley (eds.) *Partnerships for Protection: New Strategies for Planning and Management of Protected Areas*, pp. 195–204. London: IUCN and Earthscan.

Millennium Ecosystem Assessment. 2005a. *Ecosystems and Human Well Being: Wetlands and Water Synthesis*. Washington, DC: Water Resources Institute, 70 pp.

Millennium Ecosystem Assessment. 2005b. *Our Human Planet: Summary for Decision Makers*. Millennium Ecosystem Assessment, Washington, DC: Island Press, 109 pp.

Munasinghe, M. and J. McNeeley (eds.). 1994. *Linking Conservation and Sustainable Development*. Washington, DC: The World Bank.

Navid, D. 1988. Developments in the Ramsar Convention. In D. D. Cook et al. (eds.). *The Ecology and Management of Wetlands Volume 2: Management Use and Value of Wetlands*, pp. 21–27. Portland, OR: Timber Press.

Oviedo, G. and J. Brown. 1999. Building alliances with indigenous peoples to establish and maintain potential areas. In S. Stolton and N. Dudley (eds.) *Partnerships for Protection: New Strategies for Planning and Management of Protected Areas*, pp. 99–108. London: IUCN and Earthscan.

Ramsar Convention. 2000. *Handbook 5: Establishing and Strengthening Local Communities and Indigenous People's Participation in the Management of Wetlands: Annex: Case Studies on Local and Indigenous People's Involvement in Wetland Management*. Gland, Switzerland: Ramsar Convention Bureau.

Ratner, B. D., D. Than Ha, M. Kosal, A. Nissapa, and S. Champhengxay. 2004. *Undervalued and Overlooked; Sustaining Rural Livelihoods Through Better Governance of Wetlands*, CABI Publication.

Schuyt, K. and L. Brander. 2004. *The Economic Values of the World's Wetlands*. Gland, Switzerland: World Wildlife Found, 30 pp.

Smardon, R.C. (ed.). 1983. *The Future of Wetlands; Assessing Visual-Cultural Values*, Allanheld-Osmun & Co., Totowa, NY 226 pp.

Smardon, R. C. 2003. The role of nongovernmental organizations for sustaining wetland heritage values. In Gravi-Bardos, M. and S. Gichard-Anguis (eds.) *Cross-Gazes to the Heritage Concept Worldwide to the end of the 20th Century*, pp. 795–815. Paris, France: Institute de Geography, Paris IV, Sorbonne.

Smardon, R. C. and B. B. Faust. 2006. Introduction: International policy in the biosphere reserves in Mexico's Yucatan peninsula. *Landscape Urban Planning* 74: 160–192.

Smardon, R.C. 2006. Heritage values and functions of wetlands in Southern Mexico, *Landscape and Urban Planning*, 74(3–4): 296–312.

Smart, M. and K. J. Kanters. 1991. Ramsar participation and wise use. *Landscape and Urban Planning*, 20(1–3): 269–274.

Stolton, S. and N. Dudley (Eds.) 1999. *Partnerships for Protection: New Strategies for Planning and Management of Protected Areas*. London: IUCN and Earthscan.

Terborgh, J. 1999. *Requiem for Nature*. Washington, DC: Island Press.

Turner, R. K., J. C. M. van de Bergh, T. Soderqvist, A. Barendregt, J. van de Straaton, E.
    Maltby, and E. C. van Ierland. 2000. Ecological-economic analysis of wetlands: Scientific
    integration for management and policy. *Ecological Economics* 35: 7–23.
UNESCO. 1984. Action Plan for Biosphere Reserves. *Nature and Resources*, 24: 11–22.
UNESCO. 1995. The Seville strategy for biosphere reserves. *Natural Resources*, 31(2): 2–17.
Whitten, S. M. and J. Bennett. 2005. *Managing Wetlands for Private and Social Good.*
    Canberra, Australia: Edward Elger Publishing.
Woodward, R. T. and Y. S. Wui. 2001. The economic value of wetland services: A meta
    analysis. *Ecological Economics* 37: 257–270.
World Wildlife Fund. 1992. *Statewide Wetland Strategies; A Guide to Protecting and Mana-
    ging the Resource.* Washington, DC: Island Press.

Index